Intertemporal Asset Pricing

Contributions to Economics

Christoph M. Schneider
Research and Development Management: From the Soviet Union to Russia
1994. ISBN 3-7908-0757-5

Bernhard Böhm/Lionello F. Punzo (Eds.)
Economic Performance
1994. ISBN 3-7908-0811-3

Lars Olof Persson/Ulf Wiberg
Microregional Fragmentation
1995. ISBN 3-7908-0855-5

Ernesto Felli/Furio C. Rosati/ Giovanni Tria (Eds.)
The Service Sector: Productivity and Growth
1995. ISBN 3-7908-0875-X

Giuseppe Munda
Multicriteria Evaluation in Fuzzy Environment
1995. ISBN 3-7908-0892-X

Giovanni Galizzi/ Luciano Venturini (Eds.)
Economics of Innovation: The Case of Food Industry
1996. ISBN 3-7908-0911-X

David T. Johnson
Poverty, Inequality and Social Welfare in Australia
1996. ISBN 3-7908-0942-X

Rongxing Guo
Border-Regional Economics
1996. ISBN 3-7908-0943-8

Oliver Fratzscher
The Political Economy of Trade Integration
1996. ISBN 3-7908-0945-4

Ulrich Landwehr
Industrial Mobility and Public Policy
1996. ISBN 3-7908-0949-7

Arnold Picot/Ekkehard Schlicht (Eds.)
Firms, Markets, and Contracts
1996. Corr. 2nd printing 1997. ISBN 3-7908-0947-0

Thorsten Wichmann
Agricultural Technical Progress and the Development of a Dual Economy
1997. ISBN 3-7908-0960-8

Ulrich Woitek
Business Cycles
1997. ISBN 3-7908-0997-7

Michael Carlberg
International Economic Growth
1997. ISBN 3-7908-0995-0

Massimo Filippini
Elements of the Swiss Market for Electricity
1997. ISBN 3-7908-0996-9

Giuseppe Gaburro (Ed.)
Ethics and Economics
1997. ISBN 3-7908-0986-1

Frank Hoster/Heinz Welsch/ Christoph Böhringer
CO_2 Abatement and Economic Structural Change in the European Internal Market
1997. ISBN 3-7908-1020-7

Christian M. Hafner
Nonlinear Time Series Analysis with Applications to Foreign Exchange Rate Volatility
1997. ISBN 3-7908-1041-X

Sardar M.N. Islam
Mathematical Economics of Multi-Level Optimisation
1998. ISBN 3-7908-1050-9

Sven-Morten Mentzel
Real Exchange Rate Movements
1998. ISBN 3-7908-1081-9

Lei Delsen/Eelke de Jong (Eds.)
The German and Dutch Economies
1998. ISBN 3-7908-1064-9

Mark Weder
Business Cycle Models with Indeterminacy
1998. ISBN 3-7908-1078-9

Tor Rødseth (Ed.)
Models for Multispecies Management
1998. ISBN 3-7908-1001-0

Michael Carlberg
Intertemporal Macroeconomics
1998. ISBN 3-7908-1096-7

Sabine Spangenberg
The Institutionalised Transformation of the East German Economy
1998. ISBN 3-7908-1103-3

continued on page 288

Bernd Meyer

Intertemporal Asset Pricing

Evidence from Germany

With 49 Figures
and 27 Tables

Physica-Verlag
A Springer-Verlag Company

Series Editors
Werner A. Müller
Martina Bihn

Author
Dr. Bernd Meyer
Frankfurt am Main, Germany

Printed with the financial support of the
Stiftungsfond Deutsche Bank, Essen, Germany

ISBN 3-7908-1159-9 Physica-Verlag Heidelberg New York

Cataloging-in-Publication Data applied for
Die Deutsche Bibliothek – CIP-Einheitsaufnahme
Meyer, Bernd: Intertemporal asset pricing: evidence from Germany; with 27 tables / Bernd Meyer. – Heidelberg: Physica-Verl., 1999
(Contributions to economics)
ISBN 3-7908-1159-9

Softcover Design: Erich Kirchner, Heidelberg

SPIN 10694526 88/2202-5 4 3 2 1 0 – Printed on acid-free paper

Foreword

In the mid-eighties Mehra and Prescott showed that the risk premium earned by American stocks cannot reasonably be explained by conventional capital market models. Using time additive utility, the observed risk premium can only be explained by unrealistically high risk aversion parameters. This phenomenon is well known as the equity premium puzzle. Shortly afterwards it was also observed that the risk-free rate is too low relative to the observed risk premium.

This essay is the first one to analyze these puzzles in the German capital market. It starts with a thorough discussion of the available theoretical models and then goes on to perform various empirical studies on the German capital market. After discussing natural properties of the pricing kernel by which future cash flows are translated into securities prices, various multiperiod equilibrium models are investigated for their implied pricing kernels. The starting point is a representative investor who optimizes his investment and consumption policy over time. One important implication of time additive utility is the identity of relative risk aversion and the inverse intertemporal elasticity of substitution. Since this identity is at odds with reality, the essay goes on to discuss recursive preferences which violate the expected utility principle but allow to separate relative risk aversion and intertemporal elasticity of substitution.

There are two basic approaches for testing equilibrium models. The conventional approach is to estimate the parameters of the equilibrium from observed data. The other approach is to implement certain parameter values for the equilibrium model and then derive the implications for the behavior of securities prices. This is called the calibration approach and used in this essay. Based on German financial market data from 1960 to 1994, various properties of risk premia and risk-free rates are estimated. As in the USA, the risk premium puzzle also exists in Germany. Therefore recursive preferences are necessary also for Germany to explain stock market data. The essay does not stop here, but discusses various extensions of the equilibrium model to obtain a better empirical fit. Even then the observed risk premium cannot be reasonably explained with time additive preferences. Although recursive preferences remedy this problem, the essay shows that other properties of price processes cannot be explained by the equilibrium models. For example, none of the models can explain the volatility of the risk-free rate and of the risk premium. Therefore, even the more elaborate equilibrium models can explain only some properties of observed price movements. Overall, this essay represents a very successful investigation of the German capital market. It adds much to our understanding, at the same time it points to various directions for future research.

Günter Franke

Acknowledgements

This monograph was written while I was working at the Department of Economics and Statistics and at the special research program "Internationalisierung der Wirtschaft" (*Sonderforschungsbereich 178*) at the University of Konstanz. In January 1998, the Department of Economics and Statistics at the University of Konstanz accepted this monograph as a dissertation. It could not have been accomplished without the help of the following people and organisations whose contributions are hereby gratefully acknowledged.

First of all, I would like to thank my supervisor, Professor Dr. Günter Franke, for his support and advice during the various stages of my work. What I have learned from him is far beyond the contents of this work. I owe thanks to Professor Dr. Nikolaus K. A. Läufer for taking on the role as second referee and the positive critique.

Furthermore, I would like to thank my colleagues in the Department of Economics and Statistics and, especially, in the research unit for international finance. In particular, I am indebted to Dr. Axel F. A. Adam-Müller, Dr. Markus Herrmann, Dr. Dieter Hess and Mathias Külpmann for many constructive discussions and for comments on earlier versions of this monograph. Additionally, I thank the student research assistants for their numerous errands to the library and, especially, our secretary, Mrs. Gisela Laniecki, for creating an informal and harmonic atmosphere.

I had the opportunity to present parts of this monograph at the annual meetings of the European Finance Association in Milano (1995) and Oslo (1996) and at the Seventh Symposium on Money, Finance, Banking, and Insurance in Karlsruhe (1996).

I gratefully acknowledge the financial support of the *Deutsche Forschungsgemeinschaft* through the *Sonderforschungsbereich 178* "Internationalisierung der Wirtschaft" at the University of Konstanz. I also thank *Deutsche Bank AG* for financially supporting the publication of the monograph.

Finally, I am grateful to my family and friends for their support. In particular, I am indebted to Mrs. Doris Decker for her encouragement and constant support during the final stage of the work and the conscientious reading of the monograph.

Bernd Meyer

Contents

II Intertemporal Asset Pricing: Empirical Analysis 107

Chapter 1

Introduction

1.1 Subject of Analysis

Investors are concerned with the choice between risky and risk-free investments. It is consensus among economists that investors demand a reward for bearing the risks of risky investments such as stocks as compared to investments free of risk. The quantification of this tradeoff between risk and expected return is one of the central problems of financial economics and has been the focus of theoretical and empirical research for many decades. But the question of how to measure risk and how to calculate the market reward for bearing a given amount of risk is not fully answered.

The Capital Asset Pricing Model (CAPM) of Sharpe (1964), Lintner (1965), Mossin (1966) and Black (1972) and the Arbitrage Pricing Theory (APT) of Ross (1976) yield only partial answers. According to the CAPM the expected return on an asset is linear in the covariance of its return with the return on the market portfolio; i.e. an asset's expected return is measured relative to the expected return on the market portfolio. The risk-free rate and the risk premium on the market portfolio are taken as exogenously given. Similarly, the APT, assuming that the return on an asset is linear in more than one risk factor, takes the factor risk premia as exogenously given. Moreover, the CAPM and the APT are one-period models that hold in a multiperiod world only under additional assumptions. Empirical evidence indicates that the risk-free rate and the risk premium on the market portfolio are not constant but change with economic conditions. Autocorrelation in risk-free rates and risk premia implies some predictability of asset returns. Taking time-varying risk-free rates and risk premia into account, this predictability of asset returns may be consistent with market efficiency, an issue that was neglected in financial economics for many years. Today the importance of time-variation in expected returns is widely recognized.

The CAPM and the APT determine neither the average levels of the risk-free rate and the factor risk premia nor their movements over time. But knowing the determinants of the levels and movements is of crucial interest to academics as well as to investors and investment professionals.

This study deals with both issues. In particular, it focuses on the levels and movements of the risk-free rate and the equity premium, i.e. the risk premium on the stock market, and addresses these issues in intertemporal equilibrium models. While there exist numerous studies empirically investigating asset pricing models such as the CAPM and the APT for Germany, there is a lack of empirical evidence on intertemporal equilibrium asset pricing. This study attempts to fill this gap, at least partly.

Evidence from the United States suggests that explaining the empirically observed average equity premium is not an easy task. A positive average equity premium is to be expected since investors would not hold stocks unless they were compensated for bearing the additional risk by earning a higher average rate of return on stocks than on risk-free investments. But, assuming that the ex-post observed average equity premium equals the ex-ante expected one, the size of the historical average equity premium in the US appears to be far in excess of what is justified by standard intertemporal equilibrium models with reasonable levels of risk aversion. This finding is known as the "Equity Premium Puzzle" (cf. Mehra and Prescott (1985)). Jointly explaining risk-free and risky returns poses an even greater problem for intertemporal equilibrium models. Given the high level of equity premia, the risk-free rate observed in the US appears to be too low to be consistent with these models. This finding has been termed the "Risk-free Rate Puzzle" (cf. Weil (1989)). As pointed out by several authors (e.g., Kocherlakota (1996)), the failure to explain the risk-free rate and the average equity premium has broad implications for macroeconomic theory. If models widely applied in macroeconomics cannot explain risk-free rates as low as those empirically observed and equity premia as high as those empirically observed, then macroeconomic theory also cannot explain why people save, given the low risk-free rate, and why people are so averse to invest in stocks.

Before the purpose and the organization of this study are described in section 1.3, section 1.2 provides international evidence on the risk-free rate and the equity premium. This helps to clarify whether the average levels of the risk-free rate and the equity premium are similar across countries[1].

---

[1] A more thorough investigation of observed rates of return on German assets is presented in part II of the analysis.

1.2 International Evidence on the Risk-free Rate and the Equity Premium

The equity premium is defined as the expected arithmetic return[2] on equity in excess of the return on a risk-free investment. Mehra and Prescott (1985, table 1) report an average real equity premium of 6.18 percent per annum for the 90 years from 1889 to 1978 in the US. The real risk-free rate of return is the real return on three-month government treasury bills exhibiting an arithmetic average of .8 percent per annum. Siegel (1992) extends the time period analyzed by Mehra and Prescott to the period 1802 to 1990. For the periods 1802-1990, 1871-1990 and 1926-1990 he estimates arithmetic average real equity premia of 4.7 percent, 6.5 percent, and 8.0 percent per annum, respectively (Siegel 1992, tables I and II)[3]. The average real risk-free rates over the same periods are 3.1 percent, 1.8 percent and .6 percent per annum.

An important issue is whether similar findings can also be obtained for countries other than the US. Often, the comparison of historical average rates of return is diluted by investigating different time periods[4] or applying return time series that cannot be compared. A reliable comparison of average returns in different countries should therefore satisfy the following conditions. (1) The investigation periods should be equal across countries.

[2]The arithmetic average return or mean return on an asset over, for example, yearly time intervals does not measure a compound rate of return over periods consisting of several years. If one is interested in the compound rate of return to a buy-and-hold strategy the geometric average return should be used. The geometric average return is always less than the arithmetic average return, except when all returns are equal. However, the arithmetic average return can be interpreted as the expected return over the next time interval and finance theory usually applies to expected returns; risk is traded off against per period expected returns. The primary objective of this study is to analyze whether or not observed rates of return are consistent with asset pricing models, but not to measure the long-term performance of buy-and-hold strategies. Therefore, arithmetic average returns are used throughout this study.

[3]These values are the differences between the arithmetic average real returns on stocks given in the third row of table I and the arithmetic average real returns on treasury bills given in row nine of table II.

[4]The results of Stehle and Hartmond (1991), for example, illustrate the effect of different time periods. For the period 1954 to 1988 they calculate an geometric average real German stock return of 8.7 percent per annum. The average real risk-free rate is 2.2 percent per annum indicating a geometric average equity premium of 6.5 percent per annum (cf. Stehle and Hartmond (1991, table 4)). For the years 1960 to 1988, however, they only observe an average equity premium of 2.1 percent per annum. Morawietz (1994) investigates the average German equity premium in the period 1870 to 1992. Unfortunately, his study contains some data errors and exhibits several methodological problems. For example, the values of the stock price index for March 1972 and May 1972 (Morawietz 1994, p. 323) suffer from a decimal point error. Also, the real equity premium is calculated as the difference between the real return on stocks and the *nominal* return on the risk-free investment (see Morawietz (1994, p. 224)).

(2) The time series should exhibit the same data frequency and the same averages (arithmetic or geometric) should be computed. (3) The time series of stock returns should uniformly include dividends or, alternatively, dividend returns must be added. (4) The maturity and the style of the assets chosen to represent the risk-free rate should be comparable. (5) Differences in inflation and taxation should be considered. With the exception of tax adjustments, Campbell (1996) comes close to satisfying these conditions in his study comparing risk-free rates and equity premia of 12 different countries[5]. For rather short periods he observes arithmetic average real equity premia between -.31 percent per annum (Spain) and 9.54 percent per annum (Switzerland) (cf. Campbell (1996, table 5)). With 6.60 percent per annum the German equity premium does not differ considerably from the other countries' values[6].

Subsequently, international evidence on the risk-free rate and the equity premium is provided for the 35 years from 1960 to 1994, covering the G-7 countries and Belgium. The employed time series use monthly data. Except for the time series of the return on Japanese long-term bonds[7], they cover the period January 1960 to December 1994. Table 1.1 lists the arithmetic averages and the standard deviations of (1) monthly rates of inflation, (2) monthly nominal percentage changes in the stock price indices[8], (3) monthly average nominal annual dividend yields, (4) monthly averages of nominal three-month treasury bill rates and (5) monthly averages of nominal returns on long-term government bonds. All values are given in percent per annum. The means and standard deviations of the monthly rates of inflation and the monthly percentage changes in the stock price indices are annualized by taking the gross rates to the power of 12 and multiplying the standard deviation by $\sqrt{12}$.

[5] Unfortunately, the covered time period is quite short; the start dates of the quarterly time series range from 1970.1 to 1978.4 and the end dates range from 1993.3 to 1994.3.

[6] Also Solnik (1996) analyzes international investments. A comparison of average returns on international stocks, bonds and treasury bills received by US Dollar investors in the period 1971 to 1994 is given in Solnik (1996, exhibit 4.9).

[7] This time series begins in June 1966.

[8] Notice that the national stock markets are of different size, both absolutely and in proportion to national gross domestic product or gross national product. The stock market capitalization relative to the nominal gross national product in the year 1995, for example, is (in descending order) 1.52 for the United Kingdom, 1.22 for the US, .88 for Kanada, .63 for Japan, .44 for Belgium, .38 for France, .27 for Germany and .23 for Italy (Deutsche Bundesbank 1997, p. 28).

Table 1.1: Means and standard deviations (in parentheses) of monthly rates of inflation, of monthly nominal percentage changes in the stock price indices, of nominal dividend yields, of monthly averages of nominal three-month treasury bill rates and of monthly averages of nominal returns on long-term government bonds for eight industrialized countries for the period 1960 to 1994. All values are given in percent per annum.

Country	CPI[a]		Stocks[b]		Dividend[c]		3-Month TB[d]		Long Bonds[e]	
Belgium	4.56	(1.37)	4.27	(14.09)	5.70	(1.86)	7.81	(3.23)	8.40	(2.20)
Canada	5.06	(1.42)	7.34	(15.19)	3.55	(.69)	7.60	(3.62)	8.70	(2.66)
France	5.99	(1.35)	8.50	(21.25)	4.37	(1.86)	8.07	(2.67)	9.36	(2.98)
Italy	8.53	(2.27)	7.90	(23.97)	3.25	(1.04)	10.36	(4.40)	10.16	(4.35)
Japan	5.07	(2.86)	9.97	(15.82)	2.19	(1.45)	4.93	(1.35)	6.78[f]	(1.45)
United Kingdom	7.36	(2.46)	9.81	(17.18)	4.88	(1.16)	8.62	(3.26)	9.72	(2.83)
United States	4.72	(1.19)	6.56	(11.93)	3.75	(.88)	6.13	(2.80)	7.31	(2.53)
West Germany	3.39	(1.11)	5.11	(15.34)	3.68	(.88)	6.32	(2.57)	7.49	(1.24)
Average[g]	5.59	(1.75)	7.43	(16.84)	3.92	(1.23)	7.48	(2.99)	8.49	(2.53)

[a]Annualized percentage changes in the seasonally unadjusted monthly consumer price indices measuring monthly average prices (rates of inflation). Except for Belgium these indices are taken from the Citicorp database. For Belgium the index comes from the OECD Leading Indicators database.

[b]Annualized nominal percentage changes in monthly stock price indices measuring monthly average prices not including dividends. Except for Belgium these indices are taken from the Citicorp database. For Belgium the index comes from the OECD Main Economic Indicators database.

[c]Average nominal annual dividend yields taken from the OECD Financial Affairs database. For Germany, the dividend yields including the tax credit provided by the Statistisches Bundesamt are used.

[d]Monthly averages of the nominal yield on 3-month treasury bills. For Japan, the averages of the yield on 2-month financial bills and for France and Italy, the averages of the yield on short-term (3 to 5 years) bonds is used instead. These data are taken from the OECD Financial Affairs database. For Germany, the monthly averages of the 3-month money market rate published by the Deutsche Bundesbank are used.

[e]Monthly averages of the average nominal yield to maturity of long-term government bonds taken from the OECD Main Economic Indicators database, respectively the OECD Leading Indicators database (France and Italy).

[f]6/1966-12/1994

[g]Equally weighted average of the eight values.

First, consider the returns on stocks. The average nominal return on stocks is calculated as the average nominal percentage change of the stock price index plus the average nominal dividend yield. It varies between 8.79 percent per annum (Germany) and 14.69 percent per annum (United Kingdom). These differences cannot be attributed to differences in the risk measured by the

Figure 1.1: Average risk and return of the stock market of eight industrialized countries for the period 1960 to 1994. Risk is measured by the annualized standard deviation of the nominal monthly returns on stocks. The return is measured by the annualized average of nominal monthly stock returns (left graph) and, alternatively, by the annualized average of real monthly stock returns (right graph).

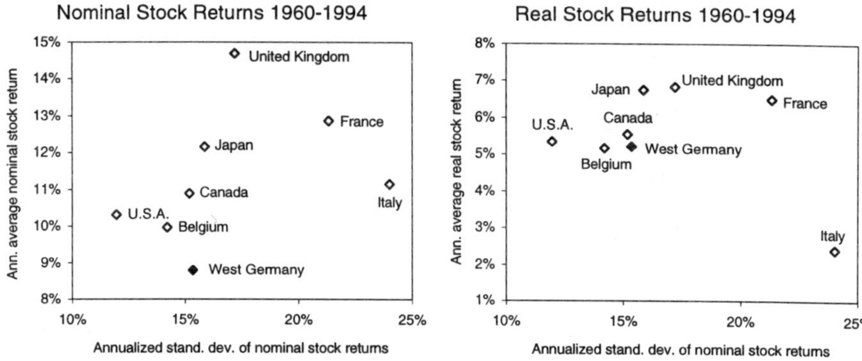

standard deviation of the nominal returns on stocks[9] as is evident from the left graph in figure 1.1. However, the average rate of inflation varies between 3.39 percent per annum (Germany) and 8.53 percent per annum (Italy), so that the findings for real returns on stocks may be different. Average real returns on stocks are calculated by dividing the average nominal gross returns on stocks by the average gross rate of inflation and then subtracting one[10]. These real returns are depicted in the right graph of figure 1.1. Except for Italy (2.41 percent per annum) the average real returns on stocks lie in a small range of 1.66 percentage points between 5.17 percent per annum (Belgium) and 6.83 percent per annum (United Kingdom). The average German real return on stocks is 5.22 percent per annum.

[9]The standard deviation of the nominal returns on stocks is calculated from the standard deviation of the nominal percentage changes in the stock price index and the standard deviation of the dividend yields assuming that both are independent.

[10]The correct calculation would have been to deflate each month's stock return with the corresponding rate of inflation and then calculate the arithmetic average of these real returns on stocks. Unfortunately, the monthly time series of dividend yields measures the average annual dividend return of stocks, rather than the dividend return in each month, so that the stock return consisting of capital gains and dividend payments cannot be calculated separately for each month. The same problem applies for the time series of average yields on treasury bills or long-term bonds. However, the bias imposed by deflating the averages instead of averaging deflated values is small. For Germany, a precise investigation is given in section 6.2.

Figure 1.2: Annualized average of monthly values of the real risk-free rate and of the real equity premium for eight industrialized countries for the period 1960 to 1994.

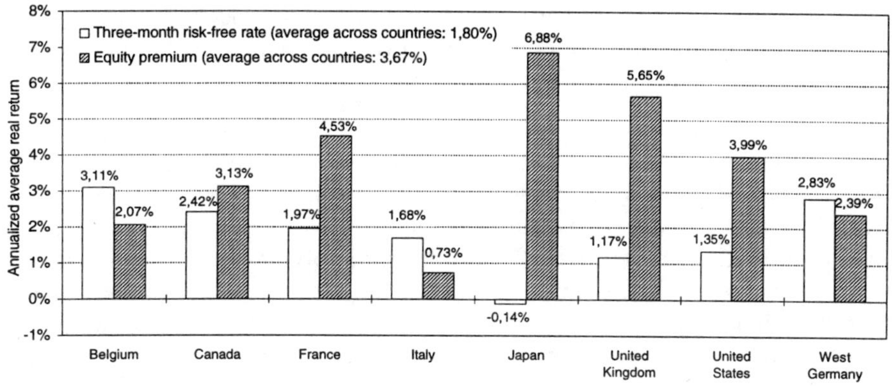

Now, consider the risk-free rate and the equity premium. The risk-free rate is represented by the return on a three-month risk-free investment. The average nominal risk-free rate varies in a range of 5.43 percentage points between 4.93 percent per annum (Japan) and 10.36 percent per annum (Italy) (see table 1.1). The average real risk-free rates are illustrated in figure 1.2. They lie in a range of 3.25 percentage points with a minimum of -.14 percent per annum (Japan) and a maximum of 3.11 percent per annum (Belgium). With 2.83 percent per annum Germany has the second largest average real risk-free rate. The average real equity premium, also depicted in figure 1.2, is calculated as the difference between the average real return on stocks and the average real risk-free rate. It is positive for each country[11]. However, the range (6.15 percentage points) is quite large. The highest average real equity premium is observed for Japan (6.88 percent per annum), the lowest for Italy (.73 percent per annum). With 2.39 percent per annum the average German real equity premium lies 1.28 percentage points below the average of the eight countries. Thus, although the average equity premia observed in Belgium and Italy are smaller than the average German equity premium, there is some evidence that the average equity premium is smaller in Germany than in most other countries. This difference is yet not significant[12].

[11]Except for the average Italian equity premium, the average real equity premia are significantly positive at a five percent level of significance on the basis of the parametric t-statistic.

[12]Separately comparing the average German equity premium with the equity premium of each other country using a parametric t-test reveals that the average German equity premium is significantly smaller than the average equity premium in Japan (one percent level of significance) and the average equity premium in the United Kingdom (five percent level of significance). However, comparing the average German equity premium with the

Alternatively representing the risk-free rate by the average real return on long-term bonds does not alter these findings. On average, long-term bonds yield .96 percent per annum in excess of the three-month treasury bills. With the exception of Italy the spread between the average return on long-term bonds and the average return on treasury bills is almost the same for all countries[13]. Therefore, the average equity premium in excess of bond returns is generally about one percentage point below that in excess of treasury bill returns.

Overall, these results suggest, that the German situation may differ from the situation studied by Mehra and Prescott (1985) (average real risk-free rate of .8 and average real equity premium of 6.18 percent per annum), but, investigating the same time periods and applying comparable time series, it does not differ significantly from the situation in other industrialized countries. It is important to notice that similar levels of risk-free rates and equity premia in different countries do not imply that they are similarly puzzling. An equity premium and a risk-free rate are only puzzling relative to a given variability of macroeconomic activity and a given covariance structure between historical rates of return and economic fundamentals such as aggregate consumption. Both may differ across time periods and countries, especially because the importance of the national capital markets is different and stocks are held by different groups of investors in these markets.

average of the equity premia of the other countries, no significant difference is found.

[13]For Italy, for which only returns on short-term bonds are available instead of treasury bill rates, the spread between the average return on long-term bonds and the average return on short-term bonds is negative.

1.3 Purpose and Outline of Analysis

The purpose of the analysis is threefold. The first issue addresses the Equity Premium Puzzle. It is examined whether the size of the empirically observed average German equity premium can be explained as a compensation for the additional risk of stocks. Therefore, the focus is on the relation between the risk and the return of the stock market portfolio. Risks and returns of individual stocks are not investigated. The second issue addresses the Risk-free Rate Puzzle. It is analyzed whether the relation between the empirically observed German average risk-free rate and the empirically observed German average equity premium can be explained. Therefore, the analysis is conceptually based on intertemporal equilibrium asset pricing models that allow for a simultaneous pricing of risky and risk-free assets. While the first and second issue relate to a long run unconditional perspective, the third issue relates to a short run conditional perspective. This issue presumes that the risk-free rate, the conditional expected equity premium and the conditional expected variance of the equity premium are time-varying. It is investigated how the short-term co-movement of the risk-free rate and the equity premium, and, hence, the co-movement of the risk-free rate and the return on stocks can be modeled and explained.

The main purpose of the analysis can thus be summarized by the question:

- Are German risk-free interest rates and equity premia consistent with intertemporal equilibrium asset pricing models?

From a practitioner's perspective the analysis addresses the interaction between the stock and the bond market. It investigates the fundamental macroeconomic sources that drive the expected returns on both markets and, thus, determine the average levels and the co-movement of the prices of stocks and bonds.

The objective of this study is not to add to the literature of statistical tests of intertemporal equilibrium asset pricing models applying methods such as the Generalized Method of Moments (GMM) estimation of Hansen and Singleton (1982). Instead, the aim is to provide an understanding of the dimensions along which these asset pricing models may fail in Germany and to guide the search for better models. Therefore, two alternative approaches for the empirical investigation of intertemporal asset pricing models are applied, the variance bound approach of Hansen and Jagannathan (1991) and the calibration and simulation approach of Kydland and Prescott (1982). Calibration exercises are widely conducted in business cycle theory and such exercises have unveiled the Equity Premium Puzzle and the Risk-free Rate

Puzzle for the US. Up to now both approaches have not been applied to German data[14].

The analysis is organized in two parts: A theoretical analysis of intertemporal asset pricing in part I (chapters 2 to 5) and an empirical analysis in part II (chapters 6 to 9).

Chapter 2 presents the central theoretical concept of the analysis, the market pricing kernel approach. This is a broad concept for intertemporal asset pricing that places few restrictions a priori on the structure of asset pricing models. The market pricing kernel gives Arrow-Debreu prices per unit probability and hence is a random variable. The knowledge of the distribution of the market pricing kernel for a specific date permits the pricing of any payoff at that date. Any parametric model of the market pricing kernel specifies a parametric intertemporal asset pricing model. The relation of the market pricing kernel approach to other characterizations of intertemporal asset pricing as well as the implications of arbitrage free and complete markets for the market pricing kernel are discussed. Finally, a general linear asset pricing model in terms of the market pricing kernel is derived. The whole analysis is based on discrete-time theory.

Chapter 3 reviews various approaches for reaching conclusions about the properties of the market pricing kernel. It particularly motivates and describes the variance bound approach of Hansen and Jagannathan (1991) that is applied in chapter 7. This approach is to place restrictions on the unconditional distribution of the market pricing kernel using unconditional sample moments of empirically observed rates of return. Satisfying these restrictions is necessary for any parametric model of the market pricing kernel to be consistent with the empirically observed rates of return.

Chapter 4 is concerned with the derivation of parametric models of the market pricing kernel, i.e. the specification of intertemporal asset pricing models. The consumption-based equilibrium asset pricing approach is used to derive these models. Alternative approaches are considered briefly. Given the models' failure to explain the historical average risk-free rate and average equity premium in the US, the usually employed time-additive expected preference framework seems to be too restrictive. Therefore, the analysis is generalized to the recursive non-expected preference framework of Kreps and Porteus (1978) and the corresponding parametrizations of Epstein and Zin (1990) and Weil (1989). These preferences allow for a separate specification of risk-aversion and elasticity of intertemporal substitution as well as for preferences over the timing of the resolution of uncertainty.

[14]For a first application of the variance bound approach, Missong (1994) is the only exception.

In chapter 5 the calibration approach for the empirical investigation of parametric models of the market pricing kernel is introduced and compared with the estimation approach. The calibration approach, which is applied in chapter 8, is to specify a model economy and an exogenous endowment process and to compare specific properties of the equilibrium rates of return implied by reasonably parametrized market pricing kernels with those of observed rates of return. An example illustrates this approach. Finally, a brief overview of the existing literature applying the calibration approach in a related context is given.

Chapter 6, the first chapter of part II, gives an overview of the empirical analysis and the data, covering the 35 years from 1960 to 1994. Real tax-adjusted rates of return on the German stock market, the bond market and risk-free investments are calculated for different taxation scenarios, different methods of return calculation and different lengths of the time intervals over which the returns are measured. Stylized facts of these rates of return are presented. Finally, the consumption data required for the derived parametric models of the market pricing kernel are described.

Variance bounds of the market pricing kernel are analyzed in chapter 7. The first part of the chapter compares the variance bounds imposed by different sets of assets and studies the sensitivity of the variance bounds to taxation, the method of return calculation, and the length of the time intervals over which returns are measured. This illustrates the information content of different sets of German return time series for analyzing intertemporal asset pricing models. Using the calculated variance bounds, the second part evaluates the parametric models of the market pricing kernel derived in chapter 4.

Chapters 8 contains the calibration exercises. The objective is to choose the parameters of the parametric models of the market pricing kernel derived in chapter 4 such that the implied unconditional equilibrium risk-free rate and equity premium match their empirically observed counterparts. Markov state-switching processes are estimated for the dynamic structure of the analyzed model economies. These processes are chosen because they permit closed-form solutions for the equilibrium rates of return (cf. Lucas (1978)), while still allowing to model time variation in the risk-free rate as well as in the equity premium. Various modifications of the standard model, e.g., the introduction of leverage and the separation of aggregate consumption and aggregate dividends, are studied. But, the assumptions of complete and fricitionless asset markets and of the rationality of the individuals are maintained throughout the calibration exercises. Monte Carlo simulations of the calibrated models are conducted to permit statistical inference.

In chapter 9 some of the parametric models of the market pricing kernel derived in chapter 4 and calibrated in chapter 8 are assessed with respect to properties of empirically observed rates of return, other than the first and second unconditional moments. For example, it is explored whether the empirically observed correlation between the risk-free rate of return and the equity premium is consistent with the equilibrium model and whether the empirically observed mean-reversion in asset prices over long horizons (cf. Fama and French (1988a) and Poterba and Summers (1988)) can be explained by the time-variation in the conditional expected returns implied by the calibrated models. Finally, implied conditional moments are compared with the corresponding empirically observed conditional moments.

Chapter 10 summarizes the results and concludes.

Part I

Intertemporal Asset Pricing: Theory

Chapter 2

The Market Pricing Kernel Approach

2.1 The Market Pricing Kernel

2.1.1 Definition

In a frictionless market[1] it follows from the Law of One Price[2] that a market pricing kernel[3] $\tilde{\phi}_{t+1}$ exists for each date t such that the ex-dividend price P_t^i of any asset i can be calculated from the equation

$$P_t^i = \mathrm{E}_t \left\{ \tilde{\phi}_{t+1} \cdot \left(\tilde{P}_{t+1}^i + \tilde{D}_{t+1}^i \right) \right\}. \tag{2.1}$$

$\tilde{P}_{t+1}^i$ denotes the stochastic ex-dividend price of the asset i at date $t+1$, $\tilde{D}_{t+1}^i$ is the stochastic amount of any dividends, interest or other payments out of the asset i at date $t+1$, and $\mathrm{E}_t\{.\}$ is the expectation taken for date $t+1$, given the information available at date t. The market pricing kernel relates expected future payoffs to present prices. It may depend on

---

[1]The market is assumed to be competitive and perfect; that is, individuals perceive prices as beyond their influence, there are no transaction costs or taxes, assets are perfectly divisible, and the full proceeds from short sales can be invested.

[2]The Law of One Price requires that assets with the same payoffs must trade at the same price (cf. Dybvig and Ross (1989, p. 58)).

[3]In the literature the market pricing kernel sometimes is called "stochastic discount factor" (e.g., Hansen and Jagannathan (1997)) or "pricing asset" (e.g., Duffie (1988)). The term "stochastic discount factor" stems from the present value model, where prices equal the expected discounted value of future cash flows. Recursively substituting equation (2.1) and using the law of iterated expectations yields the traditional present value model with stochastic discount factors: $P_t^i = \mathrm{E}_t \left\{ \sum_{\tau=t+1}^{\infty} \left(\tilde{D}_\tau^i \cdot \prod_{k=t+1}^{\tau} \tilde{\phi}_k \right) \right\}$. See Shiller (1981, p. 430) for an application of a present value model with time-varying real discount rates.

the endowments, preferences and future investment opportunities of the investors and may therefore be expressed as a function of other random variables $\tilde{x}_{t+1}$, $\tilde{\phi}_{t+1} \equiv \phi(\tilde{x}_{t+1})$, such as the market return or the growth rate of consumption. This is discussed in section 2.2 and chapter 4.

In terms of gross returns it follows equivalently that the expectation of the product of the market pricing kernel and the gross return on any asset i, given the information at date t, equals one:

$$1 = \mathrm{E}_t \left\{ \tilde{\phi}_{t+1} \cdot \tilde{R}^i_{t+1} \right\}, \qquad \text{where} \qquad \tilde{R}^i_{t+1} = \frac{\tilde{P}^i_{t+1} + \tilde{D}^i_{t+1}}{P^i_t}. \tag{2.2}$$

Equation (2.2) is an important starting point of this study.

Before equivalent characterizations of intertemporal asset pricing are discussed, the notation for a discrete state-space is introduced. Assume, that the uncertainty of the outcome at date $t+1$ can be described by a discrete number of states of nature that can occur at date $t+1$. Given the information at date t, state s, $s \in \{1, \ldots, S\}$, occurs at date $t+1$ with probability prob_s. For this discrete state-space, equation (2.1) can be written as

$$P^i_t = \sum_{s=1}^{S} \mathrm{prob}_s \; \phi_{t+1,s} \; e^i_{t+1,s} \,, \tag{2.3}$$

where $\phi_{t+1,s}$ denotes the market pricing kernel for state s and $e^i_{t+1,s}$ denotes the total payoff of asset i at date $t+1$ if state s occurs, i.e. $e^i_{t+1,s} = P^i_{t+1,s} + D^i_{t+1,s}$.

2.1.2 Equivalent Characterizations

There are equivalent ways of characterizing intertemporal asset pricing, e.g., using prices for state-contingent claims or risk-neutral probabilities. The characterizations are briefly compared for the discrete state-space.

The price of a state-contingent claim is the price of a future payoff of one unit in just one state of nature. $\pi_{t+1,s}$, for example, is the time t price of a payoff of one unit in state s at date $t+1$. The time t price of a security i, P^i_t, can then be expressed as the sum across all states of its future, state-dependent payoffs, $e^i_{t+1,s}$, multiplied by $\pi_{t+1,s}$,

$$P^i_t = \sum_{s=1}^{S} \pi_{t+1,s} \; e^i_{t+1,s} \,. \tag{2.4}$$

The prices of state-contingent claims incorporate the state-probabilities. Therefore, prices of state-contingent claims simultaneously measure two separate aspects, namely, the probability of the states, prob_s, and the relative desirability of the payoff in state s. Both aspects are not necessarily interrelated. For example, the desirability of a payoff in state s may depend on the value of some exogenously given variables in state s, independent of the probability of state s. Therefore, it may be advantageous to specify both aspects separately. A comparison of equations (2.3) and (2.4) reveals that the state-probability times the market pricing kernel equals the price of the state-contingent claim, or, under the assumption that the state has a positive probability, the market pricing kernel equals the probability-deflated price of a state-contingent claim. Thus, the market pricing kernel is equivalent to the price of a state-contingent claim per unit probability (cf. Dybvig and Ross (1989, pp. 64-5)). Hence, the market pricing kernel does not incorporate the state-probability and, therefore, it exclusively measures the desirability of the payoff in state s. This approach is especially useful for analyzing choice problems under uncertainty because under additional assumptions the market pricing kernel can be related to the marginal utility of consumption[4].

In another characterization of intertemporal asset pricing, P_t^i is the expected value of the future payoff under artificial risk-neutral state-probabilities, prob'_s, discounted at the risk-free rate. The artificial state-probabilities, prob'_s, are called risk-neutral because under this probability measure all assets yield an expected return equal to the risk-free rate (Duffie 1996, p. 5). This risk-neutral probability measure is also referred to as an equivalent martingale measure because under this measure all asset prices discounted at the risk-free rate are martingales (Duffie 1996, p. 29)[5]. Using this representation the price of asset i at date t is given by

$$P_t^i = \frac{1}{R_{t+1}^f} \sum_{s=1}^{S} \text{prob}'_s \, e_{t+1,s}^i \,, \tag{2.5}$$

where R_{t+1}^f is the gross return on a one-period risk-free asset. Comparing equation (2.5) with equation (2.4), it is easy to verify that the risk-neutral probability measure deflated by the risk-free rate equals the prices of state-contingent claims, $\text{prob}'_s / R_{t+1}^f = \pi_{t+1,s}$. Hence, the risk-neutral probability measure also measures state-probability and desirability of a state-dependent payoff simultaneously. The market pricing kernel is equivalent to the probability- and risk-free rate-deflated risk-neutral probability measure, $\phi_{t+1,s} = \text{prob}'_s / (\text{prob}_s R_{t+1}^f)$. The risk-neutral probability measure is useful for analyses that do not explicitly consider individual preferences. Another characteristic feature of the risk-neutral probability measure is that it allows to separate intertemporal and cross-sectional aspects. This may be

[4]See section 4.2 for details.

[5]A stochastic process is a martingale if its conditional expected future values are the current value, for any future and current time (Duffie 1988, p. 135).

advantageous for intertemporal problems with a stochastic risk-free rate[6].

The motivation of using the market pricing kernel approach is that the market pricing kernel does not simultaneously measure the probability of a future state and the desirability of a payoff in that state. In contrast, it characterizes the desirability of a payoff in a future state independent of the state probability. The desirability of a payoff in a future state may depend on realizations of exogenously given variables in this state, e.g., the payoff of other securities or the tastes and endowments of investors, so that the market pricing kernel may be expressed in terms of these exogenous variables (see discussion in section 2.2). Furthermore, the market pricing kernel approach allows a convenient representation of intertemporal asset pricing, especially in the case of a continuum of states, using expectations under objective probabilities (equations (2.1) and (2.2)).

2.1.3 Absence of Arbitrage Opportunities

The Law of One Price is less restrictive than the absence of arbitrage opportunities because it only requires that assets with the same payoffs must trade at the same price. The absence of arbitrage opportunities further requires that assets having nonnegative payoffs and at least one positive payoff with positive probability have positive prices. As can be seen from equation (2.3) this restriction requires that $\phi_{t+1,s} > 0 \; \forall s$ and, hence, $\tilde{\phi}_{t+1} > 0$. Besides the restriction that the market pricing kernel must be positive, the absence of arbitrage opportunities places no other restrictions on the properties of the market pricing kernel. However, while the Law of One Price is not equivalent to the absence of arbitrage opportunities, it is its immediate consequence. This becomes clear by considering Farkas' Lemma. It implies that there is no arbitrage if and only if nonnegative prices for state-contingent claims, $\pi_{t+1,s} \geq 0 \; \forall s$, exist, such that for any security i its price P_t^i equals the sum of its future, state-dependent payoffs, $e_{t+1,s}^i$, evaluated by $\pi_{t+1,s}$: $P_t^i = \sum_{s=1}^{S} \pi_{t+1,s} e_{t+1,s}^i \; \forall i$, where S denotes the state-space (Franke and Hax 1994, p. 362). Recall, that $\phi_{t+1,s} \cdot \text{prob}_s = \pi_{t+1,s}$. Hence, under the absence of arbitrage opportunities the existence of a positive market pricing kernel follows via the existence of nonnegative prices for state-contingent claims from Farkas' Lemma. Notice, that Farkas' Lemma also implies the other direction, i.e. that the absence of arbitrage opportunities follows from the existence of a positive market pricing kernel. Given a positive market pricing kernel and nonnegative state-probabilities, the prices for state-contingent claims are nonnegative and, hence, there is no arbitrage.

[6] This advantage of the risk-neutral probability measure can be transferred to the market pricing kernel approach by using the market pricing kernel in form of a forward measure, $\phi'_{t+1,s} = \phi_{t+1,s} / R_{t+1}^f$. For an application see, for example, Franke, Stapleton, and Subrahmanyam (1996).

2.1.4 Existence of a One-period Risk-free Asset

Assume that a one-period risk-free asset exists[7]. Then it follows from equation (2.2) and hence from the Law of One Price that the market pricing kernel must have a conditional expectation of one divided by the gross return on the one-period risk-free asset, R^f_{t+1},

$$\mathrm{E}_t\left\{\tilde{\phi}_{t+1}\right\} = \frac{1}{R^f_{t+1}}. \tag{2.6}$$

Using R^f_{t+1}, equation (2.2) implies for the excess return on any asset i over the one-period risk-free rate, $\tilde{p}^i_{t+1} \equiv \tilde{R}^i_{t+1} - R^f_{t+1}\ \forall i$, the equation

$$0 = \mathrm{E}_t\left\{\tilde{\phi}_{t+1} \cdot \tilde{p}^i_{t+1}\right\}. \tag{2.7}$$

Many empirical tests are based on equation (2.7). However, unless equation (2.6) is used in addition, equation (2.7) is weaker than equation (2.2), because (2.7) is equivalent to $\mathrm{E}_t\{\tilde{\phi}_{t+1} \cdot \tilde{R}^i_{t+1}\} = \Theta_{t+1}\ \forall i$, where Θ_{t+1} is an arbitrary constant, while equation (2.2) restricts Θ_{t+1} to 1.

2.1.5 Completeness of the Market

The market pricing kernel $\tilde{\phi}_{t+1}$ is not necessarily uniquely identified. For example, if two different states can occur at date $t+1$ ($s \in \{1,2\}$) and only one asset i with the state-dependent payoffs $e^i_{t+1,1} = 1$ and $e^i_{t+1,2} = 0$ is traded, then $\phi_{t+1,2}$ is not identified. The market pricing kernel is uniquely identified if and only if the market is complete (cf. Dothan (1990, p. 74)). A market is said to be complete if at least as many linearly independent securities are traded in the market as states of nature can occur. Securities are said to be linearly independent if the securities' payoffs are such that it is impossible to replicate the payoffs of one security by a portfolio of the other securities. Using these linearly independent securities primitive securities can be created for each state by forming portfolios of the existing securities. Primitive or Arrow-Debreu securities (as opposed to complex securities) have a payoff of one unit in one state of nature and zero otherwise. Hence, primitive or Arrow-Debreu securities represent state-contingent claims. Therefore, the simplest form of a complete market is a market where as many independent primitive securities are traded as states of nature can occur.

[7]In a complete market a one-period risk-free asset always exists.

2.2 Arbitrage Free Versus Equilibrium Asset Pricing

If the distribution of the market pricing kernel, or, equivalently, the market pricing kernel for each possible state and the state-probabilities are known, any asset can be priced. Therefore, the question emerges, under which circumstances the market pricing kernel is observable or how information about the market pricing kernel can be gained otherwise.

First, consider a complete market. In a complete market the prices of as many linearly independent securities are observable as states of nature can occur. Thus, the prices for state-contingent claims on all possible states are observable and unique, and any asset can be priced according to equation (2.4). The principle of arbitrage free asset pricing is based on this observability of the prices for state-contingent claims. However, for arbitrage free asset pricing the completeness of the market is not necessarily required. Only prices for claims on those states must be observable in which the security to be priced has payoffs. In other words, using arbitrage free asset pricing, only those assets can be priced that lie in the marketed space, the subset of the whole state-space that can be achieved by forming portfolios of securities with observable prices (Duffie 1988, p. 93). This property is also referred to as the "spanning property", i.e. an asset can be priced using arbitrage free asset pricing principles if it is "spanned" by (lies in the "span" of) securities with observable prices (e.g., Duffie (1996, p. 8)). Thus, arbitrage based models only yield the price of one asset relative to the prices of other assets[8].

Now, consider an incomplete market. Considering this situation seems reasonable, because "complete financial markets, while a useful abstraction, are not an everyday occurrence, however. Securities number at most a few thousand, while the relevant set of states is no doubt much larger" (Hakansson 1989, p. 139). A new security paying off in a state on which no claims have been traded before cannot be priced using arbitrage free asset pricing principles. Prices for such securities can only be derived, if the securities' payoffs in these states are related to exogenous data such as preferences and endowments of individuals. This is the equilibrium approach of asset pricing. Within this approach the market pricing kernel can, for example, be derived from explicit assumptions on the preferences of individuals and on the stochastic process that governs the exogenous endowment (see section 4.2).

[8]Notice, that as long as primitive securities are not traded the prices for state-contingent claims are not directly observable and hence must be calculated from observed prices of complex securities. For securities markets in reality this task may not be trivial (see discussion in section 3.2).

2.3 Market Pricing Kernel and Linear Asset Pricing

The Law of One Price is the only assumption necessary for the existence of a market pricing kernel. Without further assumptions a general linear valuation equation can be derived[9]. Equation (2.2) must hold for any asset i. Therefore, differences in expected returns result from the differences in the covariance with the market pricing kernel[10],

$$\mathrm{E}_t\{\tilde{R}^i_{t+1}\} = \left(\mathrm{E}_t\{\tilde{\phi}_{t+1}\}\right)^{-1} - \frac{\mathrm{cov}_t(\tilde{\phi}_{t+1}, \tilde{R}^i_{t+1})}{\mathrm{E}_t\{\tilde{\phi}_{t+1}\}}. \tag{2.8}$$

The first term, $\mathrm{E}_t\{\tilde{\phi}_{t+1}\}^{-1}$, is equal for all assets. The second term measures the asset's risk premium. Since this pricing formula is valid for any asset and any portfolio of assets, it must also be valid for an asset z that satisfies $\mathrm{cov}_t(\tilde{\phi}_{t+1}, \tilde{R}^z_{t+1}) = 0$. From equation (2.8) it follows that the conditionally expected return on this "reference asset" equals the inverse of the conditional expectation of the market pricing kernel, $\mathrm{E}_t\{\tilde{R}^z_{t+1}\} = \mathrm{E}_t\{\tilde{\phi}_{t+1}\}^{-1}$. A risk-free asset, for example, is such an asset z, because zero variance implies zero covariance with each random variable. If a one-period risk-free asset exists the return on the asset z can be replaced by the risk-free rate of return. However, the asset z is not required to be risk free, because any asset that has zero correlation with the market pricing kernel satisfies $\mathrm{cov}_t(\tilde{\phi}_{t+1}, \tilde{R}^z_{t+1}) = 0$. Using the return on asset z, equation (2.8) yields the weaker[11] restriction for the expected excess return on any asset i over the reference asset z,

$$\mathrm{E}_t\{\tilde{R}^i_{t+1} - \tilde{R}^z_{t+1}\} = -\frac{\mathrm{cov}_t(\tilde{\phi}_{t+1}, \tilde{R}^i_{t+1})}{\mathrm{E}_t\{\tilde{\phi}_{t+1}\}}. \tag{2.9}$$

Define an asset's conditional beta-coefficient to be the conditional covariance of the asset's returns with the market pricing kernel normalized by the conditional variance of the market pricing kernel, $\beta^{\tilde{\phi},\tilde{R}^i}_{t+1} = \mathrm{cov}_t(\tilde{\phi}_{t+1}, \tilde{R}^i_{t+1})/\mathrm{var}_t(\tilde{\phi}_{t+1})$. Because the beta-coefficient is zero for the reference asset z, this asset is often referred to as a "zero-beta asset" (Black 1972)[12]. The asset pricing formula (2.9) can be rewritten to establish a linear

[9]See Epstein (1992, pp. 26-7) or Duffie (1996, pp. 11-2).

[10]For two stochastic variables, $\tilde{x}$ and $\tilde{y}$, $\mathrm{E}\{\tilde{x} \cdot \tilde{y}\} = \mathrm{E}\{\tilde{x}\} \cdot \mathrm{E}\{\tilde{y}\} + \mathrm{cov}(\tilde{x}, \tilde{y})$.

[11]See the discussion in section 2.1.4.

[12]Sometimes the zero-beta asset is defined to be the asset that has the minimum variance of all assets uncorrelated with $\tilde{\phi}_{t+1}$. Minimum variance is not required here, however.

relation between the expected excess return on any asset i over the return on a zero-beta asset and its beta-coefficient:

$$\mathrm{E}_t\{\tilde{R}^i_{t+1} - \tilde{R}^z_{t+1}\} = -\beta^{\tilde{\phi},\tilde{R}^i}_{t+1} \cdot \frac{\mathrm{var}_t(\tilde{\phi}_{t+1})}{\mathrm{E}_t\{\tilde{\phi}_{t+1}\}}. \tag{2.10}$$

Equation (2.10) must hold for any portfolio m with gross return $\tilde{R}^m_{t+1}$. Solving for $\mathrm{var}_t(\tilde{\phi}_{t+1})/\mathrm{E}_t\{\tilde{\phi}_{t+1}\}$ and substituting into equation (2.10) yields

$$\mathrm{E}_t\{\tilde{R}^i_{t+1} - \tilde{R}^z_{t+1}\} = \frac{\beta^{\tilde{\phi},\tilde{R}^i}_{t+1}}{\beta^{\tilde{\phi},\tilde{R}^m}_{t+1}} \mathrm{E}_t\{\tilde{R}^m_{t+1} - \tilde{R}^z_{t+1}\}. \tag{2.11}$$

Portfolio m is a second reference asset. If there exists a portfolio with gross returns $\tilde{R}^\phi_{t+1}$ that are perfectly (positively or negatively) correlated with the market pricing kernel and if this portfolio is used as the second reference asset, then $\beta^{\tilde{\phi},\tilde{R}^i}_{t+1} / \beta^{\tilde{\phi},\tilde{R}^\phi}_{t+1}$ equals[13] $\beta^{\tilde{R}^\phi,\tilde{R}^i}_{t+1}$ and equation (2.11) can be rewritten as

$$\mathrm{E}_t\{\tilde{R}^i_{t+1} - \tilde{R}^z_{t+1}\} = \beta^{\tilde{R}^\phi,\tilde{R}^i}_{t+1} \cdot \mathrm{E}_t\{\tilde{R}^\phi_{t+1} - \tilde{R}^z_{t+1}\}, \tag{2.12}$$

or, equivalently, if a one-period risk-free asset exists and this asset is used as the reference asset z,

$$\mathrm{E}_t\{\tilde{R}^i_{t+1}\} - R^f_{t+1} = \beta^{\tilde{R}^\phi,\tilde{R}^i}_{t+1} \cdot \left(\mathrm{E}_t\{\tilde{R}^\phi_{t+1}\} - R^f_{t+1}\right). \tag{2.13}$$

This linear intertemporal valuation formula only assumes the Law of One Price. Risky assets whose returns are positively correlated with $\tilde{R}^\phi_{t+1}$ have positive risk premia over the return on a zero-beta asset, e.g., a risk-free asset. The return on a zero-beta asset is characterized by a zero covariance with the market pricing kernel, and $\tilde{R}^\phi_{t+1}$ is characterized by a perfect correlation with the market pricing kernel. Specifying the market pricing kernel through specific assumptions on its stochastic behavior, e.g., assuming a specific parametric model, leads to specific intertemporal asset pricing models. Parametric models of the market pricing kernel may be either arbitrarily chosen or derived from equilibrium considerations. This is done in chapter 4.

The intertemporal valuation formulas (2.12) and (2.13) are very similar to many one-period asset pricing models that ignore intertemporal aspects. In a one-period context the time subscripts can be dropped which yields linear formulas known as the "beta models of rates of return" (Duffie 1988, p. 94). Then, assuming a simple parametric model for the market pricing

[13]This can be shown using the fact that $\mathrm{corr}_t(\tilde{R}^\phi_{t+1}, \tilde{R}^i_{t+1}) = \mathrm{corr}_t(\tilde{\phi}_{t+1}, \tilde{R}^i_{t+1})$ if $\mathrm{corr}_t(\tilde{R}^\phi_{t+1}, \tilde{\phi}_{t+1}) = 1$ and $\mathrm{corr}_t(\tilde{R}^\phi_{t+1}, \tilde{R}^i_{t+1}) = -\mathrm{corr}_t(\tilde{\phi}_{t+1}, \tilde{R}^i_{t+1})$ if $\mathrm{corr}_t(\tilde{R}^\phi_{t+1}, \tilde{\phi}_{t+1}) = -1$.

kernel that relates the market pricing kernel to the return on the market portfolio so that both are perfectly correlated[14] yields the one-period Capital Asset Pricing Model (CAPM) of Sharpe (1964), Lintner (1965) and Mossin (1966), respectively the one-period Zero-Beta CAPM of Black (1972), that does not require the existence of a risk-free rate. These one-period models are often referred to as static models because by applying these models to a multiperiod context one must assume that all conditional moments are constant or at least perfectly predictable[15]. If this assumption is not satisfied these models are incorrect because potential changes in the future investment opportunity set are not taken into account[16].

2.4 Summary of Chapter 2

The market pricing kernel approach for intertemporal asset pricing is a broad concept that places very few restrictions a priori on the structure of asset pricing models. Different asset pricing models can be related by this approach because each model implies a specific functional form of the market pricing kernel. Therefore, the market pricing kernel approach provides an advantageous starting point for further analysis. In contrast, the literature is mainly concerned with analyzing specific asset pricing models. The market pricing kernel, however, can in general be interpreted as a stochastic variable that has certain properties under certain conditions, such as the positivity under the absence of arbitrage opportunities or the uniqueness if the market is complete. For its existence the Law of One Price, which is weaker than the absence of arbitrage opportunities, must be assumed. Without further assumptions a linear asset pricing relation can be derived that is very similar to the CAPM but more general. The application to asset pricing, however, necessitates further information about the distribution of the market pricing kernel. This information may either be revealed directly without parametrizing the market pricing kernel (chapter 3) or by analyzing specific parametric models of the market pricing kernel (chapter 4).

---

[14]In a one-period context such a parametric model of the market pricing kernel can be theoretically justified by assuming quadratic utility of the representative agent or general two fund separation (cf. Huang and Litzenberger (1988, chapter 4) or Franke and Hax (1994, pp. 378-81)).

[15]For a thorough discussion of this see section 4.1.

[16]For generalizations of the one-period CAPM that include additional factors capturing the variability of the investment opportunity set see Merton (1973) or Long (1974).

Chapter 3

Implications of Asset Prices for the Market Pricing Kernel

A vital interest in financial research is to derive information about the distribution of the market pricing kernel, about the state price density or, equivalently, about the risk-neutral probability distribution and the risk-free rate. As shown in the previous chapter this knowledge would enable the pricing of any asset. The traditional approach to gain insights into the properties of the market pricing kernel is to start by choosing a functional form for the pricing kernel and then test equation (2.1) empirically. Recently, more suggestive approaches attempt to reveal properties of the market pricing kernel without assuming an explicit functional form. These approaches first examine asset returns or asset prices and then analyze which properties the market pricing kernel must exhibit to be consistent with these data. One can distinguish theoretical and empirical approaches. The theoretical approaches make assumptions about the distribution or the stochastic process of asset returns or prices, whereas the empirical approaches use observed data. The first part of this chapter briefly surveys one of the theoretical approaches. Then, two empirical approaches, one employing cross-sectional data and one employing time series data, are compared. The approach that employs time series data to place restrictions on the moments of the unconditional distribution of the market pricing kernel is applied in chapter 7 and is therefore described in greater detail.

3.1 Implications of Specific Asset Price Processes

A purely theoretical approach is to make assumptions about either the stochastic process which the asset prices (respectively returns) or the market pricing kernel follow and to derive the properties which are then imposed on the other process. Franke, Stapleton, and Subrahmanyam (1996) show that the market pricing kernel must exhibit constant elasticity with respect to the price of the aggregate wealth portfolio if the asset prices follow a geometric Brownian motion. They further show that a market pricing kernel with declining elasticity is consistent with mean reverting asset prices and implies a higher volatility of asset prices than a market pricing kernel with constant elasticity.

In an equilibrium representative-agent framework the market pricing kernel is related to the preferences of the representative agent[1]. Constant elasticity of the market pricing kernel with respect to the price of the aggregate wealth portfolio is closely related to constant relative risk aversion of the representative agent. Thus, Franke, Stapleton, and Subrahmanyam (1996) conclude that models which assume a geometric Brownian motion for asset prices, e.g., the so-called preference-free Black-Scholes model, imply preferences with constant relative risk aversion in models with a representative agent. The result that the Black-Scholes model requires constant relative risk aversion of the representative agent in a general equilibrium model was also obtained by Bick (1987). In general, assuming a certain stochastic process for the asset prices or returns implies a certain class of preferences in a general equilibrium framework and, therefore, implies certain properties that the market pricing kernel must exhibit, given that the assumed process is generally consistent with an equilibrium (cf. Bick (1990, p. 674)). Hence, models which seemingly make assumptions only about the price or return generating process implicitly make assumptions also about the market pricing kernel and about preferences. Several other studies analyze the relation between certain price or return generating processes and the utility specification in a representative-agent framework, e.g., Franke (1984) and Stapleton and Subrahmanyam (1990) among others.

3.2 Estimating Conditional Distributions from Asset Prices

Empirical approaches use observed asset prices to reveal information about the market pricing kernel. The approach discussed in this section is to estimate or approximate the distribution of the pricing kernel, the state price density or, equivalently, the risk-neutral probability distribution and an implicit risk-free rate from the prices of traded financial securities, as firstly suggested by Banz and Miller (1978) and Breeden and Litzenberger (1978).

[1]In section 4.2 this relation will be deduced and functional forms for the market pricing kernel will be derived from specific assumptions about the preferences of the representative agent.

Within this approach the states of nature are defined by the price of an underlying asset at a given future date τ. The date τ state price density is then estimated from simultaneously observed prices of European options on the underlying asset with different exercise prices and the same exercise date τ. The obtained distribution is thus conditional on the specific date τ and based on the states defined by the price of the underlying asset. It can be used for pricing complex securities contingent on the underlying asset. Since the states are defined by the price of the underlying asset, other assets can be priced once their prices at date τ relative to the underlying asset's price at date τ have been specified.

One method within this empirical approach is to specify the distribution of the market pricing kernel in a parametric form and to invert option pricing models such as the Black-Scholes model to determine moments of this distribution (see Longstaff (1995) among others). The estimated distribution of the market pricing kernel can in turn be used to calculate the price of the underlying asset implicit in the option prices. Ignoring market frictions the implied value and the observed value should be equal in the absence of arbitrage. This restriction is known as the "martingale restriction" (cf. Longstaff (1995)). Neumann and Schlag (1996) test this martingale restriction for stock index options on the Deutscher Aktienindex (DAX) traded on the Deutsche Terminbörse (DTB). Similar to Longstaff's results, their results indicate that the price of the underlying asset implicit in option prices is higher than the underlying asset's observed price, i.e. the martingale restriction is violated. Further analyses show that the number of options used in the estimation process as well as the time to maturity and the average moneyness are significant determinants for the amount of the pricing error. A disadvantage of this method is that distributional assumptions must be made and that it presumes that the model used to infer the moments of the implied distribution is true and used by the market participants for valuing options.

Another method proposed by Rubinstein (1994) avoids this shortcoming. Recall, the former method was to assume a specific distribution of the market pricing kernel or the risk-neutral probabilities and to determine moments of this distribution. Then, comparing observed prices of the underlying asset and the derivative securities with the prices implied by the distribution reveals that not all assets are correctly priced. This is different in the method of Rubinstein (1994). Instead of assuming a specific distribution of the risk-neutral probabilities he proposes to estimate the distribution by minimizing its distance to an arbitrarily chosen prior distribution[2] under the constraint that it correctly prices the observed derivative securities and the underlying asset. It is not required that the distribution of the risk-neutral probabilities has the form of the prior distribution but that the estimated distribution

---

[2]Rubinstein (1994) uses the lognormal distribution as the prior distribution.

correctly prices the observed assets. No assumption about the functional relation between the option prices and the price of the underlying asset is made. Neumann and Schlag (1996) apply this method to stock index options on the Deutscher Aktienindex (DAX). Jackwerth and Rubinstein (1996) investigate the impact of different distance measures and propose to use a measure that maximizes the smoothness of the resulting probability distribution. Rubinstein (1994), Jackwerth and Rubinstein (1996) and Neumann and Schlag (1996) observe that the implied risk-neutral probability distribution differs systematically from the lognormal distribution assumed in the Black-Scholes model. Rubinstein (1994) observes that in comparison with the lognormal distribution call prices assign higher probabilities to prices close to the forward price of the underlying asset and to prices at the lower tail. Because the higher probability of low values was not observable before the stock market crash in 1987, Rubinstein attributes this observation to a phenomenon he calls "crash-o-phobia". Neumann and Schlag (1996) find that the shape of the distribution implied by call prices differs from that implied by put prices. They confirm the observation of a higher probability of values close to the forward price of the underlying asset implicit in call prices but do not find the high lower tail probability. The distribution implied by put prices, on the other hand, exhibits a high probability at both tails and less probability close to the forward price of the underlying asset. Jackwerth and Rubinstein (1996) additionally analyze the time pattern of the implied distribution. They find that the shape of the distribution changes dramatically over time, e.g., after the 1987 stock market crash. It is important to notice that these results refer to the risk-neutral probability distribution. The state price density and the distribution of the market pricing kernel may be additionally affected by changes in the risk-free rate.

The methods within the empirical approach surveyed in this section usually employ cross-sectional data, i.e. simultaneously observed prices of different assets, to reveal information about the distribution of the market pricing kernel, about the state price density or about the risk-neutral probability distribution and an implied risk-free rate for one specific date[3] and for states defined by the price of an underlying asset at that specific date. The empirical evidence indicates that the implied conditional risk-neutral probability distribution is not constant over time. The shape of this distribution may change subsequent to events such as a stock market crash because such events may affect the preferences of investors. Besides of preference changes, nonstationary conditional distributions of the risk-neutral probability may reflect potential changes in the future investment opportunity set. Unfortunately, analyses of changes in the implied conditional distribution are restricted to relatively short time periods due to the availability of option price data. In addition to the nonstationarities in the conditional

[3] Subsequently, the distribution implied by the asset prices is therefore referred to as the "conditional distribution".

risk-neutral probability distribution the conditional state price density and the conditional distribution of the market pricing kernel may be affected by changes in the risk-free rate. Therefore, the approach introduced in the next section employs time series data to reveal properties of the unconditional distribution of the market pricing kernel.

3.3 Placing Restrictions on the Unconditional Distribution Using Time Series of Asset Returns

3.3.1 Motivation

The results of the previous section have illustrated that the conditional distribution of the market pricing kernel may not be stationary. Besides the observed nonstationarities in the conditional distribution of the risk-neutral probabilities, further evidence arises from the widely observed autocorrelation in asset returns as well as from the observed fat tails and excess kurtosis in the distribution of asset returns. Furthermore, the ex-ante real risk-free rate is not constant (cf. Garcia and Perron (1996)). The empirically observed fluctuations in the one-period risk-free rate reflect nonstationarities in the conditional expectation of the market pricing kernel. As noted above, a conditional distribution of the market pricing kernel that varies over time reflects potential changes of the future investment opportunity set and/or changes in preferences. This shows that the unconditional moments of time series of asset prices or asset returns are unsuitable for revealing information about the conditional distribution of the market pricing kernel. However, they can be used to reveal information about the unconditional distribution of the market pricing kernel.

The empirical approach described in this section derives restrictions for the moments of the unconditional market pricing kernel using time series data. Before this approach is introduced one has to clarify under which conditions the unconditional distribution of the market pricing kernel exists and what it looks like under these conditions. Despite the observed nonstationarities in the conditional distribution, a constant unconditional distribution may exist. Consider, for example, a stochastic variable that follows a stationary and ergodic Markov state-switching process[4]. Although this process is stationary and exhibits a constant unconditional distribution of the stochastic

[4]This process is chosen as example because a Markov state-switching economy will be assumed later in this study (see section 8.1). Furthermore, this example is in line with the empirical findings of Garcia and Perron (1996). They estimate a Markov switching model for quarterly observations of the real interest rate in the US between 1961 and 1986 and find that a switching model with three states explains the data significantly

variable, the conditional distribution of this stochastic variable depends on the current state and, thus, may vary over time. The ergodicity of the process ensures that time series estimates of the moments converge to the moments of the unconditional distribution generated by the stochastic process as the sample size becomes large. Therefore, the only distributional assumption made subsequently is that the market pricing kernel and the asset returns are generated by stationary stochastic processes that are ergodic in the moments estimated from time series data. Under this assumption the asset returns as well as the market pricing kernel may exhibit varying conditional distributions and stochastic dependencies over time although a constant unconditional distribution exists. Time series moments may therefore be biased estimates of the moments of the conditional distributions, but converge to the moments of the unconditional distribution.

Now recall that in chapter 2 all equations are conditional on one point in time. However, this poses no problem, because under the assumption made above the equations (2.1), (2.2) and (2.7) also hold unconditionally (cf. Hansen and Jagannathan (1991, footnote 2)), e.g., $\mathrm{E}\{\tilde{\phi}_{t+1} \cdot \tilde{R}^i_{t+1}\} = 1$ for equation (2.2). This can be seen by invoking the law of iterated expectations and taking the unconditional expectation (cf. Burnside (1994, p. 58))[5]. Notice, that a market pricing kernel satisfying the unconditional equations may not satisfy each conditional equation, whereas a market pricing kernel which satisfies the equations conditionally on each possible state also satisfies the unconditional equations.

Based on the unconditional equations Hansen and Jagannathan (1991) derive and estimate a lower bound for the unconditional second moment of the market pricing kernel as a function of the unconditional first moment of the market pricing kernel and of the observed unconditional first and second moments of asset returns[6]. This approach is known as the variance bound

better than models with only one state. Within the states the real interest rate follows a random walk. The switches between the states can be attributed to structural events such as the sudden rise in the oil price in 1973 or the federal budget deficit in 1981. Gray (1996) develops a generalized regime-switching model of the nominal short-term interest rate which outperforms simple single-regime models in an out-of-sample forecasting experiment.

[5] To illustrate this result, consider again the Markov state-switching process with S states. $\mathrm{E}_s\{\tilde{\phi}_{t+1} \cdot \tilde{R}^i_{t+1}\} = 1$ is satisfied for each $s \in S$. Then the unconditional expectation $\mathrm{E}\{\tilde{\phi}_{t+1} \cdot \tilde{R}^i_{t+1}\}$ equals one because $\mathrm{E}\{\tilde{\phi}_{t+1} \cdot \tilde{R}^i_{t+1}\} = \sum_{s \in S} \mathrm{prob}_s \cdot \mathrm{E}_s\{\tilde{\phi}_{t+1} \cdot \tilde{R}^i_{t+1}\} = \sum_{s \in S} \mathrm{prob}_s \cdot 1 = 1$. prob_s is the unconditional probability of state s.

[6] Shiller (1982) was the first to discuss the variability of discount rates implied by asset prices. In contrast, the "excess-volatility" literature (cf. Shiller (1981) or Grossman and Shiller (1981)) derives upper variance bounds for asset prices given the observed variability of the cash flow (e.g., the dividends) and a *constant* discount factor. This literature starts with a simple rational valuation model where the asset's price equals the expected future cash flow divided by a constant discount rate and asks whether prices vary too much to be explained only by changes in expected cash flows. From this

approach[7]. The bound restricts the unconditional variance of the market pricing kernel, $\text{var}(\tilde{\phi})$. This variance may be decomposed into the unconditional variance of the conditional expectation of the market pricing kernel and its average conditional variance

$$\text{var}(\tilde{\phi}) = \text{var}(\text{E}_t\{\tilde{\phi}_{t+1}\}) + \text{E}\{\text{var}_t(\tilde{\phi}_{t+1})\}, \tag{3.1}$$

where $\text{var}_t(\tilde{\phi}_{t+1})$ is the variance of the market pricing kernel at date $t+1$ conditional on the information set available at date t[8]. Since $\text{E}_t\{\tilde{\phi}_{t+1}\}$ equals the reciprocal of the gross return on the risk-free rate from date t to date $t+1$ (see section 2.1.4) the first term of equation (3.1) captures the variation in the one-period risk-free rate. If the one-period risk-free rate is constant, this term equals zero. Given the empirical evidence that the real one-period risk-free rate varies over time (e.g., Garcia and Perron (1996)) it is assumed that an unconditional real risk-free rate cannot be observed[9]. Hence, the unconditional expectation of the market pricing kernel cannot be determined from the assets' returns. However, since the volatility of the one-period risk-free rate is small, most of the unconditional variance of the market pricing kernel should come from conditional variance (cf. Cochrane and Hansen (1992, p. 136)).

The variance bound allows us to investigate different aspects: First, the bound reflects the information about the unconditional first and second moments of the market pricing kernel contained in the unconditional means, variances and covariances of the asset returns under consideration and can therefore be used as a tool to analyze the information content of different sets of assets for intertemporal asset pricing[10]. Second, the bound implies restrictions on the parameter space of specific parametric models of the mar-

---

perspective the Hansen-Jagannathan bound can be interpreted as a lower bound for the variance of the discount factor given the variability of the asset prices and cash flows.

[7]Snow (1991) generalizes the variance bound approach of Hansen and Jagannathan (1991) to higher than second moments.

[8]See Gallant, Hansen, and Tauchen (1990) for a study applying conditioning information to improve on the variance bounds computed by Hansen and Jagannathan (1991).

[9]Notice, that in reality even a conditional real one-period risk-free rate may not be observable as a result of inflation risk.

[10]Snow (1991) applies this approach to investigate the firm-size-effect first documented by Banz (1981). He examines the information content of returns on portfolios sorted by firm size and shows that returns of small firms can provide restrictions on the market pricing kernel beyond those implied by a market proxy or the returns of large firms. Bekaert and Hodrick (1992) use US-dollar returns on domestic and international investments to check whether this information makes the bounds more restrictive. He and Modest (1995) explore the effects of market frictions on variance bounds of the market pricing kernel. Meyer (1997) applies the variance bound approach to analyze the information content of different portfolios of German stocks, bonds and interbank deposits. Using the estimated variance bounds Meyer (1997) further investigates the effects of different methods of return calculation, of using tax adjusted returns and of different lengths of the time intervals over which the returns are measured (see also section 7.1).

ket pricing kernel[11] and, hence, can be used to evaluate the plausibility of different parametric models of the market pricing kernel without developing special estimation or calibration tools[12].

In chapter 7 the variance bound approach is used twice, firstly, for examining the information about the market pricing kernel contained in different asset return data sets and, secondly, for comparing parametric models of the market pricing kernel derived from alternative assumptions about preferences. For this reason, the construction of variance bounds is now described in greater detail.

3.3.2 Variance Bounds of the Market Pricing Kernel

Before in section 3.3.2.2 a variance bound of the market pricing kernel for the general case of more than one risky asset is derived, section 3.3.2.1 illustrates the construction of a variance bound for the case of one risky asset alternatively using returns and excess returns on the asset.

3.3.2.1 Variance Bound Imposed by One Risky Asset

Assume that there exists only one risky asset, i.e. $I = 1$. Dropping the subscripts and using the time series estimates of the unconditional first and second moments of the return on the asset, $\mathrm{E}\{\tilde{R}\}$ and $\mathrm{std}(\tilde{R})$, it follows from equation (2.8) that the product of the standard deviation of the market pricing kernel, $\mathrm{std}(\tilde{\phi})$, and the correlation between the market pricing kernel and the return on the asset, $\mathrm{corr}(\tilde{\phi}, \tilde{R})$, is a constant given the expected value of the market pricing kernel $\mathrm{E}\{\tilde{\phi}\}$,

$$\mathrm{std}(\tilde{\phi}) \cdot \mathrm{corr}(\tilde{\phi}, \tilde{R}) = \frac{1 - \mathrm{E}\{\tilde{\phi}\} \cdot \mathrm{E}\{\tilde{R}\}}{\mathrm{std}(\tilde{R})} = \text{const.} \tag{3.2}$$

[11] In the consumption-based asset pricing framework with time-additive expected utility the restrictions on the parameter space determine, for example, the minimum level of risk aversion and time preference needed for the model to be consistent with the asset prices (see section 7.2.1).

[12] Hansen and Jagannathan (1991), Gallant, Hansen, and Tauchen (1990), Cochrane and Hansen (1992) and Missong (1994) use variance bounds to analyze parametric models of the market pricing kernel derived from intertemporally separable preferences and, alternatively, models derived from preferences exhibiting intertemporal dependencies such as habit formation or local durability (see section 4.2.1.2 for a short description of these dependencies). Parametric models derived from recursive non-expected utility are evaluated by Gordon, Samson, and Carmichael (1995) and Hung (1995). Ferson and Harvey (1992) examine parametric models of the market pricing kernel derived from time-separable preferences using seasonally unadjusted consumption data. A recent study of Bakshi and Naka (1997) uses variance bounds to evaluate various parametric models of the market pricing kernel for Japan.

Figure 3.1: The left graph shows the lower bound imposed on the standard deviation of the market pricing kernel by one risky asset. The right graph shows the bound imposed by three risky assets. To be consistent with the returns on the three assets the market pricing kernel must lie in the shaded area. To be consistent with the three assets and each portfolio of these assets it must lie in the area above the parabola.

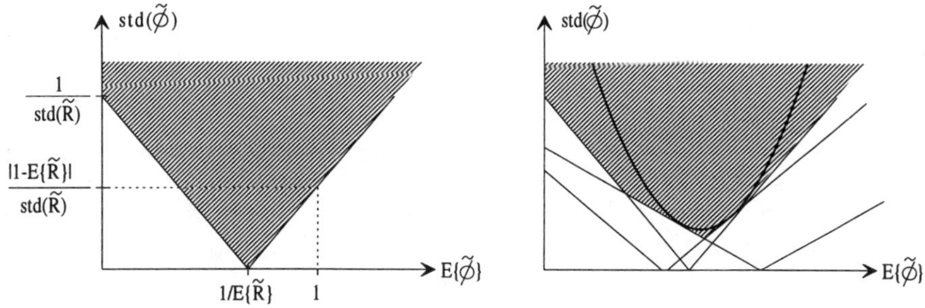

For negative [positive] $(1 - \mathrm{E}\{\tilde{\phi}\} \cdot \mathrm{E}\{\tilde{R}\})$, the market pricing kernel exhibits the lowest standard deviation if the correlation between the market pricing kernel and the return on the asset equals -1 [$+1$]. Otherwise, the standard deviation of the market pricing kernel must be larger. Hence, in the case of one asset a lower bound on the standard deviation of the market pricing kernel is given by

$$\mathrm{std}(\tilde{\phi}) \geq \frac{|1 - \mathrm{E}\{\tilde{\phi}\} \cdot \mathrm{E}\{\tilde{R}\}|}{\mathrm{std}(\tilde{R})}. \tag{3.3}$$

This bound is shown in the left graph of figure 3.1. The unconditional first and second moment of a market pricing kernel derived from a parametric model must lie in the shaded area to satisfy the restriction imposed by the variance bound. However, the variance bound is derived under the assumption of perfect correlation between the market pricing kernel and the return on the asset and, thus, does not incorporate information about the true correlation between the market pricing kernel and the return on the asset. To be consistent with the asset's returns the correlation implied by the parametric model of the market pricing kernel must satisfy equation (3.2). A low correlation requires a high standard deviation of the market pricing kernel. This is the reason why no upper bound for the variance of the market pricing kernel can be derived.

$1/\mathrm{E}\{\tilde{\phi}\}$ equals the implied unobservable, unconditional risk-free rate. Since the implied risk-free rate should be smaller than the expected return on the risky asset, the right branch of the bound shown in the left graph of figure 3.1 is more interesting than the left branch. On the right branch the correlation between the market pricing kernel and the return on the risky asset is assumed to equal -1. The slope of a straight line through the origin and any point on the branch equals the risky asset's Sharpe-ratio[13] calculated with the risk-free rate implied by the expected value of the market pricing kernel that corresponds to that point on the branch. This can be seen by rewriting equation (3.3) as $\mathrm{std}(\tilde{\phi}) \geq \mathrm{E}\{\tilde{\phi}\}\frac{|1/\mathrm{E}\{\tilde{\phi}\}-\mathrm{E}\{\tilde{R}\}|}{\mathrm{std}(\tilde{R})}$. Thus, each point of the bound equals the Sharpe-ratio of the risky asset, calculated using the implied risk-free rate, multiplied by the expected value of the market pricing kernel. Therefore, the branches become steeper, i.e. the bound becomes more restrictive, for high values of $\mathrm{E}\{\tilde{R}\}$ and low values of $\mathrm{std}(\tilde{R})$.

Alternatively, a variance bound can be derived from the excess returns on the risky asset over the observed conditional one-period risk-free rate. The conditional one-period risk-free rate varies over time and is therefore also a risky asset so that actually two risky assets exist in this case. However, since only the difference between the returns on both assets is considered this case is more closely related to the case of one risky asset than to the case of more than one risky asset. For the difference between the return on the risky asset, $\tilde{R}^i$, and the return on the one-period risk-free asset, $\tilde{R}^f$, an equation similar to equation (3.2) can be derived from equation (2.9),

$$\mathrm{std}(\tilde{\phi}) \cdot \mathrm{corr}(\tilde{\phi}, (\tilde{R}^i - \tilde{R}^f)) = \frac{-\mathrm{E}\{\tilde{\phi}\} \cdot \mathrm{E}\{\tilde{R}^i - \tilde{R}^f\}}{\mathrm{std}(\tilde{R}^i - \tilde{R}^f)} = \text{const.} \tag{3.4}$$

Hence, the lower bound for the standard deviation of the market pricing kernel is given by

$$\mathrm{std}(\tilde{\phi}) \geq \mathrm{E}\{\tilde{\phi}\}\frac{|\mathrm{E}\{\tilde{R}^i - \tilde{R}^f\}|}{\mathrm{std}(\tilde{R}^i - \tilde{R}^f)} \approx \mathrm{E}\{\tilde{\phi}\}\frac{|\mathrm{E}\{\tilde{R}^i - \tilde{R}^f\}|}{\mathrm{std}(\tilde{R}^i)}. \tag{3.5}$$

This equation shows that the bound implied by the excess returns on one risky asset i over the one-period risk-free rate equals a straight line through the origin with a slope coefficient equal to the expected excess return divided by the standard deviation of the excess return. Although the one-period risk-free rate is not assumed to be constant its standard deviation is typically small as compared to that of the risky asset. Therefore, the standard deviation of the excess return, $\mathrm{std}(\tilde{R}^i - \tilde{R}^f)$, is approximately equal to the standard deviation of the risky asset, $\mathrm{std}(\tilde{R}^i)$. Hence, the slope coefficient of the variance bound calculated from the excess returns on a risky asset over the one-period risk-free rate is approximately equal to the asset's Sharpe-ratio.

The derived variance bounds do not incorporate the positivity restriction of the market pricing kernel. Hence, the no-arbitrage condition may not be

[13] For any asset or portfolio the Sharpe-ratio is defined as the mean excess return divided by the standard deviation of return.

satisfied by a market pricing kernel in the shaded area in the left graph of figure 3.1. Using option pricing principles, Hansen and Jagannathan (1991, pp. 243-6) show how to derive a variance bound incorporating the no-arbitrage condition. Unlike the bounds derived above, this bound cannot be expressed simply as a function of the first moment of the market pricing kernel and the first and second moment of the return on the asset. Instead, the calculation of the bound may require a nonquadratic optimization problem to be solved for each value of $\mathrm{E}\{\tilde{\phi}\}$. Not surprisingly, Hansen and Jagannathan reveal that this bound is more restrictive than the bound without the no-arbitrage condition. However, both "boundaries ... coincide for points closest to the horizontal axis" (Hansen and Jagannathan (1991, p. 245)). Moreover, the findings of Cecchetti, Lam, and Mark (1994, p. 149) suggest that while the incorporation of the positivity restriction sharpens the variance bound the uncertainty associated with the location of the bound increases so that this bound is "less conclusive" than a bound that ignores the positivity restriction. Therefore, the positivity restriction is not incorporated in this study.

3.3.2.2 Variance Bound Imposed by More than One Risky Asset

Now, assume that there exist more than one risky assets, i.e. $I > 1$. In this case the market pricing kernel must not only satisfy the restrictions imposed by each individual asset, illustrated by the shaded area in the right graph of figure 3.1 for the case of three assets; moreover, it must satisfy the restriction implied by each portfolio of the I assets. This is a stronger condition, since it also takes into account the correlation between the assets. If the assets are not perfectly correlated, for each expected value of the market pricing kernel, $\mathrm{E}\{\tilde{\phi}\}$, a portfolio with a higher Sharpe-ratio can be constructed which then makes the variance bound more restrictive. For a given value of $\mathrm{E}\{\tilde{\phi}\}$ the strongest restriction is placed on the variance of the market pricing kernel by the portfolio which exhibits the maximum Sharpe-ratio. Therefore, the variance bound is constructed as follows: For each value of $\mathrm{E}\{\tilde{\phi}\}$, which implies a specific risk-free rate, the portfolio with the highest Sharpe-ratio is calculated. As will be shown in this section, this can be done by using a regression approach. The bound calculated in this way turns out to be a parabola, as depicted in the right graph of figure 3.1. A portfolio of risky assets that maximizes the Sharpe-ratio for a given risk-free rate lies on the efficient frontier for asset returns. This illustrates the duality between the variance bound of the market pricing kernel and the efficient frontier for asset returns as pointed out by Hansen and Jagannathan (1991, pp. 235-9). This duality can be used to show that an upward shift in the variance bound on the market pricing kernel implies a leftward shift in the efficient frontier for asset returns, and vice versa. Comparing the left and the right graph of this figure reveals that the variance bound imposed by a larger number of assets is more restrictive than the bound imposed by one asset[14]. The formal representation of the variance bound in the case of more than one risky asset is derived in the remaining part of this section[15].

$\tilde{\mathbf{R}}_t$ is the $[I \times 1]$ vector of the stochastic gross returns on I risky assets at date t. Then, equation (2.2) can be rewritten in unconditional form as

$$\mathbf{1} = \mathrm{E}\{\tilde{\phi}_t \cdot \tilde{\mathbf{R}}_t\}, \tag{3.6}$$

where $\mathbf{1}$ is a $[I \times 1]$ vector of ones. With $\mathrm{E}\{\tilde{\mathbf{R}}\}$, the $[I \times 1]$ vector of the unconditional expected gross returns on the I assets, and $\mathrm{E}\{\tilde{\phi}\}$, the unconditional expectation of the market pricing kernel, equation (3.6) can be rewritten as

$$\mathbf{1} = \mathrm{E}\{\tilde{\phi}\} \cdot \mathrm{E}\{\tilde{\mathbf{R}}\} + \mathrm{E}\left\{\left(\tilde{\phi}_t - \mathrm{E}\{\tilde{\phi}\}\right)\left(\tilde{\mathbf{R}}_t - \mathrm{E}\{\tilde{\mathbf{R}}\}\right)\right\}. \tag{3.7}$$

For the construction of the variance bound in the case of one risky asset (see section 3.3.2.1) the correlation between the market pricing kernel and the return on the asset was restricted to equal 1 or −1, respectively. A similar restriction is imposed by an OLS-regression of the market pricing kernel, $\tilde{\phi}_t$, on a constant and the gross return on the asset, $\tilde{R}_t$. This similarity is used for the construction of the variance bound in the case of I assets. An OLS-regression of the market pricing kernel on a constant and the vector of gross returns on the I assets, $\tilde{\mathbf{R}}_t$, is examined,

$$\tilde{\phi}_t = a + \tilde{\mathbf{R}}_t^{\mathrm{T}}\mathbf{b} + \varepsilon_t, \tag{3.8}$$

where a is a constant, $\mathbf{b}$ is the $[I \times 1]$ vector of slope coefficients, ε_t is the regression error and $\tilde{\mathbf{R}}_t^{\mathrm{T}}$ denotes the transpose of the vector $\tilde{\mathbf{R}}_t$[16]. Since the regression error and $\tilde{\mathbf{R}}_t^{\mathrm{T}}\mathbf{b}$ are uncorrelated and the variance of the regression

[14]This is the reason why explicit parametrizations of the market pricing kernel often cannot be rejected when they are tested for a single asset, whereas these models have to be rejected when they are estimated simultaneously for different assets, for example a risk-free and a risky asset (cf. Hansen and Singleton (1982), (1983) and Mehra and Prescott (1985)).

[15]This derivation of the variance bound parallels that of Cochrane and Hansen (1992) but is based on asset returns instead of asset prices. For different derivations, all using asset prices, see Hansen and Jagannathan (1991), Snow (1991) and Cecchetti, Lam, and Mark (1994).

[16]The regression (3.8) can be interpreted as a regression of the market pricing kernel on a constant and the return on a portfolio. This can be seen as follows. Defining $b = \mathbf{b}^{\mathrm{T}}\mathbf{1}$ and $\mathbf{v} = 1/b\mathbf{b}$ so that $\mathbf{v}^{\mathrm{T}}\mathbf{1} = 1$, and rewriting equation (3.8) yields $\tilde{\phi}_t = a + b\tilde{\mathbf{R}}_t^{\mathrm{T}}\mathbf{v} + \varepsilon_t = a + b\tilde{R}_t^{\mathrm{P}} + \varepsilon_t$. $\tilde{R}_t^{\mathrm{P}}$ is the return on the portfolio. $\mathbf{v}$ is the vector of the portfolio weights. The slope coefficient b and the portfolio weights are simultaneously estimated in this regression so as to simultaneously maximize the portfolio's expected return, minimize the portfolio's variance and maximize the correlation between the portfolio and the market pricing kernel. This yields the portfolio with the maximum Sharpe-ratio calculated with the implied unobservable risk-free rate $1/\mathrm{E}\{\tilde{\phi}\}$.

error is nonnegative the variance bound for the market pricing kernel is given by

$$\begin{aligned} \mathrm{var}(\tilde{\phi}) &= \mathrm{var}(\tilde{\mathbf{R}}_t^{\mathrm{T}}\mathbf{b}) + \mathrm{var}(\varepsilon_t) \\ \Rightarrow \mathrm{var}(\tilde{\phi}) &\geq \mathrm{var}(\tilde{\mathbf{R}}_t^{\mathrm{T}}\mathbf{b}). \end{aligned} \tag{3.9}$$

To obtain a tractable functional form **b** must be calculated. Using the deviation of the market pricing kernel from its mean, $(\tilde{\phi}_t - \mathrm{E}\{\tilde{\phi}\})$, instead of the market pricing kernel and the deviation of the asset returns from their means, $(\tilde{\mathbf{R}}_t - \mathrm{E}\{\tilde{\mathbf{R}}\})$, instead of the asset returns yields the following regression model

$$\left(\tilde{\phi}_t - \mathrm{E}\{\tilde{\phi}\}\right) = \left(\tilde{\mathbf{R}}_t - \mathrm{E}\{\tilde{\mathbf{R}}\}\right)^{\mathrm{T}} \mathbf{b}^* + \varepsilon_t. \tag{3.10}$$

No constant term is included in this regression. The subtraction of the expected values equals a shift of the axes and, hence, the vectors of slope coefficients **b** and $\mathbf{b}^*$ in the regression models (3.8) and (3.10) are equal. Using the standard least squares estimates for regression (3.10) the vector of slope coefficients is obtained by

$$\mathbf{b} = \Sigma_{\tilde{\mathbf{R}}}^{-1}\mathrm{E}\left\{\left(\tilde{\phi}_t - \mathrm{E}\{\tilde{\phi}\}\right)\left(\tilde{\mathbf{R}}_t - \mathrm{E}\{\tilde{\mathbf{R}}\}\right)\right\} \tag{3.11}$$

$$= \Sigma_{\tilde{\mathbf{R}}}^{-1}\left[\mathbf{1} - \mathrm{E}\{\tilde{\phi}\}\cdot\mathrm{E}\{\tilde{\mathbf{R}}\}\right], \tag{3.12}$$

where $\Sigma_{\tilde{\mathbf{R}}} = \mathrm{E}\{(\tilde{\mathbf{R}}_t - \mathrm{E}\{\tilde{\mathbf{R}}\})(\tilde{\mathbf{R}}_t - \mathrm{E}\{\tilde{\mathbf{R}}\})^{\mathrm{T}}\}$ is the covariance-matrix of the gross returns. Equation (3.12) is derived from equation (3.11) using equation (3.7). Using **b** from equation (3.12) it follows that

$$\begin{aligned} \mathrm{var}(\tilde{\mathbf{R}}_t^{\mathrm{T}}\mathbf{b}) &= \mathrm{var}\left(\tilde{\mathbf{R}}^{\mathrm{T}}\Sigma_{\tilde{\mathbf{R}}}^{-1}\left[\mathbf{1} - \mathrm{E}\{\tilde{\phi}\}\cdot\mathrm{E}\{\tilde{\mathbf{R}}\}\right]\right) \\ &= \left[\mathbf{1} - \mathrm{E}\{\tilde{\phi}\}\cdot\mathrm{E}\{\tilde{\mathbf{R}}\}\right]^{\mathrm{T}}\Sigma_{\tilde{\mathbf{R}}}^{-1\mathrm{T}}\mathrm{var}(\tilde{\mathbf{R}}^{\mathrm{T}})\Sigma_{\tilde{\mathbf{R}}}^{-1}\left[\mathbf{1} - \mathrm{E}\{\tilde{\phi}\}\cdot\mathrm{E}\{\tilde{\mathbf{R}}\}\right] \\ &= \left[\mathbf{1} - \mathrm{E}\{\tilde{\phi}\}\cdot\mathrm{E}\{\tilde{\mathbf{R}}\}\right]^{\mathrm{T}}\Sigma_{\tilde{\mathbf{R}}}^{-1\mathrm{T}}\Sigma_{\tilde{\mathbf{R}}}\Sigma_{\tilde{\mathbf{R}}}^{-1}\left[\mathbf{1} - \mathrm{E}\{\tilde{\phi}\}\cdot\mathrm{E}\{\tilde{\mathbf{R}}\}\right] \\ &= \left[\mathbf{1} - \mathrm{E}\{\tilde{\phi}\}\cdot\mathrm{E}\{\tilde{\mathbf{R}}\}\right]^{\mathrm{T}}\Sigma_{\tilde{\mathbf{R}}}^{-1}\left[\mathbf{1} - \mathrm{E}\{\tilde{\phi}\}\cdot\mathrm{E}\{\tilde{\mathbf{R}}\}\right] \end{aligned} \tag{3.13}$$

and, thus,

$$\mathrm{std}(\tilde{\phi}) \geq \sqrt{\left[\mathbf{1} - \mathrm{E}\{\tilde{\phi}\}\cdot\mathrm{E}\{\tilde{\mathbf{R}}\}\right]^{\mathrm{T}}\Sigma_{\tilde{\mathbf{R}}}^{-1}\left[\mathbf{1} - \mathrm{E}\{\tilde{\phi}\}\cdot\mathrm{E}\{\tilde{\mathbf{R}}\}\right]}. \tag{3.14}$$

Hence, a lower bound for the unconditional standard deviation of the market pricing kernel can be calculated for each unconditional expected value of the market pricing kernel given unconditional first and second moments of asset returns.

Notice, that the bound calculated from one risky asset and the stochastic one-period risk-free rate contains more information about the market pricing kernel than the bound calculated from the difference between the returns on both assets, i.e. from the excess returns on the risky asset over the one-period risk-free rate. Therefore, using the returns on one risky asset and the one-period risk-free rate in equation (3.14) yields a more restrictive bound than equation (3.5). A market pricing kernel complying with the bound imposed by the excess returns is not necessarily consistent with the returns on both assets. This result can easily be generalized to the case of I risky assets. Similar to equation (3.14) a variance bound can be derived from the excess returns on these assets over the one-period risk-free rate. A market pricing kernel complying with this bound is consistent with the excess returns paid on these assets but it is not necessarily consistent with the returns on these assets.

3.3.3 Statistical Tests of Parametrized Market Pricing Kernels Using Variance Bounds

A parametrized market pricing kernel consists of a functional form for the market pricing kernel and a complete specification of the parameter values. Most studies using variance bounds to analyze parametrized market pricing kernels are based on point estimates of the variance bounds and point estimates of the mean and the standard deviation of the parametrized market pricing kernel, e.g., Gallant, Hansen, and Tauchen (1990), Hansen and Jagannathan (1991), Bekaert and Hodrick (1992), Ferson and Harvey (1992) or Missong (1994). If the mean-standard deviation pair lies below the variance bound the parametrized market pricing kernel is rejected in this approach. An attractive feature of this approach is that the knowledge of the underlying processes of asset returns and of the market pricing kernel is not required. However, the sampling error in both the variance bound and the mean-standard deviation pair of the parametrized market pricing kernel is ignored.

Cochrane and Hansen (1992), Burnside (1994), Cecchetti, Lam, and Mark (1994), Gordon, Samson, and Carmichael (1995) and Hansen, Heaton, and Luttmer (1995) propose methods to gain statistical inference on whether a parametrized market pricing kernel is consistent with the bound or not. These tests are mainly based on the vertical distance between the mean-standard deviation pair of the parametrized market pricing kernel and the estimated variance bound. This distance is then normalized by its asymptotic standard error (cf. Cochrane and Hansen (1992), Burnside (1994) and Cecchetti, Lam, and Mark (1994)). The parametrized market pricing kernel is rejected, if its mean-standard deviation pair lies significantly below the bound. Hansen, Heaton, and Luttmer (1995) formulate a similar test that is also applicable when there are transactions costs and assets subject to short-sale constraints. Burnside (1994, p. 60) further proposes a test based on the shortest distance to the bound rather than on the vertical distance. In contrast, the approach of Gordon, Samson, and Carmichael (1995) only accounts for the uncertainty in the variance bound. Assuming i.i.d. multivariate log-normally distributed asset returns they replicate the set of asset returns 2000 times by simulation to obtain a simulated distribution of the variance bound.

The results obtained by applying these tests are controversial. The results of Burnside (1994) stress the importance of considering sampling errors to avoid an unjustified rejection of a parametrized market pricing kernel, i.e. to avoid a type I error. Cecchetti, Lam, and Mark (1994) observe that most of the variation in the vertical distance between the mean-standard deviation pair of the market pricing kernel and the variance bound is the result of uncertainty induced by the estimation of the expected value of the parametrized market pricing kernel. This observation is not surprising due to the limited observability of consumption data. However, in contrast to Burnside (1994) they find that sampling errors do not appear to explain the asset pricing puzzles observed by comparing variance bounds with parametrized market pricing kernels. A similar result is obtained by Cochrane and Hansen (1992).

A statistical test whether a parametrized market pricing kernel meets the variance bound or not allows to reject the parametrization. However, as discussed earlier, satisfying the variance bounds is a necessary but not a sufficient condition for the parametrized market pricing kernel to correctly price the assets because the market pricing kernel may not be perfectly correlated with the assets' returns. Thus, if the parametrized market pricing kernel cannot be rejected the variance bound test does not render a direct test of the pricing implications of the market pricing kernel redundant. Hence, the variance bound approach is an unsatisfactory tool for formally testing parametrized market pricing kernels. Nevertheless, it is an intuitive tool for visualizing the capability of parametric models to jointly price different assets and for understanding why a particular model might not fit the data.

3.4 Summary of Chapter 3

Recently, attempts to reveal information about the distribution of the market pricing kernel without making assumptions about the functional form of the market pricing kernel have been discussed in the literature. Theoretical investigations have shown that specific assumptions about the stochastic process which the asset prices follow imply specific properties of the stochastic process which the market pricing kernel follows, and vice versa. In an equilibrium representative-agent framework the properties of the market pricing kernel are related to the preferences of the representative investor, so that assumptions about the asset price or return generating process imply restrictions on preferences.

Empirical investigations either estimate the distribution of the market pricing kernel, respectively the risk-neutral probability distribution, conditional on one date from cross-sectional asset prices or place restrictions on the moments of the unconditional distribution of the market pricing kernel using time series data. Both methods derive properties of the market pricing kernel directly from observed data and, with the exception of one estimation method, do not require strong distributional assumptions. Estimated conditional distributions of the risk-neutral probabilities usually differ systematically from the lognormal distribution and are not stationary over time. In addition, the one-period risk-free rate may be stochastic. Hence, the conditional distribution of the market pricing kernel may not be stationary, indicating changes in the future investment opportunity set and/or changes in preferences.

The approach to estimate bounds for the first and second unconditional moment of the market pricing kernel from first and second unconditional moments of asset returns, i.e. the variance bound approach, is useful (1) for analyzing the information content of different asset price or asset return data sets for intertemporal asset pricing and (2) for comparing the potential of different parametric models of the market pricing kernel to be consistent with these data. To be consistent with these data it is necessary, but not sufficient, that a correctly specified market pricing kernel satisfies the derived variance bounds.

In chapter 7 the variance bound approach is applied to analyze the information about the market pricing kernel contained in the returns on different portfolios of German assets and to evaluate different parametric models of the market pricing kernel. Parametric models of the market pricing kernel are derived in the next chapter.

Chapter 4

Parametric Models of the Market Pricing Kernel

This chapter is concerned with the derivation of parametric models of the market pricing kernel, i.e. the specification of intertemporal asset pricing models. For this purpose the consumption-based equilibrium asset pricing approach is applied, which relates asset prices to aggregate consumption. Within this approach the market pricing kernel is derived from the optimal intertemporal consumption and investment choice of a representative agent. Alternative approaches attempt to generalize one-period models – in particular the Capital Asset Pricing Model (CAPM) – that usually relate asset prices to aggregate wealth, to the intertemporal context. These attempts suffer from problems that arise because in a multiperiod world aggregate wealth may not deterministically determine aggregate consumption, except for the last period. Although there exist theoretically founded generalizations they are of no practical use because they do not specify the relevant pricing factors. Therefore, the factors in the recently suggested Conditional CAPMs, which can be interpreted as linear multifactor models in the spirit of the theoretically founded generalizations, are more or less arbitrarily chosen. Moreover, in contrast to the consumption-based equilibrium asset pricing approach the conditional asset pricing approach takes the risk-free rate and the risk premia on the market portfolio and on a set of "factors" as exogenously given instead of determining them. This, however, is an important goal of the present analysis. This study therefore applies the consumption-based equilibrium asset pricing approach.

The chapter is organized as follows. Section 4.1 discusses the problems that arise when the one-period CAPM is applied in a multiperiod world and illustrates the conditional capital asset pricing approach. Section 4.2 gives a detailed description of the consumption-based equilibrium asset pricing approach. The behavioral implications of time-additive expected utility are discussed. Next, a more general recursive non-expected utility specification is introduced. Parametric models of the market pricing kernel are then derived for both preference specifications. A summary ends the chapter.

4.1 Conditional Capital Asset Pricing

As already stated, this analysis applies the consumption-based equilibrium asset pricing approach to derive parametric models of the market pricing kernel. However, there exists a variety of approaches for intertemporal (multiperiod) asset pricing based on the one-period CAPM. To illustrate the problems associated with these approaches they are briefly summarized in this section. First, it is illustrated that one-period models cannot simply be applied to the multiperiod case. Second, sufficient conditions are discussed that allow to apply one-period models in the multiperiod case. Third, the intuition of theoretical generalizations of the CAPM is given. Finally, the recently suggested Conditional CAPMs are discussed in the light of these facts. The general idea of these models is that the CAPM holds conditional on state variables describing the state of the economy but that it does not hold unconditionally.

Consider the one-period CAPM. As noted in section 2.3, the one-period CAPM follows from the conditional linear asset pricing relation (2.13) by dropping the time-subscripts and assuming that the market pricing kernel is perfectly correlated with the return on the market portfolio, e.g., $\tilde{\phi} = a + b\tilde{R}^M$. The CAPM, like other one-period models, thus relates asset prices to aggregate wealth; more precisely, it relates risk premia on risky assets to the expected return on the aggregate wealth portfolio (market portfolio). This is justifiable in a one-period model because at the end of the period aggregate wealth is completely consumed and therefore aggregate wealth equals aggregate consumption from which individuals actually derive utility. Asset prices should therefore be related to consumption. In a multiperiod framework, however, the relationship between aggregate wealth and aggregate consumption is not necessarily deterministic except for the final period. Only if aggregate consumption is a completely deterministic function of wealth, aggregate wealth may be substituted for aggregate consumption and the risk premia on risky assets may be related to the stochastic properties of the market portfolio as well as to those of aggregate consumption (cf. Huang and Litzenberger (1988, p. 210)).

This complicates the application of one-period models such as the CAPM in a multiperiod world. For example, at the end of an arbitrarily chosen period τ only a fraction of aggregate wealth is consumed, the rest is reinvested. The indirect utility that is derived from this wealth may therefore depend on the investment opportunities available at this time. Hence, the investment opportunities available at the end of period τ may also influence the decisions at the beginning of period τ. The same argument applies for all future periods. Thus, if a one-period model is applied in a multiperiod world then all future investment opportunity sets (FIOSs) matter. Only if the individuals have preferences such that their decisions are independent of the FIOSs[1] one-period models can be applied period by period in a multiperiod world without further assumptions.

Now assume that the FIOSs are deterministic and, moreover, constant over time. Then the investment opportunities are the same in each period so that for constant preferences individuals' decisions are the same in each period. In this case aggregate consumption can be expressed as a deterministic and constant function of aggregate wealth (cf. Huang and Litzenberger (1988, pp. 214-6)) and the risk premia on risky assets can be related to the stochastic properties of the market portfolio as well as to those of aggregate consumption. The one-period CAPM can then be applied in a multiperiod world as a sequence of identical one-period models. The same model holds for conditional as well as for unconditional expected returns[2]. Deterministic and constant FIOSs imply that all conditional moments in the equations (2.12) and (2.13) are constant, i.e. that expected returns and risks do not vary over time. Then, however, the "multiperiod economy essentially lacks any intertemporal flavor" (Huang and Litzenberger 1988, p. 210). This illustrates the static nature of the CAPM.

If the individuals' decisions are not independent of the FIOSs and if the FIOSs are not deterministic, i.e. if expected returns and risks vary stochastically over time, no deterministic function relating aggregate consumption and aggregate wealth can be given. In this case risk premia on risky as-

---

[1]This is for example the case for time-additive expected logarithmic utility, according to which the individuals generally consume a constant fraction of their wealth, independent of the future investment opportunities. Thus, for logarithmic utility a deterministic and constant relationship between aggregate consumption and aggregate wealth is generally given (see also section 4.2.2.4). This individuals' behavior is referred to as myopic behavior.

[2]A slightly weaker result is obtained, if the FIOSs are deterministic but not constant over time. Then a deterministic relation between aggregate wealth and aggregate consumption can be given for each period (cf. Huang and Litzenberger (1988, pp. 214-6)). In this case, one-period models can also be applied period by period in a multiperiod world. However, the models are not necessarily the same for each period, e.g., in the case of the CAPM the β-coefficients and/or the risk premia on the market portfolio may differ. Hence the CAPM holds conditionally, period by period, but may not hold unconditionally. Jagannathan and Wang (1996) show that if the CAPM holds conditionally, a two-factor model holds unconditionally; the average returns are jointly linear in the average market beta and a measure of "beta instability" (Jagannathan and Wang 1996, p. 5). However, under additional assumptions one-period models may also be applied period by period if the FIOSs are not deterministic. Franke (1984) shows that the period-by-period application of one-period models may be justifiable even if the FIOSs are not completely deterministic, i.e. known at the first decision date, if at each date t a preference-dependent variable is known with certainty that summarizes the FIOSs of all successive periods (Franke 1984, proposition 2).

sets cannot simply be related to the stochastic properties of the market portfolio[3]. Instead, additional factors are necessary to capture the potential changes in the FIOSs. This is shown by Merton (1973). Merton derives a continuous-time Intertemporal CAPM (ICAPM) that captures stochastic variation in the one-period risk-free rate. He obtains a multi-beta model where the conditional expected return is linear in the market return and the return on a "hedge-portfolio", which is most highly correlated with changes in the one-period risk-free rate. The hedge-portfolio enters the equilibrium model because if risk averse investors have to choose between two portfolios with the same distribution of future wealth, they will prefer the portfolio that provides the best "hedge against 'unfavorable' shifts in the investment opportunity set" (Merton 1973, p. 876). Long (1974) generalizes this result in a discrete-time model. He allows for variations beyond that in the one-period risk-free rate and obtains a model with one additional factor (hedge-portfolio) for each additional uncertainty. These intertemporal asset pricing models are purely theoretical and up to now no direct empirical investigation of these models exists because the relevant pricing factors are not specified. However, these models illustrate that under stochastic FIOSs conditional expected returns are jointly linear in more than one factor[4], one of which is the excess return on the market.

Overall it has been shown that applying the one-period CAPM to average returns in a multiperiod world requires that expected returns and risks do not vary over time (deterministic and constant FIOSs). This assumption is not likely to be satisfied in reality which may be a reason for the lack of empirical support for the CAPM[5]. It has further been noted that if the returns and risks vary over time in a perfectly predictable way (deterministic FIOSs), the CAPM holds conditionally but not unconditionally. If the returns and risks vary stochastically over time (stochastic FIOSs), the CAPM holds neither conditionally nor unconditionally in a multiperiod world. The theoretical generalizations of the CAPM for this case are of no practical use because they do not specify all pricing factors. Recently suggested so-called "Conditional CAPMs" allow for variation of moments over time through explicit models of conditional moments. This means that actually the one-period CAPM is applied in a multiperiod world, but that the moments used are conditioned on past information. Relevant conditioning information is identified through the use of instrumental variables. Usually historically ob-

[3]Notice that a similar problem arises if future consumption opportunities are uncertain. In this case no deterministic function relating aggregate consumption and aggregate wealth can be given, too (cf. Long (1974)).

[4]A this point it becomes evident that the ICAPM that is based on equilibrium arguments leads to a multifactor model comparable with the Arbitrage Pricing Theory (APT) (cf. Ross (1976)) that is based on arbitrage arguments. Both multifactor models do not specify the factors.

[5]Fama and French (1992), for example, find a flat relation between average returns and unconditional betas.

served returns or macroeconomic variables are used as instruments. The idea of these models is further supported by studies documenting that the returns and risks of individual stocks are predictable over time using technical variables such as the dividend yield or interest rates[6]. If this predictability reflects changes in required returns in equilibrium then a reliable model should consider this time variation.

In the Conditional CAPMs usually only the second moments are allowed to vary over time. This means that the conditional β-coefficient in the conditional asset pricing relations (2.12) and (2.13) is time-varying. The conditional risk premium on the market is still assumed to be constant. As a result, the expected excess return on individual stocks may vary over time. Many authors, e.g., Turtle, Buse, and Korkie (1994), use ARCH or GARCH models to model time-varying second moments. Other authors, e.g., Ferson and Schadt (1996) and Karolyi and Kho (1996), directly model the conditional β-coefficient as a linear function of instrumental variables[7]. This approach is equivalent to assuming that the market pricing kernel is a linear function of the market excess return $\tilde{p}^M_{t+1}$ (Ferson and Schadt 1996, p. 430), where the coefficients depend linearly on the instruments, i.e. $\tilde{\phi}_{t+1} = a_t + b_t\tilde{p}^M_{t+1}$, where $a_t = a(X_t)$ and $b_t = b(X_t)$ are linear functions of a set of instrumental variables X_t observable at date t. Carhart, Krail, Stevens, and Welch (1996) directly apply this approach. They model a_t and b_t directly as linear functions of instrumental variables known at date t. Hence, the linear function that relates the market pricing kernel and the excess return on the market portfolio is conditional on the instrumental variables. However, since a_t and b_t are linear in the instruments the market pricing kernel may also be interpreted as an unconditional linear function of the instrumental variables, the excess return on the market and the product of each instrumental variable and the excess return on the market. This means that the market pricing kernel can be written as $\tilde{\phi}_{t+1} = a + \sum_{k=1}^{K} b^k \tilde{R}^K_{t+1}$ leading to a multifactor asset pricing model similar to the static APT[8] [9]. Thus, models of expected stock returns that allow for variations in the β-coefficient may be replaced by multifactor models with constant coefficients. Although this simple result is obtainable only for linear functions, it corresponds to the theoretical results of Merton (1973) and Long (1974) that the introduction of stochastic FIOSs requires adding factors to the CAPM to capture the variations over time.

---

[6]Among these studies, Fama and French (1993) is probably the most cited one.

[7]Ferson and Schadt (1996) also apply this approach to the APT, i.e. they model each β-coefficient in the APT as a linear function of instrumental variables.

[8]The static Arbitrage Pricing Theory (APT) can be derived by assuming that the market pricing kernel is a linear combination of the return on K factors (benchmark portfolios), $\tilde{\phi}_{t+1} = a + \sum_{k=1}^{K} b^k \tilde{R}^K_{t+1}$ and additionally assuming that the risk premia on each factor are constant over time (cf. Missong (1996)).

[9]See also the interpretation of Ferson and Schadt (1996, p. 430).

All these models reduce asset valuation to the pricing of a number of factors. Since both the number of factors and the factors themselves are not specified, empirical investigations usually choose factors so as to maximize the fit of one of the equations (2.2), (2.12) or (2.13). The chosen factors are not theoretically justified and may be without explicit reference to the fundamental sources of risk. The estimated models may therefore overfit the data and may capture irregularities in the data. From a theoretical point of view relating the risk premia on risky assets to aggregate consumption is more satisfying in a multiperiod world. Moreover, the consumption-based equilibrium asset pricing approach yields theoretically founded models of the market pricing kernel that are fully specified. Therefore, the ensuing section employs this approach. However, implementing the derived models requires consumption data that is not available at high frequencies and burdened with measurement problems[10].

4.2 Consumption-based Equilibrium Asset Pricing

In the consumption-based equilibrium asset pricing framework asset prices are related to consumption. In equilibrium asset prices and consumption are such that no individual can improve his utility by intertemporally substituting consumption or by purchasing or selling assets. Under specific assumptions the individuals in the economy can be aggregated to a representative agent. In equilibrium the representative agent must realize his optimal intertemporal consumption and investment choice. This optimal choice is used to derive the market pricing kernel. Therefore, first, the optimal intertemporal consumption and investment choice of a representative agent is derived for a general class of recursive preferences without further specifying these preferences (section 4.2.1). Second, the parametric models of the market pricing kernel are derived for three utility specifications from this class of recursive preferences (sections 4.2.2 and 4.2.3).

4.2.1 Intertemporal Consumption and Investment Choice of a Representative Agent

The analysis starts at the level of individuals. Section 4.2.1.1 specifies the intertemporal choice problem of individuals. Before the optimal choice can be characterized it must be clarified how the individuals' preferences over

[10]To overcome this shortcoming, Campbell (1993) presents an approach to substitute consumption out of the model.

consumption paths can be represented. This is done in section 4.2.1.2. Conditions for the individuals' optimal intertemporal consumption and investment decisions are derived in section 4.2.1.3. Section 4.2.1.4 analyzes the conditions under which the aggregate behavior of individuals can be represented by the behavior of a representative agent. A parametric model of the market pricing kernel is obtained from the conditions for the optimal intertemporal investment and consumption choice of the representative agent.

4.2.1.1 Intertemporal Consumption and Investment Choice of Individuals

Consumption-based asset pricing models assume that individuals choose consumption and investment policies that maximize their utility over all possible lifetime consumption paths. Subsequently, a discrete-time multiperiod model is assumed where individuals make consumption and investment decisions at fixed points in time[11]. The individuals are assumed to live for T periods, T may increase to infinity, and to have no bequest motive. At decision date t, $t \in \{0, 1, \ldots, T\}$, the individual j owns wealth $w_t \geq 0$[12]. The individual decides to consume a proportion $c_t \geq 0$ of his wealth w_t, $c_t \leq w_t \forall t \in \{0, 1, \ldots, T\}$, (consumption choice) and to invest the remainder $(w_t - c_t)$ in a portfolio chosen from a set of I different one-period investment opportunities[13] indexed by i (investment choice).

The random gross rate of return on investment i, $i \in \{1, 2, \ldots, I\}$, from date t to date $t+1$ is denoted by $\tilde{R}^i_{t+1}$. v^i_t represents the fraction of the portfolio invested in the investment opportunity i at date t, so that $\sum_{i=1}^{I} v^i_t = 1$. The random gross return on the individual's portfolio from date t to date $t+1$ can then be written as $\tilde{R}^{\mathrm{P}}_{t+1} = \sum_{i=1}^{I} v^i_t \tilde{R}^i_{t+1}$. It is assumed that labor income is fully diversifiable, i.e. a market exists where the individuals trade claims on human capital. Under this assumption human capital, i.e. the potential to work and earn labor income, can be included into the individual's total wealth w_t and the return on human capital is part of the return on the

[11] Breeden (1989, p. 183) notes "It is unlikely that an individual would choose only to revise at fixed dates in time, regardless of what happens in between, so ... [discrete-time models] initially cause concern. However, that concern is alleviated somewhat by the fact that the qualitative properties of optimal policies in many models are unaffected by the choice of updating interval.".

[12] To keep the notation as simple as possible no indices for individual j are given. Henceforth until specified otherwise the analysis refers to individual j. However, the same analysis applies to each individual in the economy.

[13] This assumption does not mean that no multiperiod investment opportunities exist. It just requires that each investment opportunity can be traded at each date, i.e. that a secondary market exists where multiperiod investment opportunities are traded each date over their maturity.

individual's portfolio. Then the individual's random wealth at date $t+1$ is given by $\tilde{w}_{t+1} = (w_t - c_t)\tilde{R}^{\text{P}}_{t+1}\ \forall t \in \{0, \ldots, T-1\}$.

Starting with initial wealth w_t at the decision date t the individual can realize different consumption paths $(c_t, \tilde{c}_{t+1}, \ldots, \tilde{c}_T)$ through his intertemporal consumption and investment choice. This intertemporal consumption and investment choice is a dynamic choice problem, where a number of decisions are made through time[14]. At each decision date the individual looks at that date and into the future, makes plans for the entire future, but carries out only that part of the plan referring to the current data. The individual reconsiders the optimality of the initially chosen plan at subsequent dates. Comparing the consumption and investment policies planned for any date τ at date t, $\tau > t$, with those realized at date τ shows whether the individual sticks to the initially chosen plan or not, i.e. whether the choices are intertemporally consistent or not. Intertemporally consistent choices can formally be described as follows. Consider two consumption plans $\tilde{c} = (\tilde{c}_t, \tilde{c}_{t+1}, \ldots, \tilde{c}_{\tau-1}, \tilde{c}_\tau, \ldots, \tilde{c}_T)$ and $\tilde{c}' = (\tilde{c}_t, \tilde{c}_{t+1}, \ldots, \tilde{c}_{\tau-1}, \tilde{c}'_\tau, \ldots, \tilde{c}'_T)$. If $\tilde{c}$ is strictly preferred to $\tilde{c}'$ at date t then intertemporal consistency requires that the continuation $(\tilde{c}_\tau, \ldots, \tilde{c}_T)$ is strictly preferred to $(\tilde{c}'_\tau, \ldots, \tilde{c}'_T)$ at date τ (Epstein 1992, p. 17). Hence, intertemporally consistent behavior means that the individual plans his optimal consumption and investment choice at the first decision date for all subsequent periods and realizes this plan without changing it at subsequent decision dates. One reason, of course, why observed behavior might be intertemporally inconsistent is that the individual may have incorrect or incomplete information about the future. Then he would not, in general, stick to the initially chosen plan but take into account any new information (Deaton and Muellbauer 1980, p. 341). Under certainty, however, intertemporal consistency is a widely accepted requirement of rational decision making[15].

The choice between different consumption paths requires that the individual has preferences over different consumption paths. The next section discusses how these preferences can be represented.

4.2.1.2 Preferences over Consumption Paths

First, consider the case of consumption paths with certain future consumption levels c_τ, $\tau \in \{t, \ldots, T\}$. Although, in general, the future consumption

---

[14]In contrast, a static choice problem is characterized by only one decision. In the considered intertemporal choice problem consumption sequences are the objects of choice and source of utility. In sequential choice problems as a second form of dynamic choice the terminal wealth is the source of utility (cf. Epstein (1992, p. 2)).

[15]For decision making under uncertainty intertemporal consistency of choices as a requirement of rational behavior is controversially discussed in the literature (see, for example, Machina (1989)).

levels are uncertain and preference specifications for certain and uncertain consumption levels may measure different aspects of preferences (cf. Wilhelm (1986)), neglecting uncertainty at the beginning can provide some basic insights (Deaton 1992, p. 6). Assume that the individual has a binary, complete and transitive preference relation over different consumption paths so that, comparing any two consumption paths, he either strictly prefers one to the other or is indifferent between both[16]. Then the individual's preference relation can be represented by a utility function $U_t(c_t, \ldots, c_T|\Omega_t)$, which yields the utility the individual would have from a specific consumption path (Huang and Litzenberger 1988, p. 5). The subscript t denotes that utility is measured at time t, given the information available to the individual at date t, denoted by Ω_t. The representation of the preference relation by the utility function is defined as follows. If the utility of two consumption paths is equal, the individual is indifferent between them. The utility of a consumption path strictly preferred to another one is larger than the utility of that consumption path[17]. Ω_t includes the information about the previous levels of consumption, so that the utility of current and future consumption may not only depend on the level of current and future consumption but also on past consumption. This raises the question of the intertemporal separability of preferences.

Intertemporally Separable Preferences A preference function is said to be intertemporally separable or time-additively separable, if the preference function can be expressed as the sum over the discounted utility of consumption in each period $U_t(c_t, \ldots, c_T|\Omega_t) = \sum_{\tau=t}^{T} \beta_\tau u_\tau(c_\tau)$, where $\beta_\tau > 0$ is a period-dependent discount factor and $u_\tau(c_\tau)$ is a time-dependent function giving the time τ utility of consumption at date τ, solely depending on

[16]In the case of an infinite number of consumption paths three additional assumptions have to be made. First, the preference relation must be order-preserving, i.e. if a consumption path a is preferred over b then between convex combinations of a and b the one with the largest proportion of a is preferred; second, if a consumption path a is preferred over b and b is preferred over c then a unique convex combination of a and c must exist that is exactly as good as b; and, third, the set of possible consumption paths must be bounded, i.e. a least preferred element and a most preferred element must exist (cf. Jarrow (1988, pp. 8-11)).

[17]Comparing the utility of two consumption paths thus reveals which consumption path is preferred. This is true for ordinal and cardinal preferences. For ordinal preferences, however, this does not provide information up to which extent the chosen consumption path is preferred. Monotonically increasing transformations of the preference functions do not affect ordinal preferences. In contrast, cardinal preferences allow to rank differences in the utilities gained from different consumption paths. Cardinal preferences are unaffected by linear increasing transformations. For the existence of a cardinal utility function in the case of an infinite number of consumption paths strong independence of the consumption paths must be assumed in addition to the assumptions necessary for the existence of an ordinal utility function (cf. Jarrow (1988, pp. 73-5)). For the present analysis the assumption of ordinal preferences is sufficient.

the consumption at date τ (Ahlbrecht 1996, pp. 27-8, 44)[18]. Intertemporal separability implies complete time independence of the desirability of consumption levels in different periods, i.e. the consumption of an individual at one time has no effect on his desire to consume at any other time. This is not the case for intertemporally non-separable preferences.

Intertemporal non-separability occurs, for example, if the utility of consumption in a period τ depends on the consumption levels in periods before τ, e.g., $u_\tau(c_\tau|\Omega_\tau) = u_\tau(c_\tau, c_{\tau-1}, \dots c_{\tau-\gamma})$, where γ is the number of lags to be included. Considering the nature of the dependence of the utility of current consumption on past consumption levels, one distinguishes habit formation and local durability. Habit formation captures the notion that an individual gets used to consuming a certain amount and that the habit level in turn reduces the utility of current consumption. Local durability means that consumption in one period also yields satisfaction in subsequent periods. In this case past consumption increases the utility of current consumption.

Recall that the goal is to derive the optimal intertemporal consumption and investment choice, i.e. the individual must consume and invest such that the realized consumption plan yields the highest utility among all possible plans. The optimization is less complicated for intertemporally separable preferences than for intertemporally non-separable preferences. Despite the inconvenience, the subsequent analysis is not restricted to intertemporally separable preferences. In general, dynamic programming methods are applied to find the best consumption and investment policy. For the use of these methods it is assumed that the remaining utility of lifetime consumption can be divided into two parts. At date t the first part is the utility of the current consumption measured by time t utility, $u_t(c_t|\Omega_t)$. The second part is the utility of consumption for all subsequent periods, measured by time $t+1$ utility. Both parts are aggregated by an aggregator function $W_t[\cdot,\cdot]$ so that the preference function can be rewritten recursively as

$$U_t(c_t, \dots, c_T|\Omega_t) = W_t[u_t(c_t|\Omega_t), U_{t+1}(c_{t+1}, \dots, c_T|\Omega_{t+1})]. \qquad (4.1)$$

Since the tradeoff between utility of current consumption expressed as time t utility and utility of future consumption expressed as time $t+1$ utility is determined by the individual's time preference, the individual's time preference is one determinant of the aggregator function. Preferences

[18]Notice, that ordinal time-multiplicative preferences may also be intertemporally separable. Consider the case where $U_t(c_t, \dots, c_T|\Omega_t) = \prod_{\tau=t}^{T} u_\tau^*(c_\tau)$. For example, if each $u_\tau^*(c_\tau)$ can be written in exponential form $u_\tau^*(c_\tau) = \exp(\beta_\tau u_\tau(c_\tau))$, then the preference function is given by $U_t(c_t, \dots, c_T|\Omega_t) = \exp(\sum_{\tau=t}^{T} \beta_\tau u_\tau(c_\tau))$. Taking the logarithm of $U_t(c_t, \dots, c_T|\Omega_t)$ yields $U_t^*(c_t, \dots, c_T|\Omega_t) = \sum_{\tau=t}^{T} \beta_\tau u_\tau(c_\tau)$. The logarithm is a monotonically increasing transformation that does not affect ordinal preferences. Therefore ordinal time-multiplicative preferences are intertemporally separable under specific assumptions on the functions $u_\tau^*(c_\tau)$.

of this form are intertemporally separable if (1) the utility of current consumption is independent of past consumption levels, i.e. $u_t(c_t|\Omega_t) = u_t(c_t)$, and (2) the aggregator functions are linear in their second arguments, i.e. $\partial W_t[\cdot,\cdot]/\partial U_{t+1}(\cdot) = \gamma_{t+1}$, e.g., $W_t[x, y] = x + \gamma_{t+1}y$[19]. The preference function (4.1) can then be rewritten as $U_t(c_t, \ldots, c_T|\Omega_t) = \sum_{\tau=t}^{T} u_\tau(c_\tau) \prod_{k=t+1}^{\tau} \gamma_k = \sum_{\tau=t}^{T} u_\tau(c_\tau)\beta_\tau$, where $\beta_\tau = \prod_{k=t+1}^{\tau} \gamma_k$, and therefore the condition for intertemporal separability is satisfied.

Preference functions in the recursive structure (4.1) are discussed in detail at the end of this section, assuming that both the aggregator function and the utility function of current consumption are identical over time (time-independent), i.e. $W_\tau[\cdot,\cdot] = W[\cdot,\cdot]$ and $u_\tau(c_\tau|\Omega_\tau) = u(c_\tau|\Omega_\tau)\ \forall \tau \in \{t, \ldots, T\}$, and that the utility function of current consumption is independent of past consumption, i.e. $u(c_\tau|\Omega_\tau) = u(c_\tau)$. A specific parametrization is given in section 4.2.3.1. First, however, the assumption of certain future consumption levels is relaxed, which means that a consumption plan now is a sequence of random variables $(c_t, \tilde{c}_{t+1}, \ldots, \tilde{c}_T)$. Because of convenience, expected utility theory is usually applied in the case of random consumption paths. It is shown below, that for dynamic choice problems expected utility theory, however, implies quite restrictive behavioral assumptions. Therefore a more general class of recursive non-expected utility functions is introduced. First, expected utility and its shortcomings are discussed.

Expected Utility To analyze the individual's preferences over random consumption paths consider first the case of a static choice situation at date t over different distributions of consumption at date τ. Assume that the number of possible realizations of the consumption level at date τ is finite. The individual thus expresses his preferences on single-stage lotteries whose prizes are consumption levels at date τ. Assume further that the individual has a complete, binary and transitive preference relation over such lotteries so that, comparing any two lotteries, he either strictly prefers one to the other or he is indifferent between both[20]. Additionally, assume that the satisfaction gained from one realization of the consumption level at date τ does not depend on what the consumption would have been if another realization had occurred. This assumption is known as the "independence" or "substitution axiom" (Huang and Litzenberger 1988, p. 8). Finally assume that if the individual prefers a distribution of consumption $\tilde{c}_\tau$ over a

[19]For cardinal preferences these conditions are necessary (Ahlbrecht 1996, p. 202). Under specific assumptions for the function $u_t(c_t)$ ordinal preferences are also intertemporally separable if the aggregator functions are of the form $W_t[x, y] = x \cdot y$. This is the case of intertemporally separable time-multiplicative preferences (see footnote 18 within this chapter).

[20]Notice that at the beginning of the section the individual was assumed to have a complete, binary and transitive preference relation over deterministic consumption paths.

distribution $\tilde{c}'_\tau$ and $\tilde{c}'_\tau$ over a distribution $\tilde{c}''_\tau$ then a unique convex combination of $\tilde{c}_\tau$ and $\tilde{c}''_\tau$ exists that is exactly as good as $\tilde{c}'_\tau$. This assumption is known as the "Archimedean axiom" (Huang and Litzenberger 1988, p. 8). These assumptions are necessary and sufficient for the existence of a "von Neumann-Morgenstern utility function" $U_t(\cdot)$[21] giving the utility of any possible level of consumption at date τ such that the individual strictly prefers $\tilde{c}_\tau$ to $\tilde{c}'_\tau$ if and only if $\mathrm{E}_t\{U_t(\tilde{c}_\tau)\} > \mathrm{E}_t\{U_t(\tilde{c}'_\tau)\}$ (Huang and Litzenberger 1988, p. 12). E_t denotes the expectation at time t, conditional on the information available at time t. Thus, the expected utility hypothesis asserts that utility theory under uncertainty can be viewed as a two-stage process, first calculating the utility of all possible realizations and then taking the expectation of these utility values using the probabilities of the outcomes (Jarrow 1988, p. 77). Hence, the utility of a random consumption level is assumed to equal the probability-weighted average of the utility of each possible outcome. The linearity in the probabilities is the characteristic feature of expected utility (Machina 1989, pp. 1626-7). The probabilities are assumed to be objective and exogenously given[22].

Second, consider the case where consumption occurs at more than one date, e.g., at times $\tau = t, \ldots, T$. Then the individual expresses his preferences over lotteries whose prizes are consumption vectors $(c_t, \ldots, c_T)$. Usually these lotteries are compound lotteries with resolution of uncertainty over time rather than single-stage lotteries. Compound lotteries are lotteries whose outcomes are also lotteries. The left-hand lottery in figure 4.1 illustrates a compound lottery. A direct consequence of the independence or substitution axiom of expected utility theory is that individuals are indifferent between any compound lottery and its probabilistically equivalent single-stage lottery, an assumption known as the reduction of compound lotteries axiom (Machina 1989, p. 1626). The right-hand lottery in figure 4.1 represents the probabilistically equivalent single-stage lottery of the left-hand compound lottery. Expected utility maximizers are therefore indifferent between the left-hand lottery and the right-hand lottery. The expected utility of both lotteries is given by $\mathrm{E}_0\{U_0(c_o, \tilde{c}_1, \tilde{c}_2, \tilde{c}_3|\Omega_0)\} = pqU_0(c,c,c,c|\Omega_0) + p(1-q)U_0(c,c,c,c'''|\Omega_0) + (1-p)vU_0(c,c',c',c'|\Omega_0) + (1-p)(1-v)U_0(c,c',c'',c''|\Omega_0)$[23].

Thus, applying expected utility theory to random consumption paths means that at date t, first, the utility of each possible realization of the consumption

[21] This term stems from the seminal work of von Neumann and Morgenstern (1953). Notice that the von Neumann-Morgenstern utility function must be a cardinal utility function (cf. Machina (1987)).

[22] Therefore, the examined decision problem is a decision problem under risk. In contrast, in a decision problem under uncertainty the probabilities are not exogenously given. Then the probability assessments could be assumed to be interrelated with the individual's preferences. Within this approach taken by Savage (1972) the probabilities are subjective. For a good survey of different approaches for decision problems under risk and uncertainty see Kischka and Puppe (1992).

[23] Another illustrative example is given in Kreps (1988, pp. 50-2).

Figure 4.1: The left-hand lottery illustrates a compound lottery. The right-hand lottery represents the probabilistically equivalent single-stage lottery. Expected utility maximizers are indifferent between both.

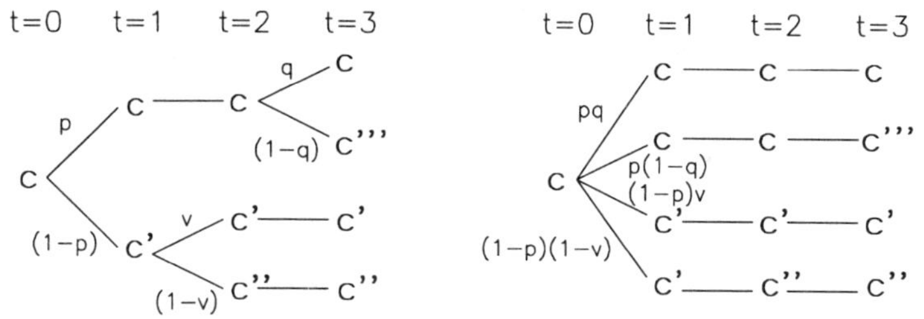

path is evaluated by a utility function for deterministic consumption paths, and, second, the expected utility is measured by the probability-weighted sum of the utility of all realizations. No assumption has been made about the utility function $U_t(c_t, \ldots c_T|\Omega_t)$. It may therefore also be recursively represented as $U_t(c_t, \ldots, c_T|\Omega_t) = W_t[u_t(c_t|\Omega_t), U_{t+1}(c_{t+1}, \ldots, c_T|\Omega_{t+1})]$. If intertemporal separability for $U_t(c_t, \ldots c_T|\Omega_t)$ is assumed (see page 50), the resulting preferences are separable across realizations and time and can be written as $\mathrm{E}_t\{U_t(c_t, \tilde{c}_{t+1}, \ldots \tilde{c}_T|\Omega_t)\} = \sum_{s=1}^{S}\sum_{\tau=t}^{T} \mathrm{prob}_s \beta_\tau u_\tau(c_{\tau,s})$, where S denotes the finite number of realizations and prob_s denotes the objective probability of realization s. Hence, expected utility and intertemporal separability together induce a doubly additive (state- and time-additive) form of preferences. Additionally assuming time-independence of the one-period discount factor and of the utility of current consumption, i.e. $\beta_\tau = \beta^{\tau-t}$ and $u_\tau(c_{\tau,s}) = u(c_{\tau,s}) \forall \tau \in \{t, \ldots, T\}$, the widely used intertemporally consistent preference function $\mathrm{E}_t\{U_t(c_t, \tilde{c}_{t+1}, \ldots \tilde{c}_T|\Omega_t)\} = \sum_{s=1}^{S}\sum_{\tau=t}^{T} \mathrm{prob}_s \beta^{\tau-t} u(c_{\tau,s})$ is obtained. This preference function is usually referred to as "time-additive expected utility". This illustrates that the use of time-additive expected utility requires strong assumptions about preferences[24].

Time of the Resolution of Uncertainty Among the axioms of expected utility theory the independence axiom is often violated in experiments. The most famous violation is the Allais Paradox[25]. Furthermore, it was illustrated that, as a consequence of the independence axiom, expected utility maximizing individuals are indifferent between a compound lottery and its probabilistically equivalent single-stage lottery. For example, consider the lotteries in figure 4.2. These lotteries are probabilistically

[24]Deaton (1992, pp. 19) comments "The double additivity ...puts a great deal of structure on preferences over states and consumption levels, perhaps too much".

[25]This Paradox and other violations of expected utility are extensively discussed in Machina (1987), (1989).

Figure 4.2: Probabilistically equivalent lotteries differing in the time of the resolution of uncertainty.

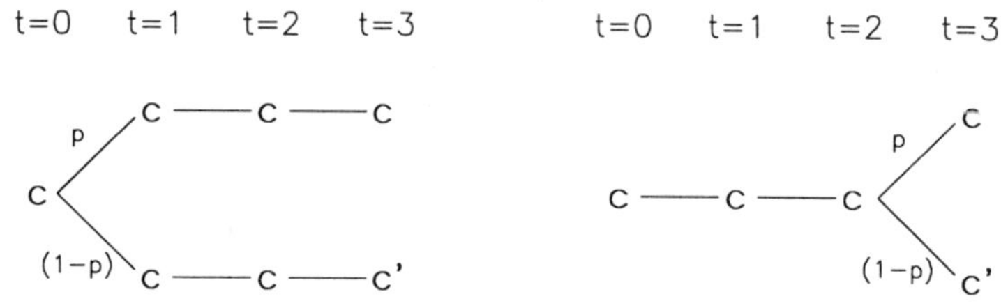

equivalent, so that expected utility maximizing individuals are indifferent between both. However, in the left-hand lottery uncertainty is resolved at time $t = 1$, whereas the consumption path is uncertain until $t = 3$ in the right-hand lottery. This implies that individuals with expected utility are indifferent to the timing of the resolution of uncertainty in multiperiod consumption lotteries. In general, expected utility maximizing individuals are indifferent to the timing of the resolution of uncertainty in the absence of choice opportunities, i.e. the consumption path cannot be rearranged, even under full information. If an early resolution of uncertainty improves planning, e.g., if the lotteries are formulated in terms of income, an early resolution of uncertainty provides more room for choice. This is also true of expected utility maximizing individuals (cf. Kreps and Porteus (1979)).

The question is whether it is plausible to assume that individuals are indifferent between the lotteries in figure 4.2 or, equivalently, whether individuals are indifferent to the timing of the resolution of uncertainty in the absence of choice opportunities. Even if these opportunities do not exist, individuals may prefer early or late resolution of uncertainty. Individuals may prefer an early resolution of uncertainty because they dislike uncertainty, or they may prefer a late resolution of uncertainty because then they can live with the hope for or illusion of a favorable outcome. Both effects, possible planning benefits of an early resolution of uncertainty and psychological costs or gains associated with the timing of the resolution of uncertainty, should be taken into consideration.

Experimental research in decision making has shown that individuals express preferences over the timing of the resolution of uncertainty even in the absence of planning benefits, i.e. individuals are not indifferent towards the timing of the resolution of uncertainty as the expected utility theory assumes. Furthermore, the preferences over an early or late resolution of uncertainty seem to depend on the decision problem and the probability and nature (good or bad) of the outcomes. In his study questioning 267 graduate students to choose between different lotteries, Ahlbrecht (1996) finds strong preferences for an early resolution of uncertainty[26]. However, in the case of a loss with high probability or a gain with low probability a significant fraction of individuals preferred a late resolution of uncertainty. It seems, as if in the case of a very likely loss, these individuals prefer to live with the hope of no loss to knowing the outcome. Similarly, in the case of a very unlikely gain, these individuals prefer to live with the hope of a gain to knowing the outcome. Comparable results are found by Chew and Ho (1994) in a study questioning 110 undergraduate students to decide in hypothetical situations representing daily decisions. They further observe that the "degree of hopefulness" does not depend on the individual's risk aversion. Thus, in everyday situations individuals seem to prefer an early resolution of uncertainty resulting from their aversion to living with uncertainty. In situations where good outcomes are very unlikely, they prefer late resolution of uncertainty because they prefer to live with the hope of a good outcome.

Non-expected Recursive Utility To capture the findings of the experimental research a more general preference representation is needed, which relaxes the reduction of compound lotteries axiom (independence axiom). Because not all axioms of expected utility theory are satisfied, this representation is called "non-expected utility". Comparable to the recursive preference representation for deterministic consumption paths, $U_t(c_t, \ldots, c_T|\Omega_t) = W_t[u_t(c_t|\Omega_t), U_{t+1}(c_{t+1}, \ldots, c_T|\Omega_{t+1})]$ (equation 4.1), Kreps and Porteus (1978) derive an intertemporally consistent recursive preference representation for random consumption paths that allows to express preferences over the timing of the resolution of uncertainty and that includes time-additive expected utility as a special case. Kreps and Porteus (1978) assume, first, that the aggregator function and the utility function of current consumption are time-independent, i.e. $W_\tau[\cdot,\cdot] = W[\cdot,\cdot]$ and $u_\tau(c_\tau|\Omega_\tau) = u(c_\tau|\Omega_\tau)\ \forall\tau \in \{t, \ldots, T\}$, and, second, that the utility function of current consumption is independent of past consumption, i.e. $u(c_\tau|\Omega_\tau) = u(c_\tau)$[27]. Now, combine the aggregator function and the utility function of current consumption to obtain a new aggregator function $V[c_t, \cdot] = W[u(c_t), \cdot]$. If $V[\cdot,\cdot]$ is linear in its second argument, the preferences are intertemporally separable (see page 50). Up to now, $U_t(c_t, \ldots, c_T|\Omega_t) =$

[26] See also Ahlbrecht and Weber (1996).

[27] This assumption rules out effects such as local durability or habit formation (cf. section 4.2.1.2). Kreps and Porteus (1978) call this "payoff history independence". In the literature the recursive Kreps and Porteus preferences are usually interpreted as a multiperiod generalization of the Ordinal Certainty Equivalence (OCE) preferences suggested by Selden (1978) since OCE-preferences, originally referring to decision problems with two points in time, yield intertemporally inconsistent preferences in a multiperiod setting (cf. Epstein and Zin (1989) and Attanasio and Weber (1989)). However, the interpretation as a generalization is not completely correct since OCE-preferences allow effects such as habit formation or local durability.

$V[c_t, U_{t+1}(c_{t+1}, \ldots, c_T|\Omega_{t+1})]$ only represents preferences over deterministic consumption paths. Kreps and Porteus (1978) extend these preferences to the case of uncertainty by replacing the second argument of the aggregator function by its expected value,

$$U_t(c_t, \tilde{c}_{t+1} \ldots, \tilde{c}_T|\Omega_t) = V[c_t, \mathrm{E}_t\{U_{t+1}(\tilde{c}_{t+1}, \ldots, \tilde{c}_T|\Omega_{t+1})\}]. \tag{4.2}$$

Thus, the expected value of the utility of different paths is calculated in each step during the backward recursion. This is in contrast to the approach of expected utility, where the expected value of the utility of all possible realizations of the whole consumption path is evaluated once at the beginning. Only if $V[\cdot,\cdot]$ is linear in its second argument, i.e. if the preferences are intertemporally separable, both approaches are equivalent and individuals exhibit timing indifference. If the aggregator function is not linear in its second argument, expected utility and recursive non-expected utility are not equivalent. Kreps and Porteus (1978, theorem 3) show that an early [late] resolution of uncertainty is preferred, if $V[\cdot,\cdot]$ is convex [concave] in its second argument.

However, timing indifference is not the only restrictive assumption of expected utility. As shown in the remainder of this section, time-additive expected utility does not permit to separately specify risk aversion and aversion to intertemporal fluctuations in deterministic consumption paths, too. Specific parametrizations of the recursive non-expected utility allow to overcome this shortcoming (see section 4.2.3.1).

Risk Aversion and Intertemporal Substitution Risk aversion and the willingness to substitute consumption over time correspond to two different aspects of preferences. Risk aversion describes the aversion to the variation of consumption across states of nature at one date while the elasticity of intertemporal substitution is concerned with the variation of consumption across time in a deterministic consumption path. Individuals who prefer a smooth consumption path, i.e. individuals who are averse to intertemporal fluctuations of consumption, exhibit a low elasticity of intertemporal substitution. A high elasticity of intertemporal substitution, on the other hand, means a low aversion to intertemporal fluctuations of consumption. Since risk aversion and elasticity of intertemporal substitution correspond to different aspects of preferences, it may be advantageous to model them separately, so that preference rankings of deterministic consumption paths are not affected by changes in the degree of risk aversion.

However, in widely used preference specifications both aspects cannot be modelled separately. Time-additive expected utility, for example, implies that the degree of intertemporal substitutability is inversely related to the degree of risk-aversion (cf. Hall (1988)). This can be shown as follows. Consider time-additive expected utility

$$U_t(c_t, \tilde{c}_{t+1}, \ldots, \tilde{c}_T) = \sum_{\tau=t}^{T} \beta^{\tau-t} \mathrm{E}\{u(\tilde{c}_\tau)\}. \tag{4.3}$$

Risk aversion is measured by the parameter of relative risk aversion, α, defined as

$$\alpha = -c_\tau \frac{u''(c_\tau)}{u'(c_\tau)}. \tag{4.4}$$

The elasticity of intertemporal substitution, σ, is defined as the relative change of consumption at date τ over consumption at date $\tau+1$ divided by the relative change in the intertemporal marginal rate of substitution, $IMRS$[28].

$$\sigma = \frac{\frac{d(c_\tau/c_{\tau+1})}{(c_\tau/c_{\tau+1})}}{\frac{dIMRS_{\tau+1}}{IMRS_{\tau+1}}} = \frac{d\frac{c_\tau}{c_{\tau+1}}}{dIMRS_{\tau+1}} \frac{IMRS_{\tau+1}}{\frac{c_\tau}{c_{\tau+1}}} \tag{4.5}$$

$$IMRS_{\tau+1} = \frac{U_t'(c_{\tau+1})}{U_t'(c_\tau)} = \beta \frac{u'(c_{\tau+1})}{u'(c_\tau)}. \tag{4.6}$$

The intertemporal marginal rate of substitution is a measure of the slope of the indifference curve whereas the elasticity of intertemporal substitution measures the curvature of the indifference curve. The total differentials of $c_\tau/c_{\tau+1}$ and $IMRS_{\tau+1}$ are

$$d\frac{c_\tau}{c_{\tau+1}} = \frac{1}{c_{\tau+1}} dc_\tau - \frac{c_\tau}{c_{\tau+1}^2} dc_{\tau+1} \tag{4.7}$$

$$dIMRS_{\tau+1} = -\beta \frac{u'(c_{\tau+1})}{(u'(c_\tau))^2} u''(c_\tau) dc_\tau + \beta \frac{1}{u'(c_\tau)} u''(c_{\tau+1}) dc_{\tau+1}. \tag{4.8}$$

Substituting (4.6), (4.7) and (4.8) in equation (4.5) yields

$$\sigma = \frac{\frac{1}{c_\tau} dc_\tau - \frac{1}{c_{\tau+1}} dc_{\tau+1}}{\frac{u''(c_{\tau+1})}{u'(c_{\tau+1})} dc_{\tau+1} - \frac{u''(c_\tau)}{u'(c_\tau)} dc_\tau}. \tag{4.9}$$

For $dc_\tau = 0$ or $dc_{\tau+1} = 0$ it follows that the elasticity of intertemporal substitution is the inverse of the coefficient of relative risk aversion (which may depend on c),

$$\sigma = \frac{1}{\alpha}. \tag{4.10}$$

[28] The elasticity of intertemporal substitution is concerned with intertemporal fluctuations in deterministic consumption paths and is not well-defined for random consumption paths (Weil 1990, footnote 11). Therefore the symbols "$\sim$" denoting random variables are dropped in the following equations.

This restriction in time-additive expected utility implies, for example, that a risk neutral individual is indifferent between a smooth consumption path and one with intertemporal fluctuations, or that highly risk averse individuals are extremely unwilling to substitute consumption over time. Another disadvantage of this restriction is that the consequences of increased risk aversion cannot be determined separately (Epstein 1992, p. 14) because a change in risk aversion simultaneously changes the elasticity of intertemporal substitution. Empirical evidence against the restriction is provided by Giovannini and Weil (1989), Bufman and Leiderman (1990), Epstein and Zin (1991) and Hardouvelis, Kim, and Wizman (1996). Therefore, a more general preference function which permits a clear separation of risk aversion and elasticity of intertemporal substitution is preferable. The Ordinal Certainty Equivalent (OCE) preferences suggested by Selden (1978) provide a complete separation between substitution and risk aversion in a two-date framework. OCE-preferences allow to change the parameter of risk aversion without affecting the ranking of deterministic consumption paths. In a setting with more than two dates parametrizations of the recursive Kreps and Porteus (1978) preferences (cf. equation (4.2)) suggested by Epstein and Zin (1989) and Weil (1989), (1990) allow a separate specification of risk aversion and intertemporal substitution[29].

Summary Expected-utility theory implies that individuals are indifferent to the timing of the resolution of uncertainty if no planning benefits exist. Furthermore, time-additive expected utility functions imply that risk aversion and aversion to intertemporal fluctuations in deterministic consumption paths are intertwined, although both correspond to different aspects of preferences. Empirical evidence indicates that both the presumed timing indifference and the presumed interrelation between risk aversion and elasticity of intertemporal substitution may not be satisfied in reality. The introduced recursive non-expected utility framework (equation (4.2)) allows to overcome these shortcomings and includes time-additive expected utility as a special case. Recall that for the derivation of a parametric model of the market pricing kernel the optimal intertemporal consumption and investment decision of the individuals is central. Up to this point the preferences of individuals can be characterized by the general utility specification introduced above. Given these preferences, conditions for the optimal intertemporal consumption and investment decision are derived in the next section.

[29] A detailed discussion of these parametrizations follows in section 4.2.3.1.

4.2.1.3 Optimal Intertemporal Consumption and Investment Choice of Individuals

Since the recursive Kreps and Porteus preferences provide a more general preference concept, the conditions for the optimal intertemporal consumption and investment decision are derived for this class of preferences. These conditions are also valid for the special case of time-additive expected utility. The maximization problem faced by an individual with recursive Kreps and Porteus preferences (see equation 4.2) at date t is to choose portfolio weights v_τ^i and consumption c_τ, $\forall i \in \{1, \ldots, I\}$ and $\forall \tau \in \{t, \ldots, T-1\}$, so as to maximize the utility of his remaining lifetime consumption $U_t(c_t, \tilde{c}_{t+1}, \ldots, \tilde{c}_T | \Omega_t) = V[c_t, \mathrm{E}_t\{U_{t+1}(\tilde{c}_{t+1}, \ldots, \tilde{c}_T | \Omega_{t+1})\}]$,

$$\max_{\{\{v_\tau^i\}_{i=1}^I, c_\tau\}_{\tau=t}^T} V[c_t, \mathrm{E}_t\{U_{t+1}(\tilde{c}_{t+1}, \ldots, \tilde{c}_T | \Omega_{t+1})\}] \equiv J_t(w_t | \Omega_t),$$

subject to the sequence of budget constraints $\tilde{w}_{\tau+1} = (w_\tau - c_\tau) \sum_{i=1}^I v_\tau^i \tilde{R}_{\tau+1}^i$, the sequence of portfolio constraints $\sum_{i=1}^I v_\tau^i = 1$ and the constraints $0 \leq c_\tau \leq w_\tau$. The value function or indirect utility function of wealth at date t, $J_t(w_t | \Omega_t)$, denotes the utility from the optimal investment and consumption policy from date t to date T, given the time t information Ω_t and wealth w_t.

This maximization problem can be solved using dynamic programming[30]. The technique of dynamic programming is based on Bellman's principle of optimality. That is, whatever the decision at date t, the subsequent decisions should proceed optimally for the subproblem starting at date $t+1$. Therefore, the idea of dynamic programming is to solve the problem from the end. At date T there is no future, and since the individual has no bequest motive it is optimal to consume total wealth, $c_T = w_T$. The optimization starts by solving for the optimal strategy for the last decision date, i.e. date $T-1$ given w_{T-1} and Ω_{T-1}. The dependence on previous decisions is given by the wealth w_{T-1} and the information set Ω_{T-1}. This yields the optimal investment and consumption policy from date $T-1$ to date T, $J_{T-1}(w_{T-1} | \Omega_{T-1})$. The optimization proceeds by going back one period, i.e. to date $T-2$, and solving the optimal investment and consumption decision at this date, given that an optimal strategy will be used thereafter, $\max_{\{\{v_{T-2}^i\}_{i=1}^I, c_{T-2}\}} V[c_{T-2}, \mathrm{E}_{T-2}\{J_{T-1}(\tilde{w}_{T-1} | \Omega_{T-1})\}]$. This yields the optimal investment and consumption policy from date $T-2$ to date T, $J_{T-2}(w_{T-2} | \Omega_{T-2})$. By successively going back period by period, the optimal investment and consumption policies are derived starting from each date until the end, where the optimal contingent policy starting at date $\tau + 1$,

---

30 For the technique of dynamic programming see, for example, Dixit (1990, pp. 162-80) or Huang and Litzenberger (1988, pp. 193-5). Kreps (1990, pp. 791-815) presents a comprehensive example.

summarized by $J_{\tau+1}(w_{\tau+1}|\Omega_{\tau+1})$, is used to find the optimal strategy at date τ. Finally the optimal investment and consumption policy from date t to date T, $J_t(w_t|\Omega_t)$, is obtained. In general, the optimization problem at date t, given that an optimal strategy will be used thereafter, can be written as:

$$J_t(w_t|\Omega_t) = \max_{\{\{v_t^i\}_{i=1}^I, c_t\}} V[c_t, \mathrm{E}_t\{J_{t+1}(\tilde{w}_{t+1}|\Omega_{t+1})\}]. \tag{4.11}$$

The analysis will be restricted to homothetic parametrizations of the recursive utility functions[31]. A utility function is homothetic if and only if it is a monotonically increasing transformation of a function that is homogeneous of degree one in its arguments, here, wealth and current consumption (Deaton and Muellbauer 1980, p. 143). For homothetic utility functions the intertemporal marginal rate of substitution is independent of the level of wealth. Therefore, the optimal consumption decision (consumption versus savings) and the optimal investment decision (portfolio allocation of the savings) are also independent of wealth, so that both decisions can be separated (Giovannini and Weil 1989, p. 8)[32]. It is further assumed that the individual strictly prefers more consumption to less and is risk averse. This ensures that the first order necessary conditions for the optimal decisions are also sufficient. The first order conditions are now derived for (1) the optimal consumption decision and (2) the optimal investment decision.

(1) Optimal Consumption Decision The maximization problem for the optimal consumption policy is:

$$\max_{c_t} V[c_t, \mathrm{E}_t\{J_{t+1}(\tilde{w}_{t+1}|\Omega_{t+1})\}] \qquad \forall t < T \tag{4.12}$$

$$\text{s.t.} \qquad \tilde{w}_{t+1} = (w_t - c_t)\tilde{R}_{t+1}^{\mathrm{P}_t^*},$$
$$0 \leq c_t \leq w_t,$$
$$w_T = c_T.$$

$\tilde{R}_{t+1}^{\mathrm{P}_t^*}$ is the gross return from date t to date $t+1$ on the portfolio obtained from the optimal investment decision in date t.

[31] Assuming homotheticity is a strong restriction. However, it keeps the calculation tractable and allows to better justify the aggregation of individual behavior to the behavior of a representative agent (see section 4.2.1.4).

[32] Homothetic utility functions have further advantages. First, because the marginal rates of substitution are wealth-independent, homotheticity implies that for constant return distributions the risk premia do not change over time as the aggregate wealth of the economy increases. Second, if all individuals in the economy have the same homothetic utility function, then aggregate behavior may be represented by a representative individual with the same utility function, even if the individuals have different levels of wealth (cf. section 4.2.1.4).

From the maximization problem in equation (4.12) it follows:

$$0 \stackrel{!}{=} \frac{\partial V}{\partial c_t} + \frac{\partial V}{\partial \mathrm{E}_t\{\cdot\}} \mathrm{E}_t \left\{ \frac{\partial J_{t+1}}{\partial \tilde{w}_{t+1}} \frac{\partial \tilde{w}_{t+1}}{\partial c_t} \right\}. \tag{4.13}$$

The budget constraint of the maximization problem yields $\partial \tilde{w}_{t+1} / \partial c_t = -\tilde{R}^{\mathrm{P}_t^*}_{t+1}$. Let $V_{x,t}$ denote the partial derivative of V with respect to its xth argument at date t. Then equation (4.13) yields the first order condition

$$V_{1,t} \stackrel{!}{=} V_{2,t} \mathrm{E}_t \left\{ \frac{\partial J_{t+1}}{\partial \tilde{w}_{t+1}} \tilde{R}^{\mathrm{P}_t^*}_{t+1} \right\} \tag{4.14}$$

$$\Leftrightarrow V_{1,t} \stackrel{!}{=} V_{2,t} \mathrm{E}_t \left\{ V_{1,t+1} \tilde{R}^{\mathrm{P}_t^*}_{t+1} \right\}. \tag{4.15}$$

Equation (4.15) follows from equation (4.14) using the Envelope-Theorem, which implies that in the optimum the marginal indirect utility of wealth equals the marginal utility of consumption, i.e. $\frac{\partial J_{t+1}}{\partial \tilde{w}_{t+1}} = V_{1,t+1}$ (see Huang and Litzenberger (1988, p. 196)). This can be seen by differentiating the indirect utility function of wealth, J_t, with respect to wealth w_t:

$$\frac{\partial J_t}{\partial w_t} = \frac{\partial V}{\partial c_t} \underbrace{\frac{\partial c_t}{\partial w_t}}_{1} + \underbrace{\frac{\partial V}{\partial \mathrm{E}_t\{\cdot\}} \mathrm{E}_t \left\{ \frac{\partial J_{t+1}}{\partial \tilde{w}_{t+1}} \underbrace{\frac{\partial \tilde{w}_{t+1}}{\partial w_t}}_{\tilde{R}^{\mathrm{P}_t^*}_{t+1}} + \frac{\partial J_{t+1}}{\partial \tilde{w}_{t+1}} \underbrace{\frac{\partial \tilde{w}_{t+1}}{\partial c_t}}_{-\tilde{R}^{\mathrm{P}_t^*}_{t+1}} \underbrace{\frac{\partial c_t}{\partial w_t}}_{1} \right\}}_{0}.$$

The optimality condition in equation (4.15) can be interpreted as follows: An additional unit of wealth at date t can be consumed or invested. An additional unit of consumption at date t yields $V_{1,t}$ additional units of time t utility. Investing the unit of wealth yields $\tilde{R}^{\mathrm{P}_t^*}_{t+1}$ additional units of wealth at date $t+1$. Consuming this additional wealth at date $t+1$ yields $(V_{1,t+1} \tilde{R}^{\mathrm{P}_t^*}_{t+1})$ additional units of time $t+1$ utility. The expected time t utility of this additional time $t+1$ utility is $V_{2,t} \mathrm{E}_t \{ V_{1,t+1} \tilde{R}^{\mathrm{P}_t^*}_{t+1} \}$. In the optimum the utility from both opportunities must be equal; for small amounts of wealth the investor is indifferent between consumption and investment. Rewriting equation (4.15) yields

$$1 \stackrel{!}{=} \mathrm{E}_t \left\{ \frac{V_{1,t+1}}{V_{1,t}} V_{2,t} \tilde{R}^{\mathrm{P}_t^*}_{t+1} \right\}, \tag{4.16}$$

which will be used later in this section. $(V_{1,t+1} V_{2,t})$ is the time t marginal utility of consumption at date $t+1$ and $V_{1,t}$ is the time t marginal utility of consumption at date t. $\frac{V_{1,t+1}}{V_{1,t}} V_{2,t}$ therefore is the individual's intertemporal marginal rate of substitution between consumption at date t and consumption at date $t+1$, $IMRS_{t+1}$[33].

[33]For a given level of utility the $IMRS_{t+1}$ measures the slope of the indifference curve, $IMRS_{t+1} = -dc_t / dc_{t+1}$.

(2) Optimal Investment Decision The maximization problem for the optimal portfolio allocation is:

$$\max_{\{v_t^i\}_{i=1}^I} V[c_t, \mathrm{E}_t\{J_{t+1}(\tilde{w}_{t+1}|\Omega_t)\}] \qquad \forall t < T \tag{4.17}$$

$$\text{s.t.} \qquad \tilde{w}_{t+1} = (w_t - c_t)\sum_{i=1}^{I} v_t^i \tilde{R}_{t+1}^i,$$

$$\sum_{i=1}^{I} v_t^i = 1.$$

Arbitrarily choosing asset k, $k \in \{1, \ldots, I\}$, as reference and combining the constraints yields

$$\tilde{w}_{t+1} = (w_t - c_t)(\tilde{R}_{t+1}^k + \sum_{\substack{i=1\\i\neq k}}^{I} v_t^i(\tilde{R}_{t+1}^i - \tilde{R}_{t+1}^k)), \qquad k \in \{1, 2, \ldots, I\}. \tag{4.18}$$

From the maximization problem (4.17) it follows:

$$0 \stackrel{!}{=} \frac{\partial V}{\partial \mathrm{E}_t\{\cdot\}} \mathrm{E}_t \left\{ \frac{\partial J_{t+1}}{\partial \tilde{w}_{t+1}} \frac{\partial \tilde{w}_{t+1}}{\partial v_t^i} \right\}. \tag{4.19}$$

Equation (4.18) yields $\partial\tilde{w}_{t+1}/\partial v_t^i = (w_t - c_t)(\tilde{R}_{t+1}^i - \tilde{R}_{t+1}^k)$ for $i \neq k$. Let $V_{x,t}$ again denote the partial derivative of V with respect to its xth argument at date t, then, again applying the Envelope-Theorem, equation (4.19) can be rewritten to obtain the first order condition

$$0 \stackrel{!}{=} \underbrace{(w_t - c_t)}_{\geq 0} \underbrace{V_{2,t}}_{>0} \mathrm{E}_t \left\{ V_{1,t+1}(\tilde{R}_{t+1}^i - \tilde{R}_{t+1}^k) \right\} \tag{4.20}$$

$$\stackrel{\text{if}(w_t - c_t)>0}{\Leftrightarrow} \qquad \mathrm{E}_t \left\{ V_{1,t+1} \tilde{R}_{t+1}^k \right\} \stackrel{!}{=} \mathrm{E}_t \left\{ V_{1,t+1} \tilde{R}_{t+1}^i \right\}. \tag{4.21}$$

This optimality condition can be interpreted as follows: As long as the right hand side of equation (4.20) is positive (negative), utility can be raised by raising (lowering) v_t^i in relation to v_t^k. In the optimum utility cannot be increased by reducing the fraction of one asset and increasing the fraction of another. The individual is then indifferent to selling a small amount of one asset and purchasing another asset instead (see equation (4.21)). This must hold for any two distinct assets i and k which are held by the individual and, in particular, for the individual's optimal portfolio at date t.

Optimal Consumption and Investment Decision and the Market Pricing Kernel The first order condition for the optimal consumption policy (equation (4.16)) and the first order condition for the optimal investment policy (equation (4.21)) can be combined to yield the first order condition for the optimal consumption and investment decision (Euler equation):

$$1 \stackrel{!}{=} \mathrm{E}_t \left\{ \underbrace{\frac{V_{1,t+1}}{V_{1,t}} V_{2,t}}_{IMRS_{t+1}} \tilde{R}^i_{t+1} \right\} \qquad \forall i. \tag{4.22}$$

This equation derived for individual j looks similar to equation (2.2) which suggests that the intertemporal marginal rate of substitution of any individual j, $IMRS_{t+1}$, may be equal to the market pricing kernel, $\tilde{\phi}_{t+1}$. In equilibrium, this is true, at least for complete and frictionless markets. The individuals trade in these markets until each individual realizes his optimal consumption and investment decision, that is until the marginal rates of substitution are equal across individuals (Huang and Litzenberger 1988, pp. 119-26). Thus, in equilibrium the marginal rate of substitution of any individual can be used to describe the equilibrium. However, the individuals are heterogeneous, i.e. they differ in their preferences, their endowments and expectations. Therefore, it would be convenient to determine the equilibrium by the marginal rate of substitution of some representative individual, usually called the "representative agent", that is endowed with aggregate wealth and consumes the aggregate consumption goods[34]. In the ensuing section it will be discussed that this is possible for complete markets, and, that under certain homogeneity assumptions regarding the individuals' preferences and expectations this may also be possible for incomplete markets. The question then is, what the utility function of the representative agent looks like.

Throughout the subsequent analysis, it is assumed that the behavior of the individuals can be represented by a representative agent and that the representative agent has preferences of the recursive non-expected utility form discussed in the previous sections. In equilibrium the representative agent consumes the aggregate consumption and holds the aggregate wealth of the economy (market portfolio). The return on the optimal portfolio held by the representative agent thus equals the return on the aggregate wealth portfolio, i.e. the market return, $\tilde{R}^{P^*_t}_{t+1} = \tilde{R}^M_{t+1}$. As a consequence of the assumed homotheticity of the recursive preferences, the existence of such a representative agent is theoretically justifiable in the present case, at least if the individuals are identical with exception of wealth[35]. In this case equation (4.22) determines the equilibrium rates of return for a given intertemporal marginal rate of substitution of the representative agent. For widely used preferences the intertemporal marginal rate of substitution is a function of

[34] In particular for empirical investigations it is necessary to replace individual-specific choice information by exogenous parameters of the economy, e.g., to replace the consumption of any particular individual by aggregate consumption.

[35] See discussion in the ensuing section 4.2.1.4.

the preference parameters and the growth rate of aggregate consumption[36]. A comparison of equations (2.2) and (4.22) reveals that the representative agent's intertemporal marginal rate of substitution for a given preference specification yields a theoretically founded parametric model of the market pricing kernel:

$$\ddot{\phi}_{t+1} = IMRS_{t+1} = \frac{V_{1,t+1}}{V_{1,t}} V_{2,t} \,. \tag{4.23}$$

Since marginal utilities, $V_{1,t}$ and $V_{1,t+1}$, as well as the discount factor, $V_{2,t}$, are assumed to be positive, the intertemporal marginal rate of substitution is nonnegative. Therefore, the market pricing kernel implied by consumption-based asset pricing models is nonnegative and hence satisfies the no-arbitrage condition.

4.2.1.4 Aggregation of Individual Behavior

This section discusses under which assumptions it is theoretically justifiable to assume that individuals can be aggregated to a single representative agent, an assumption common in empirical research.

If there is only one individual in the economy, the question of how to aggregate the individuals is trivial. The representative agent equals the single individual, i.e. has the same preferences, consumes all available consumption goods and holds the market portfolio. However, in reality an economy consists of more than one individual so that it has to be clarified whether and how the aggregate preferences and the aggregate consumption and investment decision, i.e. the aggregate behavior, can be described by a representative agent[37]. Usually the utility function in question, i.e. the utility function assumed for the individuals, is simply hypothesized for the representative agent, too. However, the existence of the representative agent and the utility function of the representative agent should ideally be derived from individual behavior using an appropriate aggregation theorem. Under specific conditions aggregate behavior can be modeled as the behavior of one representative agent whose choices coincide with the aggregate choices of the individuals.

---

[36]Therefore, taking the rates of return as given, equation (4.22) becomes a model for aggregate consumption behavior, as in modern permanent income models.

[37]Notice that if all or at least a considerable number of individuals have monotonic preferences, i.e. that these individuals strictly prefer more to less, then the market is arbitrage free (see Dybvig and Ross (1989, p. 60) or Jarrow (1988, pp. 27-9)). Hence a market pricing kernel exists (see section 2.1). The absence of arbitrage is a weaker condition than equilibrium since only relatively few individuals preferring more to less are needed to "bid away arbitrage opportunities, even in the presence of a sea of agents driven by 'animal spirits'" (Dybvig and Ross 1989, p. 57). Hence, the existence of individuals that strictly prefer more to less yields a utility theory based support for the assumption of the nonexistence of arbitrage opportunities in section 2.1.

First, suppose that all individuals are identical, i.e. the individuals have homogeneous preferences, endowments and expectations. Homogenous expectations means that conditional upon a particular future state of nature all individuals agree on the payoff to all the assets that are traded and all individuals agree on the probabilities of all future states. In this trivial case aggregate behavior can be represented by the behavior of a representative agent having the same preferences, the same endowment and the same expectations as each individual, "so that equilibrium rates of return are determined as if there exists only this" representative agent (Rubinstein 1974, p. 231). It is important to notice that in this case the representative agent is only endowed with a fraction of aggregate wealth and not with aggregate wealth itself. Similarly, he consumes only a fraction of aggregate consumption and not aggregate consumption itself.

However, the assumption of identical individuals is very strong and the aggregation problem can be solved even if individuals are allowed to differ in certain respects. The assumption of identical wealth, for example, can be relaxed, if all investors have identical homothetic utility functions, because these utility functions ensure that the optimality conditions are independent of wealth and therefore equal for all individuals[38]. In this case the representative agent can be assumed to be endowed with aggregate wealth and to consume aggregate consumption.

But identical homothetic utility functions are still a strong requirement. Less restrictive requirements for representing aggregate behavior by a representative agent, at least in equilibrium, can be made when the markets are complete. For complete markets lemma 1 in Constantinides (1982) states that heterogeneous individuals may be replaced by a representative agent, endowed with the aggregate endowment, leaving consumption and equilibrium prices unchanged. The intuition of this result is that after trading in complete markets, individuals become marginally homogeneous even though they are initially heterogeneous. The utility function of the representative agent turns out to be an optimized weighted average of the utility functions of the individuals (see also Duffie (1988, p. 23)). The weights are determined by the distribution of the initial endowments of the individuals, so that the utility function of the representative agent and therefore the prices in the economy will be affected by the distribution of the initial endowments across individuals (Huang and Litzenberger 1988, p. 144). If, in addition, (1) all individuals have homogeneous expectations and if (2) the individuals have time-additive expected utility functions that are increasing, strictly concave, and differentiable, then the representative agent also has time-additive expected utility (cf. Constantinides (1982, lemma 2) or Huang and Litzenberger (1988, p. 141)). But the utility function of the rep-

[38]This corresponds to the homothetic recursive preferences dealt with in the present analysis.

resentative agent still depends on the distribution of the initial endowments of the individuals.

Under specific conditions the equilibrium prices can also be determined independently of the distribution of the initial endowments[39]. If (1) all individuals have homogeneous expectations, (2) all individuals have time-additive expected utility, $U_t^j(c_t^j, \ldots, c_T^j) = \sum_{\tau=t}^{T} \beta_j^{T-\tau} u^j(c_\tau^j)$, where the utility function $u^j(c_\tau^j)$ is of the hyperbolic absolute risk aversion (HARA) class[40], and (3) all individuals j have the same preference parameter B^j and the same one-period discount factor β_j or (3') all individuals j have the preference parameter $B^j = 0$ (exponential utility), then the equilibrium rates of return are determined as if there exists only one representative individual with time-additive expected utility of the HARA-class, endowed with the arithmetic average wealth of the individuals (cf. Rubinstein (1974, aggregation theorems (ii)+(iii)) or Huang and Litzenberger (1988, pp. 144-8)). The assumption of homogenous expectations can be relaxed, if (1) $A^j = 0$ and $B^j = 1$ for all individuals, or (2) the market is complete, $B^j = 0$ for all individuals and all individuals have the same one-period discount factor β_j, or (3) the market is complete, $A^j = 0$ and $B^j = 1$ for all individuals and all individuals have the same one-period discount factor β_j and the same initial wealth (Rubinstein 1974, aggregation theorems (iv)-(vi)).

The discussed conditions for representing aggregate preferences of the individuals by the preferences of a representative agent are sufficient but not necessary. There may well exist weaker conditions that allow to represent aggregate behavior by the behavior of a representative agent (see discussion in Lewbel (1989)). However, the behavior of the representative agent, i.e. the behavior of the aggregate, may differ considerably from the behavior of the individuals (cf. Kirman (1992)). "The representative [agent] ...is a purely mathematical result and need not have economic content" (Lewbel 1989, p. 622). Especially, it is difficult to aggregate recursive preferences across individuals. In the case of homogeneous expectations of the individuals and the existence of complete markets, the utility function of the representative agent is generally not even weakly recursive (Epstein 1992, pp. 24-5). Only for identical individuals with exception of wealth the aggregation of homo-

[39]This property is known as the "aggregation property" (Huang and Litzenberger 1988, p. 144).

[40]HARA-class utility is defined as follows:

$$\begin{aligned} u^j(c_\tau^j) &= \frac{1}{B^j - 1}(A^j + B^j c_\tau^j)^{\frac{B^j-1}{B^j}} \quad \text{with} \quad A^j + B^j c_\tau^j > 0 \\ \text{and} \quad u^j(c_\tau^j) &= \ln(A^j + c_\tau^j) \quad \text{for} \quad B^j = 1 \\ \text{and} \quad u^j(c_\tau^j) &= -\exp(-c_\tau^j / A^j) \quad \text{for} \quad B^j = 0. \end{aligned}$$

Most commonly-used utility functions such as the quadratic, constant absolute risk aversion or constant relative risk aversion utility function can be obtained as special cases of the HARA-class by choosing particular values of A^j and B^j.

thetic recursive preferences yields a similar recursive utility function for the representative agent.

So, what can be done to model aggregate behavior, if the individuals do not satisfy this homogeneity conditions? The "current lack of a superior alternative" (Epstein 1992, p. 25) and the argument, that a representative agent "vastly simplifies a great deal of macro work and thought" (Lewbel 1989, p. 631) leads back to assuming the existence of the representative agent with a specific utility function, ignoring valid aggregation concerns. A consequence of this approach is that, whenever a representative agent model is tested, a joint hypothesis is tested: the validity of the particular model of interest and the hypothesis that the choices of the aggregate can indeed be described as the choices of a representative agent (see Kirman (1992, p. 125)). A particular parametric model of the market pricing kernel derived from the intertemporal consumption and investment choice of a representative agent may therefore be rejected because the representation of all individuals by the representative agent is rejected and not the parametric model derived from individual preferences.

4.2.1.5 Summary

In the consumption-based asset pricing framework, the market pricing kernel is derived from the optimal intertemporal consumption and investment decision of a representative agent. In this framework the market pricing kernel equals the intertemporal marginal rate of substitution of the representative agent. The intertemporal marginal rate of substitution is assumed to be positive so that the market pricing kernel implied by consumption-based asset pricing models is nonnegative and therefore satisfies the no-arbitrage condition. Equation (2.2) can be interpreted as the representative agent's Euler equation, the first order condition for the optimal portfolio-consumption choice. The derived parametric model of the market pricing kernel depends on the preference specification.

Time-additive expected utility has some important shortcomings. First, it does not allow to express preferences concerning the timing of the resolution of uncertainty, and, second, risk aversion and time preference are not separately specifiable. To overcome these shortcomings, the more general recursive non-expected utility class suggested by Kreps and Porteus (1978) is introduced and the conditions for the optimal intertemporal consumption and investment choice are derived. The aggregation of individual behavior to the behavior of a representative agent is theoretically justifiable only under strong assumptions on preferences. Therefore, one could argue that the existence of a representative agent is an unrealistic assumption. However, the theoretically founded functional form of the market pricing kernel derived from the consumption-based equilibrium asset pricing approach may hold even if a representative agent does not exist. It should therefore be seen as an alternative which is at least as good as ad-hoc approaches. Moreover, as a consequence of its foundation on individual preferences the derived model of the market pricing kernel is economically interpretable and the influence of policy changes may be analyzed. The next sections use the concept of recursive preference to derive alternative functional forms of the market pricing kernel from alternative parametrizations of the recursive preferences. Section 4.2.2 deals with the special case of time-additive expected utility. 4.2.3 uses a parametrization of the recursive preferences independently suggested by Epstein and Zin (1989) and Weil (1989).

4.2.2 Representative Agent with Time-additive Expected Utility

Section 4.2.2.1 introduces time-additive expected utility with constant relative risk aversion as a parametrization of the recursive preferences given in equation (4.2). Using the optimality condition derived in the previous section, section 4.2.2.2 derives the market pricing kernel for this utility specification. Section 4.2.2.3 employs the obtained parametric model of the market pricing kernel to derive the Consumption-based CAPM (C-CAPM). Finally, section 4.2.2.4 illustrates that for time-additive expected logarithmic utility which is a special case of time-additive expected utility with constant unitary relative risk aversion, the C-CAPM is closely related to the static one-period CAPM.

4.2.2.1 Time-additive Expected Utility with Constant Relative Risk Aversion

Assume that the representative agent's preferences can be represented by time-additive expected utility[41]

$$U_t(c_t, \tilde{c}_{t+1}, \ldots) = \sum_{\tau=t}^{\infty} \beta^{\tau-t} \mathrm{E}\{u(\tilde{c}_\tau)\} \qquad 0 < \beta < 1, \tag{4.24}$$

where $u(c)$ is the constant relative risk aversion utility function

$$\begin{aligned} u(c) &= \frac{c^{1-\alpha}}{1-\alpha} \quad \text{for} \quad 0 \leq \alpha < \infty \quad , \quad \alpha \neq 1 \quad \text{and} \\ u(c) &= \ln(c) \quad \text{for} \quad \alpha = 1. \end{aligned}$$

[41] Time-additive expected utility with constant relative risk aversion is a special case of the HARA-class utility. Therefore, the assumed existence of a representative agent is theoretically justified by an appropriate aggregation theorem (see section 4.2.1.4). The representative agent is assumed to live for ever.

α is the parameter of relative risk aversion and β is the discount factor. The discount factor should be less than 1.

Rewriting time-additive expected utility with constant relative risk aversion in the recursive structure of Kreps and Porteus (1978) given in equation (4.2) yields

$$\begin{aligned} U_t(c_t, \tilde{c}_{t+1}, \ldots) &= V[c_t, \mathrm{E}_t\{U_{t+1}(\tilde{c}_{t+1}, \ldots)\}] \\ &= \frac{1}{1-\alpha} c_t^{1-\alpha} + \beta \mathrm{E}_t\{U_{t+1}(\tilde{c}_{t+1}, \ldots)\}. \end{aligned} \quad (4.25)$$

Notice, that the aggregator function $V[\cdot, \cdot]$ is linear in the second argument.

4.2.2.2 Deriving the Market Pricing Kernel

The partial derivatives necessary for the derivation of the market pricing kernel (see equation (4.23)) are

$$V_{1,t} = c_t^{-\alpha}, \qquad V_{1,t+1} = \tilde{c}_{t+1}^{-\alpha} \qquad \text{and} \qquad V_{2,t} = \beta,$$

so that the intertemporal marginal rate of substitution and hence the market pricing kernel is given by

$$\tilde{\phi}_{t+1} = IMRS_{t+1} = \beta \left(\frac{\tilde{c}_{t+1}}{c_t} \right)^{-\alpha}. \quad (4.26)$$

The discussion in section 4.2.1.2 has shown that time-additive expected utility implies that the elasticity of intertemporal substitution, σ, is equal to the reciprocal of the parameter of relative risk aversion, α. This can easily be illustrated in the present case of constant relative risk aversion. Substituting equation (4.26) in the definition of the elasticity of intertemporal substitution (equation (4.5)) yields

$$\begin{aligned} \sigma &= \frac{d \frac{c_t}{\tilde{c}_{t+1}}}{dIMRS_{t+1}} \frac{IMRS_{t+1}}{\frac{c_t}{\tilde{c}_{t+1}}} \\ &= \frac{1}{\alpha\beta} \left(\frac{c_t}{\tilde{c}_{t+1}} \right)^{1-\alpha} \beta \left(\frac{c_t}{\tilde{c}_{t+1}} \right)^{\alpha} \left(\frac{c_t}{\tilde{c}_{t+1}} \right)^{-1} = \frac{1}{\alpha}. \end{aligned} \quad (4.27)$$

4.2.2.3 The Consumption-based CAPM

The derived parametric model of the market pricing kernel (equation (4.26)) yields a Consumption-based CAPM (C-CAPM). Substituting equation (4.26) into the linear asset pricing model in terms of the market pricing kernel given in equation (2.9) yields

$$\mathrm{E}_t\left\{\tilde{R}^i_{t+1} - \tilde{R}^z_{t+1}\right\} = -\left(\mathrm{E}_t\left\{\left(\frac{\tilde{c}_{t+1}}{c_t}\right)^{-\alpha}\right\}\right)^{-1} \cdot \mathrm{cov}_t\left(\left(\frac{\tilde{c}_{t+1}}{c_t}\right)^{-\alpha}, \tilde{R}^i_{t+1}\right). \tag{4.28}$$

In this C-CAPM the systematic risk of an asset is measured by the covariance of its return with a function of consumption growth (cf. Grossman and Shiller (1981)), whereas in the static CAPM the systematic risk is measured by the covariance of the asset's return with the return on the market.

The implied asset pricing relation is non-linear and therefore not easy to analyze with traditional linear estimation techniques[42]. These techniques can only be applied if a linearized version of the C-CAPM is used, where the excess return on each asset i is given by a linear function of $\mathrm{cov}_t((\tilde{c}_{t+1}/c_t), \tilde{R}^i_{t+1})$. Breeden (1979) derives a linearized version in a continuous-time framework. In discrete-time a linearized version can be derived by assuming conditional joint lognormality of consumption growth and asset returns (cf. Breeden and Litzenberger (1978) and Hansen and Singleton (1983)). Most studies estimate the C-CAPM in discrete-time. Therefore, the derivation of the linearized C-CAPM assuming conditional joint lognormality is illustrated in the following excursus.

Excursus: Consumption-based CAPM Assuming Joint Lognormality Assume that the conditional joint distribution of $\tilde{\lambda}_{t+1} = \tilde{c}_{t+1}/c_t$ and $\tilde{R}^i_{t+1}$ is lognormal. Substituting equation (4.26) in equation (4.22) and taking logs yields[43]

$$\begin{aligned} 0 &\stackrel{!}{=} \ln \mathrm{E}_t\{\beta\tilde{\lambda}^{-\alpha}_{t+1}\tilde{R}^i_{t+1}\} \\ &= \ln\beta - \alpha\mathrm{E}_t\{\ln\tilde{\lambda}_{t+1}\} + \mathrm{E}_t\{\ln\tilde{R}^i_{t+1}\} \\ &\quad + \frac{1}{2}\left(\alpha^2\mathrm{var}_t(\ln\tilde{\lambda}_{t+1}) + \mathrm{var}_t(\ln\tilde{R}^i_{t+1})\right) \\ &\quad - \alpha\mathrm{cov}_t(\ln\tilde{\lambda}_{t+1}, \ln\tilde{R}^i_{t+1}). \end{aligned} \tag{4.29}$$

[42] Singleton (1990), (1994) surveys different estimation techniques for intertemporal asset pricing models. Besides these estimation techniques there exist other methods such as the variance bound approach (see section 3.3.2) or the calibration approach (see chapter 5) for the empirical investigation of the C-CAPM. The differences between the estimation and the calibration approach are discussed in section 5.1.

[43] See appendix A.1 for a derivation of the expected value of the product of two random variables that are jointly lognormally distributed. For a derivation of the linearized C-CAPM similar to that presented here see Campbell (1996, pp. 14-5).

For a risk-free asset f it follows

$$0 \stackrel{!}{=} \ln\beta - \alpha \mathrm{E}_t\{\ln\tilde{\lambda}_{t+1}\} + \ln R^f_{t+1} + \frac{1}{2}\alpha^2 \mathrm{var}_t(\ln\tilde{\lambda}_{t+1}). \tag{4.30}$$

Solving (4.29) for $\mathrm{E}_t\{\ln\tilde{R}^i_{t+1}\}$ and (4.30) for $\ln R^f_{t+1}$ and calculating the difference yields

$$\begin{aligned} \mathrm{E}_t\{\ln\tilde{R}^i_{t+1}\} - \ln R^f_{t+1} &= \alpha \mathrm{cov}_t(\ln\tilde{\lambda}_{t+1}, \ln\tilde{R}^i_{t+1}) - \frac{1}{2}\mathrm{var}_t(\ln\tilde{R}^i_{t+1}) \\ \Leftrightarrow \quad \ln \mathrm{E}_t\{\tilde{R}^i_{t+1}\} - \ln R^f_{t+1} &= \alpha \mathrm{cov}_t(\ln\tilde{\lambda}_{t+1}, \ln\tilde{R}^i_{t+1}) \\ \Leftrightarrow \quad \ln \mathrm{E}_t\{\tilde{R}^i_{t+1}/R^f_{t+1}\} &= \alpha \mathrm{cov}_t(\ln\frac{\tilde{c}_{t+1}}{c_t}, \ln\tilde{R}^i_{t+1}). \end{aligned} \tag{4.31}$$

The left-hand side of (4.31) can be interpreted as a logarithmic risk premium on asset i over the return on a risk-free asset. This risk premium equals the coefficient of relative risk aversion times the covariance of the logarithmic return on asset i with logarithmic consumption growth. This formulation obtained by assuming conditional joint lognormality of asset returns and consumption growth is economically similar to equation (17') in Breeden (1979, p. 275) which is derived in a continuous-time environment. Define the conditional consumption beta of asset i to be $\beta^{\tilde{c},\tilde{R}^i}_{t+1} = \mathrm{cov}_t(\ln(\tilde{c}_{t+1}/c_t), \ln\tilde{R}^i_{t+1})/\mathrm{var}_t(\ln(\tilde{c}_{t+1}/c_t))$. Equation (4.31) must hold for any portfolio with gross returns $\tilde{R}^c_{t+1}$. Using this portfolio as a second reference asset equation (4.31) yields[44]

$$\ln \mathrm{E}_t\{\tilde{R}^i_{t+1}/R^f_{t+1}\} = \beta^{\tilde{c},\tilde{R}^i}_{t+1}/\beta^{\tilde{c},\tilde{R}^c}_{t+1} \ln \mathrm{E}_t\{\tilde{R}^c_{t+1}/R^f_{t+1}\}. \tag{4.32}$$

If there exists a portfolio with logarithmic returns $\ln\tilde{R}^{c*}_{t+1}$ that are perfectly correlated with logarithmic consumption growth, it follows that $\beta^{\tilde{c},\tilde{R}^i}_{t+1}/\beta^{\tilde{c},\tilde{R}^{c*}}_{t+1} = \beta^{\tilde{R}^{c*},\tilde{R}^i}_{t+1}$ and equation (4.32) can be rewritten as

$$\ln \mathrm{E}_t\{\tilde{R}^i_{t+1}/R^f_{t+1}\} = \beta^{\tilde{R}^{c*},\tilde{R}^i}_{t+1} \ln \mathrm{E}_t\{\tilde{R}^{c*}_{t+1}/R^f_{t+1}\}. \tag{4.33}$$

"Thus, the relevant risk of a security's return may be summarized by a single beta with respect to consumption" (Breeden 1979, p. 276).

Empirical Evidence The parametric model of the market pricing kernel (4.26) will be applied in the calibration exercises in the second part of this analysis. Therefore the main results of a number of studies estimating the

[44]See section 2.3 for a detailed description of similar steps.

related C-CAPM are now briefly presented; first, results of studies applying the traditional linear estimation techniques, second, results of studies applying more recent non-linear estimation techniques. Empirical evidence from calibration exercises using time-additive expected utility with constant relative risk aversion is discussed in section 5.3.

Hansen and Singleton (1983) estimate the C-CAPM assuming joint lognormality of consumption growth and asset returns using US monthly time series data. For a value-weighted stock index only they cannot reject the model, but using Treasury bill returns only or stock index returns and Treasury bill returns together the model is rejected. Mankiw, Rotemberg, and Summers (1985) also reject the linearized model using quarterly US Treasury bill returns. The linearized C-CAPM derived in continuous-time is rejected by Grossman, Melino, and Shiller (1987) using US time series data. Bossard (1988) and Missong (1994) apply the linearized discrete-time model to German data. While Bossard (1988) cannot reject the model, Missong (1994) finds no evidence for it. Notice that only Hansen and Singleton (1983) estimate the model simultaneously for more than one asset.

Brown and Gibbons (1985) criticize that the estimation results may not be robust to departures from the assumed distributions and that, since the advantage of utility-based pricing theories is the lack of strong distributional assumptions regarding security returns, empirical investigations of these theories should not require distributional assumptions. They suggest a nonparametric estimator for the parameter of relative risk aversion that is directly applicable to the Euler equation so that no linearization has to be made. Applying this estimator to monthly stock returns they show that $\alpha = 1$ (logarithmic utility) is consistent with the data over the period 1926-1981. The distribution-free Generalized Method of Moment (GMM) estimation technique introduced by Hansen and Singleton (1982) is also directly applicable to the Euler equation[45] and thus overcomes the shortcoming of strong distributional assumptions. This method can be interpreted as a generalization of Brown and Gibbons's nonparametric estimator. Hansen and Singleton's application of the GMM technique using monthly US data for the period 1959 to 1978 provides mixed results for an equally weighted stock index, whereas the model cannot be rejected using a value weighted stock index[46]. Using more than one return time series simultaneously they find strong evidence against the C-CAPM. Singleton (1994) analyzes possible

[45]Hansen and Singleton (1982) show how to use the fact that $\mathrm{E}_t\{\tilde{u}_{t+1}\} = 0$, where $\tilde{u}_{t+1}$ is the error term $\tilde{u}_{t+1} = \beta(\tilde{c}_{t+1}/c_t)^{-\alpha}\tilde{R}^i_{t+1} - 1$, to construct instrumental variables estimators of the unknown parameters and to test overidentifying restrictions. The comparison of the calibration and the estimation approach in section 5.1 provides a more detailed description of the GMM estimation technique.

[46]The results originally obtained by Hansen and Singleton (1982) are quite favorable to the C-CAPM but Hansen and Singleton later corrected their results due to an error discovered in their consumption series (see Hansen and Singleton (1984)).

sources of mismeasurement such as the length of the chosen data interval or autocorrelation in data that may lead to the rejection of the model. However, he concludes "that the more important factors underlying previous findings are likely to be misspecification of preferences or violation of the conditions for aggregation across consumers" (Singleton 1994, p. 290). Another study estimating the C-CAPM and accounting for measurement problems associated with reported consumption is provided by Breeden, Gibbons, and Litzenberger (1989). A more recent study by Hardouvelis, Kim, and Wizman (1996) estimates the C-CAPM (among other models) simultaneously for 10 portfolios of US stocks sorted by company size in the period 1959 to 1991 using the GMM technique. The overidentifying restrictions are rejected but in contrast to previous studies a significantly positive parameter of relative risk aversion of 122.4 is estimated (Hardouvelis, Kim, and Wizman 1996, p. 291). Bossard (1988) and Missong (1994) apply the GMM technique to German data. Both studies find that the model is generally not supported or that the estimated parameters are not significant. Similar findings are made by Lund and Engsted (1996) for the Danish, the German, the Swedish and the UK stock market. Although they cannot reject the C-CAPM in general, the estimates of the parameter of relative risk aversion are mostly implausible, i.e. negative, and imprecise, i.e. exhibit a large standard deviation.

Overall, empirical evidence indicates that either the C-CAPM does not hold or, considering the estimation problems, that the applied estimation techniques are insufficient. Especially jointly explaining the returns on different assets, e.g., risk-free and risky returns, seems to pose a problem to the C-CAPM.

4.2.2.4 Time-additive Expected Logarithmic Utility and the Static CAPM

Time-additive expected logarithmic utility is the special case of time-additive expected utility with constant relative risk aversion for $\alpha = 1$ (cf. equation (4.24)). The parametric model of the market pricing kernel derived from time-additive expected logarithmic utility is then given by (cf. equation (4.26))

$$\tilde{\phi}_{t+1} = \beta \frac{c_t}{\tilde{c}_{t+1}}. \tag{4.34}$$

As noted in section 4.1, it can be shown (see, for example, Giovannini and Weil (1989, appendix B.1)) that in the case of time-additive expected logarithmic utility optimal consumption is proportional to wealth with a constant proportionality factor, which means that a constant fraction of wealth is consumed independent of expectations of future variables. Hence, time-additive expected logarithmic utility implies myopia in consumption and savings decisions[47]. For an infinite planning horizon[48] the proportionality factor equals $1-\beta$, i.e. $c_t = (1-\beta)w_t$. In the case of the representative agent this proportionality holds for aggregate consumption and aggregate wealth. The budget constraint for the representative agent is $\tilde{w}_{t+1} = (w_t - c_t)\tilde{R}^M_{t+1}$, where $\tilde{R}^M_{t+1}$ denotes the gross return on the aggregate wealth portfolio (market portfolio). Combining both equations yields $\tilde{c}_{t+1} = \beta c_t \tilde{R}^M_{t+1}$. Substituting this expression in equation (4.34) shows that for logarithmic utility the market pricing kernel is equal to the reciprocal of the gross return on the aggregate wealth portfolio, a result first noted by Rubinstein (1976):

$$\tilde{\phi}_{t+1} = (\tilde{R}^M_{t+1})^{-1}. \tag{4.35}$$

Approximate $(1/\tilde{R}^M_{t+1})$ by $(2 - \tilde{R}^M_{t+1})$[49]. Then $\tilde{R}^M_{t+1}$ is perfectly negatively correlated with the market pricing kernel so that $\tilde{R}^\phi_{t+1}$ in equation (2.13) can be replaced by the return on the market portfolio, $\tilde{R}^M_{t+1}$. This yields a conditional asset pricing model equivalent to the static one-period CAPM[50]

$$\mathrm{E}_t\{\tilde{R}^i_{t+1}\} - R^f_{t+1} = \beta^{\tilde{R}^M,\tilde{R}^i}_{t+1} \cdot \left(\mathrm{E}_t\{\tilde{R}^M_{t+1}\} - R^f_{t+1}\right). \tag{4.36}$$

The equivalence of a period-by-period application of the static one-period CAPM and the intertemporal consumption-based CAPM for time-additive expected logarithmic utility is a consequence of the fact that this utility specification does not only imply myopia in consumption and savings decisions but also implies myopia in investment decisions. This means that the individuals ignore stochastic variation in the investment opportunity set in their portfolio allocation. In general, myopic valuation means that intertemporal asset pricing equals the period-by-period application of one-period pricing models that are independent of the future investment opportunity set (cf. Franke (1984)).

4.2.3 Representative Agent with Recursive Non-expected Utility

Having presented the time-additive expected utility with constant relative risk aversion, the Epstein and Zin (1989) parametrization of the recursive Kreps and Porteus (1978) preferences given in equation (4.2) is analyzed next. Section 4.2.3.1 introduces this parametrization. It illustrates that risk

---

[47]It is important to notice that myopia in consumption and savings decisions does not imply myopia in investment decisions. Investment decisions are said to be myopic if the portfolio allocation is independent of expectations of future variables.

[48]This condition is satisfied, because the representative agent is assumed to live for ever.

[49]This approximation is exact in continuous-time (cf. Missong (1996)).

[50]See also the discussion in section 4.1.

aversion and elasticity of intertemporal substitution are separately specified and how this parametrization allows to model preferences over the timing of the resolution of uncertainty. Using the previously derived optimality condition for the representative agent, section 4.2.3.2 derives the market pricing kernel for this utility specification. Section 4.2.3.3 employs the obtained parametric model of the market pricing kernel to derive a two-factor CAPM.

4.2.3.1 Recursive Non-expected Utility

Recall, Kreps and Porteus define an aggregator function $V[c_t, \mathrm{E}_t\{U_{t+1}(\tilde{c}_{t+1}, \ldots)\}]$ that aggregates current consumption and the expected value of the utility of future consumption to the total utility of the consumption path (see equation (4.2)). In contrast, Epstein and Zin use an aggregator $\mathcal{V}[c_t, \mu_t(\mathcal{U}_{t+1}(\tilde{c}_{t+1}, \ldots))]$ where the second argument is not the expected value but a certainty equivalent of random future utility. This modification allows for a more intuitive interpretation because the aggregator function $\mathcal{V}[\cdot, \cdot]$ exclusively captures the time preference and the certainty equivalent function $\mu_t(\cdot)$ exclusively captures the risk aversion. Risk aversion and time preference are therefore separately modeled. For the derivation of the market pricing kernel it will be shown that the parametrization of Epstein and Zin can be transformed into the preference structure of Kreps and Porteus via a monotonically increasing transformation that does not change the ordinal preferences.

Epstein and Zin (1989) use a certainty equivalent function $\mu_t(\cdot)$ that exhibits constant relative risk aversion, i.e. $\mu_t(\cdot) = (\mathrm{E}_t\{(\cdot)^{1-\alpha}\})^{1/(1-\alpha)}$ for $\alpha \geq 0$ and $\alpha \neq 1$, respectively $\mu_t(\cdot) = \exp(\mathrm{E}_t\{\ln(\cdot)\})$ for $\alpha = 1$. α again denotes the parameter of relative risk aversion and $\mathrm{E}_t\{\cdot\}$ denotes the expectation operator given the information available at date t. An example of how this certainty equivalent function works is given in figure 4.3. In this example it is assumed that at date $t+1$ two different states can occur with equal probability. Consider consumption path M. For this path the utility at date $t+1$, $\mathcal{U}_{t+1}(c_{t+1}, \tilde{c}_{t+2} \ldots)$, equals 5 in one state and 15 in the other. The other three consumption paths exhibit either a lower risk or a higher risk than consumption path M. The right graph in the figure presents the certainty equivalents $\mu_t(\mathcal{U}_{t+1}(\tilde{c}_{t+1}, \ldots))$ for different levels of risk aversion. The certainty equivalent decreases if the risk of the consumption path increases or if the degree of risk aversion increases. Since risk aversion is exclusively captured by the certainty equivalent function and since the form of this function ensures that the certainty equivalent of a deterministic value is the value itself, the preference ranking of deterministic consumption paths is not affected by changes in the degree of risk aversion within this class of utility functions.

Figure 4.3: Certainty equivalents of random consumption paths differing exclusively in the risk. The certainty equivalents are depicted as functions of risk aversion α.

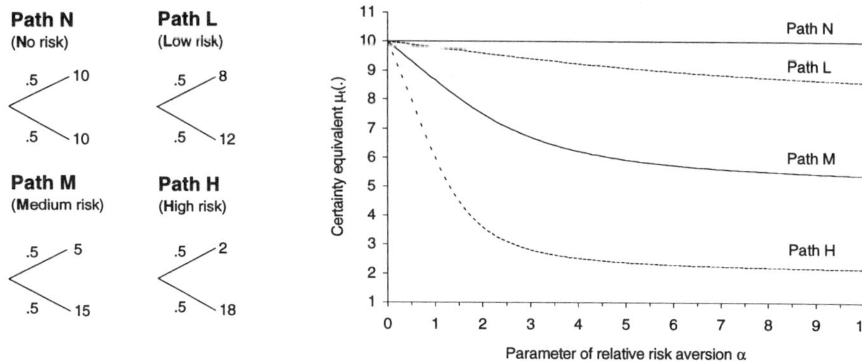

The aggregator function $\mathcal{V}[c_t, \mu_t(\cdot)] = [c_t^{1-\rho} + \beta\mu_t(\cdot)^{1-\rho}]^{\frac{1}{1-\rho}}$ for $\rho \geq 0$ and $\rho \neq 1$, respectively $\mathcal{V}[c_t, \mu_t(\cdot)] = \exp(\ln c_t + \beta \ln \mu_t(\cdot))$ for $\rho = 1$, has CES structure (constant elasticity of substitution) with the elasticity of intertemporal substitution σ equal to $1/\rho$. The value of σ can be anywhere between 0 and ∞. The larger σ, the greater the substitutability between current consumption and the certainty equivalent of random future utility. The limiting case of $\sigma = 0$ is where both must be in a fixed relation as complements to each other. The other limiting case, where σ goes to infinity is where both are perfect substitutes for each other. A high elasticity of intertemporal substitution σ, i.e. a low value of ρ, therefore characterizes the willingness to substitute consumption over time which results in a low aversion to intertemporal fluctuations of consumption. On the other hand, a low elasticity of intertemporal substitution σ, i.e. a high value of ρ, reflects a high aversion to intertemporal fluctuations. Thus, ρ can be interpreted as the aversion to intertemporal fluctuations and σ as the preference for a smooth consumption path. β is the discount factor. The discount factor measures the preference of an early or late consumption. Hence, the aggregator function fully captures the time preference, consisting of the discount factor and the elasticity of intertemporal substitution, and is not affected by changes in the degree of risk aversion. Figure 4.4 presents an example for the preference of a smooth consumption path, or, in other words, for the aversion to intertemporal fluctuations[51]. Because the discount factor β is set to one, the consumption paths differ exclusively in their smoothness. The right graph in figure 4.4 depicts the utility derived from these consumption paths as a function of the elasticity of intertemporal substitution. Utility

[51]Notice that for the sake of clarity of the illustration the transformed aggregator function $\mathcal{V}[c_t, \mu_t(\cdot)] = [.5 \cdot c_t^{1-\rho} + .5 \cdot \beta\mu_t(\cdot)^{1-\rho}]^{1/(1-\rho)}$ is used in this example.

Figure 4.4: Utility of deterministic consumption paths differing exclusively in the smoothness. Utility is depicted as a function of the elasticity of intertemporal substitution σ. The discount factor is $\beta = 1$.

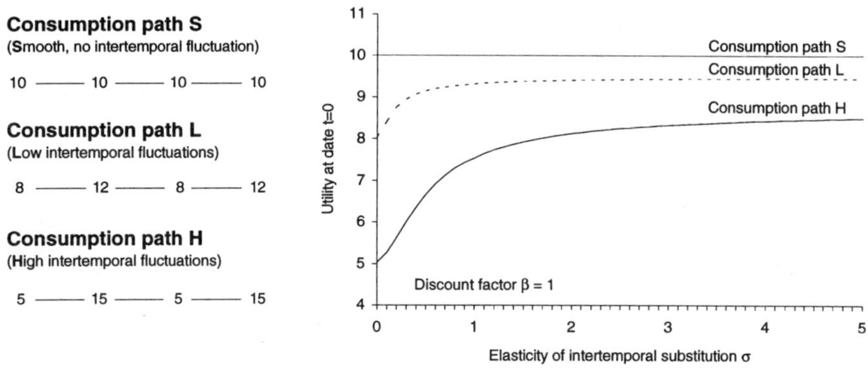

increases, if the elasticity of intertemporal substitution increases, i.e. if the aversion to intertemporal fluctuations decreases, and if the intertemporal variability of consumption decreases.

Combining the aggregator function and the certainty equivalent function yields[52]

$$
\begin{aligned}
\mathcal{U}_t(c_t, \tilde{c}_{t+1}, \ldots) &= \mathcal{V}[c_t, \mu_t(\mathcal{U}_{t+1}(\tilde{c}_{t+1}, \ldots))] \qquad (4.37)\\
&= \left[c_t^{1-\rho} + \beta\mu_t(\mathcal{U}_{t+1}(\tilde{c}_{t+1}, \ldots))^{1-\rho}\right]^{\frac{1}{1-\rho}}\\
&= \left[c_t^{1-\rho} + \beta\left(\mathrm{E}_t\left\{\mathcal{U}_{t+1}(\tilde{c}_{t+1}, \ldots)^{1-\alpha}\right\}\right)^{\frac{1-\rho}{1-\alpha}}\right]^{\frac{1}{1-\rho}}\\
\text{for} \quad & 0 < \beta \leq 1,\ \rho \geq 0,\ \rho \neq 1,\ \alpha \geq 0,\ \alpha \neq 1.
\end{aligned}
$$

In contrast to the time-additive expected utility specification this recursive non-expected utility representation clearly distinguishes between time and risk preference because the elasticity of intertemporal substitution is not restricted to equal the reciprocal of the parameter of risk aversion. However, if the elasticity of intertemporal substitution is restricted to equal the reciprocal of the parameter of risk aversion, $\sigma = 1/\alpha = 1/\rho$, the recursive structure (4.37) is equivalent to time-additive expected utility with constant relative risk aversion.

[52]For $\rho = 1$, the recursive utility function is $\mathcal{U}_t(\cdot) = c_t(\mathrm{E}_t\{\mathcal{U}_{t+1}(\cdot)^{1-\alpha}\})^{\frac{\beta}{1-\alpha}}$. Using L'Hôpital's rule, it can be shown that this is the limit of equation (4.37) for $\rho \to 1$. For $\alpha = 1$ the utility function equals $\mathcal{U}_t(\cdot) = [c_t^{1-\rho} + \beta(\exp(\mathrm{E}_t\{\ln \mathcal{U}_{t+1}(\cdot)\}))^{1-\rho}]^{\frac{1}{1-\rho}}$ and for $\alpha = 1$ and $\rho = 1$, $\mathcal{U}_t(\cdot) = c_t(\exp(\mathrm{E}_t\{\ln \mathcal{U}_{t+1}(\cdot)\}))^{\beta}$.

Figure 4.5: Utility of random consumption paths differing exclusively in the time of the resolution of uncertainty. Utility is depicted as a function of risk aversion α. The elasticity of intertemporal substitution is $\sigma = .5$, the discount factor is $\beta = .95$.

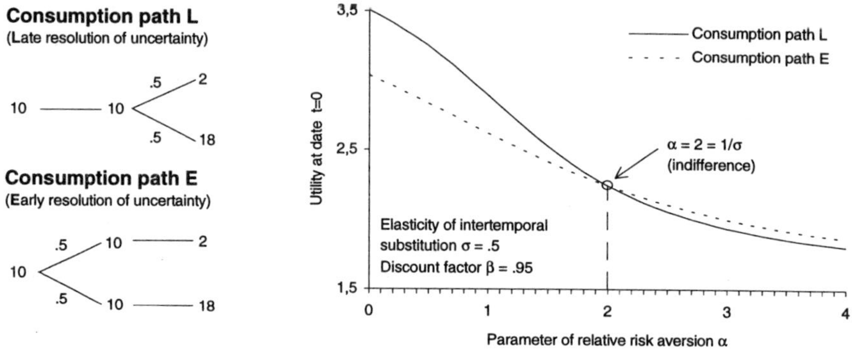

As illustrated in section 4.2.1.2, the recursive Kreps and Porteus preferences allow to represent preferences over the timing of the resolution of uncertainty. An early [late] resolution of uncertainty is preferred, if the aggregator function is convex [concave] in its second argument. Notice, that the second argument was the expected value of the utility of future consumption. In the given parametrization an early [late] resolution of uncertainty is preferred if $\alpha >$ [<]ρ (Epstein and Zin 1989, p. 952)[53]. One can now raise the question of how to interpret the link between the relation between risk aversion α and elasticity of intertemporal substitution $\sigma = 1/\rho$ and the preference over the timing of the resolution of uncertainty. Recall that ρ is a measure for the aversion to intertemporal fluctuations. If $\alpha > \rho$, i.e. if risk aversion is larger than the aversion to intertemporal fluctuations, an early resolution of uncertainty is preferred, even if it leads to deterministic consumption paths exhibiting fluctuations with certainty. If, however, the individual is more averse to these fluctuations than to risk, i.e. $\alpha < \rho$, then a late resolution of uncertainty is preferred. In general, there is a trade-off between safety and intertemporal stability of utility. If $\alpha = \rho$, i.e. in the case where recursive non-expected utility equals time-additive expected utility, the individual is indifferent to the resolution of uncertainty. This is illustrated in the example in figure 4.5. The two random consumption paths L and E differ exclusively in the time of the resolution of uncertainty. The right graph presents the utility derived from both consumption paths for different levels of risk aversion α and constant elasticity of intertemporal

[53]In the ensuing section an ordinally equivalent parametrization in the Kreps and Porteus (1978) structure is derived (equation (4.38)). Differentiating (4.38) twice with respect to $\mathrm{E}_t\{U_{t+1}(\tilde{c}_{t+1},\ldots)\}$ the condition $\alpha >$ [<]ρ for convexity [concavity] can be obtained.

substitution $\sigma = .5$. If $\alpha = 2$, i.e. if α is the reciprocal of σ, both consumption paths yield the same utility. However, if α is larger [smaller] than 2, the consumption path E [L], i.e. an early [late] resolution of uncertainty, is preferred.

4.2.3.2 Deriving the Market Pricing Kernel

A parametric model of the market pricing kernel is derived from the Epstein and Zin (1989) parametrization of the recursive preferences given in equation (4.37). This parametrization is based on an aggregator function, $\mathcal{V}[\cdot,\cdot]$, that has a certainty equivalent of random future utility as the second argument. The optimality conditions derived in section 4.2.1.3, however, apply to the recursive structure of Kreps and Porteus (1978) given in equation (4.2), where the second argument of the aggregator function, $V[\cdot,\cdot]$, is the expected value of random future utility. To permit the application of these optimality conditions the Epstein and Zin parametrization is transformed into the recursive structure of Kreps and Porteus by the monotonically increasing transformation[54] $U_t(\cdot) = \frac{1}{1-\alpha}(\mathcal{U}_t(\cdot))^{1-\alpha}$,

$$\begin{aligned} U_t(c_t, \tilde{c}_{t+1}, \ldots) &= V[c_t, \mathrm{E}_t\{U_{t+1}(\tilde{c}_{t+1}, \ldots)\}] \qquad (4.38) \\ &= \frac{1}{1-\alpha}\left[c_t^{1-\rho} + \beta\left(\mathrm{E}_t\left\{(1-\alpha)\cdot U_{t+1}(\tilde{c}_{t+1}, \ldots)\right\}\right)^{\frac{1-\rho}{1-\alpha}}\right]^{\frac{1-\alpha}{1-\rho}} \\ &= \left[\theta c_t^{1-\rho} + \beta\left(\mathrm{E}_t\left\{U_{t+1}(\tilde{c}_{t+1}, \ldots)\right\}\right)^{\frac{1-\rho}{1-\alpha}}\right]^{\frac{1-\alpha}{1-\rho}}. \end{aligned}$$

Here, θ denotes $(1-\alpha)^{-\frac{1-\rho}{1-\alpha}}$.

Let $J_t(w_t)$ again denote the indirect utility of wealth, when the optimal consumption and investment strategy is realized:

$$J_t(w_t) \equiv \max_{\{\{v_t^i\}_{i=1}^I, c_t\}} \left[\theta c_t^{1-\rho} + \beta\left(\mathrm{E}_t\left\{J_{t+1}(\tilde{w}_{t+1})\right\}\right)^{\frac{1-\rho}{1-\alpha}}\right]^{\frac{1-\alpha}{1-\rho}}. \qquad (4.39)$$

The optimal consumption and investment policy for this preference function is equal to that for the preference function defined in equation (4.37) because both are ordinally equivalent. Now, the intertemporal marginal rate of substitution and hence the market pricing kernel for this utility specification is derived[55]. The partial derivatives necessary for the derivation of

[54] A monotonically increasing transformation guarantees the ordinal equivalence of the original and the transformed preference function.

[55] Although the utility function has a different form for $\alpha = 1$ or $\rho = 1$ (see footnote 52 in this chapter), the derived intertemporal marginal rate of substitution is also valid for $\alpha = 1$. For $\rho = 1$ the intertemporal marginal rate of substitution must be derived differently (see Giovannini and Weil (1989, pp. 11-2)).

the market pricing kernel (see equation (4.23)) are

$$\begin{aligned} V_{1,t} &= \theta\,(1-\alpha)\,c_t^{-\rho}\,J_t(w_t)^{1-\frac{1-\rho}{1-\alpha}} \\ V_{1,t+1} &= \theta\,(1-\alpha)\,\tilde{c}_{t+1}^{-\rho}\,J_{t+1}(\tilde{w}_{t+1})^{1-\frac{1-\rho}{1-\alpha}} \\ V_{2,t} &= \beta\,\mathrm{E}_t\left\{J_{t+1}(\tilde{w}_{t+1})\right\}^{\frac{1-\rho}{1-\alpha}-1}\,J_t(w_t)^{1-\frac{1-\rho}{1-\alpha}} \end{aligned}$$

so that the intertemporal marginal rate of substitution and hence the market pricing kernel is given by

$$\tilde{\phi}_{t+1} = IMRS_{t+1} = \beta\left(\frac{\tilde{c}_{t+1}}{c_t}\right)^{-\rho}\left(\frac{\mathrm{E}_t\left\{J_{t+1}(\tilde{w}_{t+1})\right\}}{J_{t+1}(\tilde{w}_{t+1})}\right)^{\frac{1-\rho}{1-\alpha}-1}. \tag{4.40}$$

This equation illustrates that, given the intertemporal non-separability of the preference specification when $\alpha \neq \rho$, the intertemporal marginal rate of substitution depends not only upon the consumption levels at dates t and $t+1$ but also upon the future intertemporal utility (cf. Epstein (1992, p. 32)). To obtain a tractable form of the intertemporal marginal rate of substitution the second factor must be calculated for the optimum. Recall that the utility function is assumed to be homothetic (see section 4.2.1.3), i.e. in the optimum a change in w_t does not affect the proportion of w_t consumed at date t, c_t/w_t. Considering this, it can be seen from equation (4.39) that the indirect utility of wealth is homogeneous of degree $1-\alpha$ in wealth[56]. Hence, the indirect utility of wealth at date t can be written as

$$J_t(w_t) = \Theta_t w_t^{1-\alpha}, \tag{4.41}$$

where Θ_t denotes a proportionality factor depending on the information available at date t. Since at date t the information available at date $t+1$ is unknown, $\tilde{\Theta}_{t+1}$ is random. Using (4.41) and the budget constraint of the representative agent, $J_{t+1}(\tilde{w}_{t+1})$ can be written as

$$J_{t+1}(\tilde{w}_{t+1}) = \tilde{\Theta}_{t+1}(w_t - c_t)^{1-\alpha}(\tilde{R}^M_{t+1})^{1-\alpha}. \tag{4.42}$$

[56] For the final period, where $w_T = c_T$, it is easy to verify that $J_T(w_T)$ is homogeneous of degree $1-\alpha$. It is now shown that $J_t(w_t)$ is homogeneous of degree $1-\alpha$ if $J_{t+1}(\tilde{w}_{t+1})$ is homogeneous of degree $1-\alpha$:

$$\begin{aligned} J_t(k\cdot w_t) &= \max_{\{\{v_t^i\}_{i=1}^I, c_t\}}\left[\theta(k\cdot c_t)^{1-\rho} + \beta\left(\mathrm{E}_t\left\{J_{t+1}(k\cdot\tilde{w}_{t+1})\right\}\right)^{\frac{1-\rho}{1-\alpha}}\right]^{\frac{1-\alpha}{1-\rho}} \\ &= \max_{\{\{v_t^i\}_{i=1}^I, c_t\}}\left[\theta\cdot k^{1-\rho}\cdot c_t^{1-\rho} + \beta\left(\mathrm{E}_t\left\{k^{1-\alpha}\cdot J_{t+1}(\tilde{w}_{t+1})\right\}\right)^{\frac{1-\rho}{1-\alpha}}\right]^{\frac{1-\alpha}{1-\rho}} \\ &= k^{1-\alpha}\cdot J_t(w_t). \end{aligned}$$

Substituting this expression in equation (4.40) yields:

$$IMRS_{t+1} = \beta \left(\frac{\tilde{c}_{t+1}}{c_t} \right)^{-\rho} \left(\frac{\mathrm{E}_t \left\{ \tilde{\Theta}_{t+1} (\tilde{R}^M_{t+1})^{1-\alpha} \right\}}{\tilde{\Theta}_{t+1} (\tilde{R}^M_{t+1})^{1-\alpha}} \right)^{\frac{1-\rho}{1-\alpha} - 1} . \tag{4.43}$$

What remains is to solve for Θ in the optimum. Substituting equation (4.42) in equation (4.39) and then taking the derivative of the indirect utility of wealth at date t with respect to c_t yields

$$0 \stackrel{!}{=} \underbrace{(1-\alpha)}_{\neq 0} \underbrace{[J_t(w_t)]^{1-\frac{1-\rho}{1-\alpha}}}_{\neq 0} \left[\theta c_t^{-\rho} - \beta (w_t - c_t)^{-\rho} \mathrm{E}_t \left\{ \tilde{\Theta}_{t+1} (\tilde{R}^M_{t+1})^{1-\alpha} \right\}^{\frac{1-\rho}{1-\alpha}} \right]$$

$$\Leftrightarrow \qquad \mathrm{E}_t \left\{ \tilde{\Theta}_{t+1} (\tilde{R}^M_{t+1})^{1-\alpha} \right\}^{\frac{1-\rho}{1-\alpha}} \stackrel{!}{=} \frac{\theta}{\beta} \left(\frac{c_t}{w_t - c_t} \right)^{-\rho} . \tag{4.44}$$

After substituting equations (4.41) and (4.42) and the optimality condition (4.44) in equation (4.39), it can be solved for Θ_t:

$$\begin{aligned} \Theta_t w_t^{1-\alpha} &= \left[\theta c_t^{1-\rho} + \beta (w_t - c_t)^{1-\rho} \frac{\theta}{\beta} \left(\frac{c_t}{w_t - c_t} \right)^{-\rho} \right]^{\frac{1-\alpha}{1-\rho}} \\ \Leftrightarrow \qquad \Theta_t &= \left[\theta \left(\frac{c_t}{w_t} \right)^{-\rho} \right]^{\frac{1-\alpha}{1-\rho}} . \end{aligned} \tag{4.45}$$

Finally, substituting (4.44) and (4.45) in equation (4.43) leads to the intertemporal marginal rate of substitution and, hence, to the market pricing kernel

$$\tilde{\phi}_{t+1} = IMRS_{t+1} = \left[\beta \left(\frac{\tilde{c}_{t+1}}{c_t} \right)^{-\rho} \right]^{\frac{1-\alpha}{1-\rho}} (\tilde{R}^M_{t+1})^{\frac{1-\alpha}{1-\rho} - 1} . \tag{4.46}$$

This intertemporal marginal rate of substitution has independently been derived by Epstein and Zin (1989) and Weil (1989). It is a geometrically weighted average of the intertemporal marginal rate of substitution for time-additive expected utility with constant relative risk aversion (equation (4.26)) and the intertemporal marginal rate of substitution for time-additive expected logarithmic utility (equation (4.35)). It therefore depends on the random growth rate of consumption, $\tilde{c}_{t+1}/c_t$, and on the random return on the market portfolio, $\tilde{R}^M_{t+1}$[57]. Comparing the equations (4.46) and (4.40) shows that the return on the market, $\tilde{R}^M_{t+1}$, enters the intertemporal marginal rate of substitution as a measure of the innovation in the future

[57] The inclusion of the return on the market portfolio makes this model susceptible to the Roll (1977) critique of the static CAPM. He argues that the market portfolio including nontraded assets can never be empirically identified and that therefore models including the market portfolio are not testable.

equilibrium utility. If $\alpha = \rho$, the intertemporal marginal rate of substitution for time-additive expected utility with constant relative risk aversion is obtained, leading to the consumption-based CAPM (cf. section 4.2.2.3). If $\alpha = 1$, the intertemporal marginal rate of substitution for time-additive expected logarithmic utility is obtained, leading to the static CAPM (cf. section 4.2.2.4)[58].

4.2.3.3 A Two-factor CAPM

Firstly recall that time-additive expected utility with constant relative risk aversion leads to the C-CAPM, a one-factor asset pricing model that measures an asset's systematic risk by the covariance of its return with a function of consumption growth. Secondly recall that time-additive expected logarithmic utility leads to the static CAPM, a one-factor asset pricing model that measures an asset's systematic risk by the covariance of its return with the return on the market. As shown in the previous section, the intertemporal marginal rate of substitution for recursive non-expected utility is a geometrically weighted average of that for time-additive expected utility with constant relative risk aversion and that for time-additive expected logarithmic utility. Therefore a two-factor CAPM may be derived from recursive non-expected utility that measures an asset's systematic risk by the covariance of its return with a function of consumption growth and the covariance of its return with the return on the market. This can be seen by substituting the parametric model of the market pricing kernel (equation (4.46)) into the linear asset pricing model in terms of the market pricing kernel given in equation (2.9) and approximating the geometric average by an arithmetic average:

$$\begin{aligned} \mathrm{E}_t\left\{\tilde{R}^i_{t+1} - \tilde{R}^z_{t+1}\right\} &= -\left(\mathrm{E}_t\left\{\left(\frac{1-\alpha}{1-\rho}\right)\beta\left(\frac{\tilde{c}_{t+1}}{c_t}\right)^{-\rho}\right.\right. \\ &\quad \left.\left. + \left(1-\frac{1-\alpha}{1-\rho}\right)(\tilde{R}^M_{t+1})^{-1}\right\}\right)^{-1} \\ &\quad \cdot\left[\left(\frac{1-\alpha}{1-\rho}\right)\beta\,\mathrm{cov}_t\left(\left(\frac{\tilde{c}_{t+1}}{c_t}\right)^{-\rho}, \tilde{R}^i_{t+1}\right)\right. \\ &\quad \left. + \left(1-\frac{1-\alpha}{1-\rho}\right)\mathrm{cov}_t\left((\tilde{R}^M_{t+1})^{-1}, \tilde{R}^i_{t+1}\right)\right]. \end{aligned}$$

[58] To obtain the static CAPM, ρ is not required to equal one in this more general framework. Giovannini and Weil (1989) show that in this general framework $\alpha = 1$ implies a myopia in investment decisions independent of ρ, whereas $\rho = 1$, i.e. $\sigma = 1$, implies a myopia in consumption and savings decisions independent of the value of α. Hence, if $\alpha = \rho = 1$, myopic behavior in both consumption and savings decisions and investment decisions is implied. This is exactly the case of time-additive expected logarithmic utility (cf. section 4.2.2.4).

As in the case of the C-CAPM this asset pricing relation is non-linear. Empirical estimation is much easier if the model is linearized, i.e. if the excess return on each asset i is given by a linear function of $\mathrm{cov}_t((\tilde{c}_{t+1}/c_t), \tilde{R}^i_{t+1})$ and $\mathrm{cov}_t(\tilde{R}^M_{t+1}, \tilde{R}^i_{t+1})$. Linearized versions can again be derived in a continuous-time framework (cf. Duffie and Epstein (1992)) or in discrete-time by assuming conditional joint lognormality of consumption growth, the market return and asset returns[59] (cf. Giovannini and Weil (1989) or Epstein and Zin (1991)). For illustration the discrete-time linearization of the two-factor CAPM assuming conditional joint lognormality of consumption growth, market return and asset returns is derived in the following excursus.

Excursus: Two-factor CAPM Assuming Joint Lognormality Assume that the conditional joint distribution of $\tilde{\lambda}_{t+1} = \tilde{c}_{t+1}/c_t$, $\tilde{R}^M_{t+1}$ and $\tilde{R}^i_{t+1}$ is lognormal. Substituting equation (4.46) in equation (4.22), taking logs and replacing $(1-\alpha)/(a-\rho)$ by θ yields[60]

$$\begin{aligned} 0 &\stackrel{!}{=} \ln \mathrm{E}_t\{\beta^\theta \tilde{\lambda}_{t+1}^{-\rho\theta}(\tilde{R}^M_{t+1})^{\theta-1}\tilde{R}^i_{t+1}\} \\ &= \theta \ln \beta - \rho\theta \mathrm{E}_t\{\ln \tilde{\lambda}_{t+1}\} + (\theta-1)\mathrm{E}_t\{\ln \tilde{R}^M_{t+1}\} + \mathrm{E}_t\{\ln \tilde{R}^i_{t+1}\} \\ &\quad + \frac{1}{2}\left(\rho^2\theta^2 \mathrm{var}_t(\ln \tilde{\lambda}_{t+1}) + (\theta-1)^2 \mathrm{var}_t(\ln \tilde{R}^M_{t+1}) + \mathrm{var}_t(\ln \tilde{R}^i_{t+1})\right) \\ &\quad - \rho\theta(\theta-1)\mathrm{cov}_t(\ln \tilde{\lambda}_{t+1}, \ln \tilde{R}^M_{t+1}) - \rho\theta \mathrm{cov}_t(\ln \tilde{\lambda}_{t+1}, \ln \tilde{R}^i_{t+1}) \\ &\quad + (\theta-1)\mathrm{cov}_t(\ln \tilde{R}^M_{t+1}, \ln \tilde{R}^i_{t+1}). \end{aligned} \tag{4.47}$$

For a risk-free asset f it follows

$$\begin{aligned} 0 &\stackrel{!}{=} \theta \ln \beta - \rho\theta \mathrm{E}_t\{\ln \tilde{\lambda}_{t+1}\} + (\theta-1)\mathrm{E}_t\{\ln \tilde{R}^M_{t+1}\} + \ln R^f_{t+1} \\ &\quad + \frac{1}{2}\left(\rho^2\theta^2 \mathrm{var}_t(\ln \tilde{\lambda}_{t+1}) + (\theta-1)^2 \mathrm{var}_t(\ln \tilde{R}^M_{t+1})\right) \\ &\quad - \rho\theta(\theta-1)\mathrm{cov}_t(\ln \tilde{\lambda}_{t+1}, \ln \tilde{R}^M_{t+1}). \end{aligned} \tag{4.48}$$

[59]This assumption of joint lognormality is inconsistent because the market return is a weighted average of the asset returns, and a linear combination of lognormal variables is not lognormal. However, Jorion and Giovannini (1993) argue, that in continuous-time joint lognormality can hold exactly and that joint lognormality is a good approximation in discrete-time because gross returns do not deviate much from unity.

[60]See appendix A.1 for a derivation of the expected value of the product of three random variables that are jointly lognormally distributed.

Solving (4.47) for $\mathrm{E}_t\{\ln \tilde{R}^i_{t+1}\}$ and (4.48) for $\ln R^f_{t+1}$ and calculating the difference yields

$$\begin{aligned}
\mathrm{E}_t\{\ln \tilde{R}^i_{t+1}\} - \ln R^f_{t+1} &= \rho\theta \mathrm{cov}_t(\ln \tilde{\lambda}_{t+1}, \ln \tilde{R}^i_{t+1}) \\
&\quad -(\theta - 1)\mathrm{cov}_t(\ln \tilde{R}^M_{t+1}, \ln \tilde{R}^i_{t+1}) \\
&\quad -\frac{1}{2}\mathrm{var}_t(\ln \tilde{R}^i_{t+1}) \\
\Leftrightarrow \ln \mathrm{E}_t\{\tilde{R}^i_{t+1}/R^f_{t+1}\} &= \frac{\rho(1-\alpha)}{1-\rho}\mathrm{cov}_t\left(\ln \frac{\tilde{c}_{t+1}}{c_t}, \ln \tilde{R}^i_{t+1}\right) \\
&\quad +\frac{\alpha-\rho}{1-\rho}\mathrm{cov}_t(\ln \tilde{R}^M_{t+1}, \ln \tilde{R}^i_{t+1}). \qquad (4.49)
\end{aligned}$$

The left-hand side of this equation can be interpreted as a logarithmic risk premium on asset i over the return on a risk-free asset. The equation states that this risk premium is a linear combination of the conditional covariance of the logarithmic return on asset i with logarithmic consumption growth (first term) and the conditional covariance of the logarithmic return on asset i with the logarithmic return on the market portfolio (second term). The weights only depend on the parameter of relative risk aversion α and the elasticity of intertemporal substitution $\sigma = 1/\rho$. The consumption-based CAPM follows if $\alpha = \rho$ (see equation (4.31)) and the static CAPM follows if $\alpha = 1$, but in general a linear combination of the two common asset pricing models is obtained.

Empirical Evidence The parametric model of the market pricing kernel (4.46) will be applied in the calibration exercises in the second part of this analysis. Therefore, the main results of a number of studies estimating the related two-factor CAPM are now briefly presented. Empirical evidence from calibration exercises using this preference specification is discussed in section 5.3.

Giovannini and Weil (1989) and Jorion and Giovannini (1993) estimate the linearized two-factor model using cross-sectional data for the US. Giovannini and Weil (1989) find evidence for a unit coefficient of relative risk aversion and an elasticity of intertemporal substitution different from one. A contrary result is found by Jorion and Giovannini (1993). They cannot reject the hypothesis that α and ρ are equal. Alternatively applying the distribution-free GMM estimation technique suggested by Hansen and Singleton (1982) their results do not change. They conclude that the two-factor model does not help to better predict the cross-sectional variation in expected returns. Epstein and Zin (1991) provide evidence in favor of the two-factor model using monthly US time series data. Applying the GMM technique the expected utility hypothesis, i.e. the equality of α and ρ, is strongly rejected. In general they find that the elasticity of intertemporal substitution is lower than one and the parameter of relative risk aversion is close to one. Their results are therefore comparable with those of Giovannini and Weil (1989) for cross-sectional data. Also applying the GMM technique Bufman and Leiderman (1990) estimate the model using Israeli data. They also reject the time-additive expected utility model, but, in contrast to Giovannini and Weil (1989) and Epstein and Zin (1991) they find that both relative risk aversion and the elasticity of intertemporal substitution are larger than one. The study of Hardouvelis, Kim, and Wizman (1996) based on ten size-portfolios of US stocks in the period 1959 to 1991 provides additional evidence for $\alpha \neq \rho$. In the GMM-estimated model both factors, the market factor and the consumption factor, are positively priced and statistically significant. However, the overidentifying restrictions are rejected. As in the case of the C-CAPM their estimation yields a high value of relative risk aversion of 276.3 (Hardouvelis, Kim, and Wizman 1996, p. 291).

The empirical research of Sauer and Murphy (1992) using German data can be interpreted in the light of the two-factor model. They conduct a comparative test that directly determines which asset pricing model, the C-CAPM or the static CAPM, better explains the data by simultaneously regressing average returns on the returns predicted by both models in a multiple regression model. The two-factor model derived from recursive non-expected utility provides a theoretical justification of this ad hoc approach. The results of Sauer and Murphy (1992) indicate that the static CAPM explains risk-return relationships in Germany significantly better than the C-CAPM. A direct estimation of the two-factor model using German data, alternatively applying the linearized model or the GMM estimation technique, is given in Missong (1994). However, no clear results are obtained. For example, the GMM estimates of α and ρ are not significant.

Overall, the evidence suggests that the market covariance is more important, but that both the market factor and the consumption factor are statistically significant.

4.3 Summary of Chapter 4

Parametric models of the market pricing kernel are derived using the consumption-based equilibrium asset pricing approach. This approach is chosen for two reasons. First, applying wealth-based one-period models such as the CAPM in a multiperiod world restricts the expected returns and risks to be constant over time, or at least to be perfectly predictable. The empirical application of theoretically founded generalizations allowing for stochastic variations in expected returns and risks (ICAPM) is prevented by the fact that not all relevant factors are specified within these models. The factors chosen in recent applications of Conditional CAPMs allowing for variations in expected returns and risks have therefore no theoretical justification. Second, the consumption-based equilibrium asset pricing approach determines the risk-free rate and the risk premium on the market, whereas the conditional asset pricing approach assumes that both are exogenously given. However, one purpose of the analysis is to investigate whether the observed risk-free rate and equity premium are consistent with theoretical models of asset pricing.

Within the consumption-based equilibrium asset pricing approach the market pricing kernel is derived as the representative agent's intertemporal marginal rate of substitution between consumption at successive dates. Time-additive expected utility implies strong behavioral restrictions such as the indifference to the timing of the resolution of uncertainty and the interrelation between risk aversion and aversion to intertemporal fluctuations in consumption. Empirical evidence exists that these restrictions may not be satisfied. Therefore, the intertemporal marginal rate of substitution is derived for a class of less restrictive recursive preferences that captures time-additive expected utility as a special case. Finally, parametric models of the market pricing kernel are derived for three alternative parametrizations of the recursive preferences, namely, time-additive expected utility with constant relative risk aversion, time-additive expected logarithmic utility and recursive non-expected utility. Using the first parametrization, the Consumption-based CAPM (C-CAPM) is derived. As a consequence of the myopic behavior of individuals with logarithmic utility, the second parametrization yields an asset pricing model closely related to the static CAPM. The third parametrization allows to separate risk aversion and the elasticity of intertemporal substitution as well as to express preferences over the timing of the resolution of uncertainty. From this parametrization a two-factor CAPM is derived. The empirical evidence on these models is in favor of the two-factor CAPM. The estimated parameter of risk aversion and elasticity of intertemporal substitution do not satisfy the interrelation presumed by the C-CAPM. Furthermore, jointly explaining the returns on different assets seems to pose a problem especially to the C-CAPM.

Until now the empirical evidence stems from studies that apply estimation techniques, mainly the GMM technique. As is argued in the ensuing chapter, these estimation techniques, although econometrically rigorous, may not be able to provide strong evidence for or against the underlying theoretical models so that alternative approaches such as the variance bound approach or the calibration approach should also be used when empirically investigating the models.

Chapter 5

The Calibration Approach for Empirically Investigating Parametric Models of the Market Pricing Kernel

Section 5.1 introduces the calibration approach and compares it with the more common estimation approach for the empirical investigation of parametric models of the market pricing kernel. Both approaches have strengths and weaknesses as has been discussed in the literature. This discussion is briefly surveyed and the use of the calibration approach for the present analysis is motivated. Section 5.2 illustrates the calibration approach for a simple model economy with i.i.d. production growth, using both the parametric model of the market pricing kernel derived from time-additive expected utility with constant relative risk aversion and that derived from recursive non-expected utility in turns. Section 5.3 reviews studies applying the calibration approach. A summary ends the chapter.

5.1 Calibration Versus Estimation

Besides the variance bound approach introduced in section 3.3 there exist several other approaches for the empirical investigation of parametric models of the market pricing kernel. Basically, estimation and testing approaches can be distinguished from calibration and simulation approaches. To distinguish both approaches consider the parametric model of the market pricing kernel derived from time-additive expected utility with constant relative risk aversion (equation (4.26)). It is a function of the growth rate of consumption $\tilde{\lambda}_{t+1}$, the parameter of risk aversion α, and the discount factor β: $\tilde{\phi}_{t+1} = \phi(\tilde{\lambda}_{t+1}, \alpha, \beta)$. In this case the estimation approach is to take the observed rates of return and the observed aggregate consumption and to estimate the parameters of the preference function so as to optimize the fit of equation (2.2), $\mathrm{E}_t\{\tilde{\phi}_{t+1} \cdot \tilde{R}^i_{t+1}\} = 1 \, \forall i, t$, according to some statistical criterion. In contrast, calibration is a two-step procedure. First, one specifies a model economy (general equilibrium model) that, given a stochastic process for the exogenous endowment, in the present case the growth rate of aggregate consumption $\tilde{\lambda}_{t+1}$, and values of the preference parameters, allows to generate a time series of the endogenous equilibrium rates of return. The endowment process may either be estimated directly from the observed time series of aggregate consumption or chosen so as to mimic specific properties of this time series. Second, by varying the preference parameters one tries to match specific properties of the implied equilibrium rates of return with those of the observed rates of return. Thus, there is no claim that the model explains all the characteristics of the observed rates of return.

Subsequently, the estimation and the calibration approach are characterized using the Euler equation from intertemporal asset pricing, $\mathrm{E}_t\{\tilde{\phi}_{t+1} \cdot \tilde{R}^i_{t+1}\} = 1 \, \forall i, t$. The market pricing kernel $\tilde{\phi}_{t+1}$ derived from the optimal consumption and investment choice of a representative agent[1] is a function of the growth rate of aggregate consumption $\tilde{\lambda}_{t+1}$, the market return $\tilde{R}^M_{t+1}$ and a vector of preference parameters θ: $\tilde{\phi}_{t+1} = \phi(\tilde{\lambda}_{t+1}, \tilde{R}^M_{t+1}, \theta)$.

Estimation Given time series for $\tilde{\lambda}_{t+1}$, $\tilde{R}^M_{t+1}$ and $\tilde{R}^i_{t+1}$, estimation means finding the parameters θ so as to optimize the fit of the Euler equation. Usually, standard econometric techniques are applied in estimation[2] and the validity of the specification is examined by testing restrictions, by general goodness of fit tests or by comparing the fit of two nested models (cf. Canova (1994)). As previously discussed in section 4.2.2.3 and section 4.2.3.3, estimation is complicated by the non-linearity of the derived market pricing kernel functions. To apply linear estimation techniques such as OLS- or GLS-regressions, the asset pricing models must be linearized. For the linearization strong distributional assumptions have to be made, which might not be innocuous (cf. Brown and Gibbons (1985) and section 4.2.2.3). Given the possibility of large shocks, the non-linear nature of the models is potentially important. To overcome the need for linearization, Hansen and Singleton (1982) (HS) propose the distribution-free Generalized Method of Moments (GMM) estimation technique for directly estimating the parameter vector θ from the Euler equation of intertemporal asset pricing models. The GMM technique is now the most widely used estimation method in this context[3]. Therefore, it is briefly summarized.

---

[1] See equations (4.26), (4.35) and (4.46).

[2] Singleton (1990), (1994) provide a survey of different methods for the estimation of intertemporal asset pricing models.

[3] It is interesting to notice that the GMM technique and the calibration approach have

Define the forecast error to be $\tilde{u}^i_{t+1} = \phi(\tilde{\lambda}_{t+1}, \tilde{R}^M_{t+1}, \theta) \cdot \tilde{R}^i_{t+1} - 1$ (HS, p. 1273). The Euler equation implies that $\mathrm{E}_t\{\tilde{u}^i_{t+1}\} = \mathrm{E}\{\tilde{u}^i_{t+1}|\Omega_t\} = 0\ \forall i$, given the information available at date t, Ω_t. A set of X observable instrumental variables is used to represent the information available to the representative agent at date t, Ω_t. The forecast error is additive and uncorrelated with any information available to the agent at date t, i.e. with each instrumental variable $\tilde{x}_t \in \Omega_t$. This "orthogonality condition" implies $\mathrm{E}\{\tilde{u}^i_{t+1}\tilde{x}_t\} = 0$ for each asset and each instrumental variable (HS, equation (3.3)). For a correctly specified model the $I \cdot X$ sample orthogonality conditions $\mu^i_x = 1/T \sum_{t=1}^{T} \tilde{u}^i_{t+1}\tilde{x}_t$ should be close to zero for large values of T (HS, p. 1275). These sample orthogonality conditions are used to construct a criterion function whose minimizer is the estimate of the parameter vector θ. Basically, one minimizes a weighted sum of squared μ^i_x. The construction of the criterion function (HS, equation (3.5)) guarantees that the parameter estimator is consistent, asymptotically normal, and has an asymptotic covariance matrix that can be estimated consistently (HS, p. 1274). If the number of orthogonality conditions, i.e. the number of assets times the number of instruments $(I \cdot X)$, is larger than the number of parameters l, $(I \cdot X - l)$ orthogonality conditions exist that are not used in estimation, but that should be close to zero if the model is true. These overidentifying restrictions can be tested using a chi square statistic (HS, p. 1278).

Applying estimation techniques such as GMM usually relies on the assumption of the "axiom of correct specification" (Pagan 1994, p. S8). This may cause problems for the analyzed equilibrium models that are based on strong simplifying assumptions. If the model is not correctly specified, i.e. if the observations do not lie in the set of possible solutions of the model, the estimates may be biased. In addition, serially correlated but stationary error terms and measurement problems may strongly bias the estimation of Euler equations using the GMM estimation technique (cf. Kugler (1988)). Hence, the correct specification of the model is an important issue in using the GMM technique.

Another important issue in using the GMM technique is the selection of the instruments. First, it embodies assumptions regarding the information available to the individuals and, second, it affects the efficiency of the estimators and the identification of the parameters. The intuition for the use of instrumental variables is that the instruments allow to predict the conditional expectations and conditional covariances of the variables in the Euler equation, especially of consumption growth. If the instruments applied do not help to predict these variables, the preference parameters may not be identifiable. A thorough application of the GMM technique should therefore investigate whether or not the applied instruments have explanatory power

---

originally been introduced in articles published in the same journal and the same year, namely by Hansen and Singleton (1982) and Kydland and Prescott (1982), respectively.

for the conditional expectations and conditional covariances of the variables in the Euler equation.

The robustness of the GMM estimation procedure to alternative specifications of the intertemporal asset pricing model and different sets of instrumental variables is investigated by Neely and Whiteman (1996). They find that estimates of the parameter of risk aversion vary widely across data sets and attribute this to the failure of identification due to the unpredictability of consumption growth, at least using typical instruments such as lagged consumption growth and lagged return on the market (see, for example, Hansen and Singleton (1982) or Epstein and Zin (1991)). "Thus, while Hansen and Singleton attempted to exploit additional information in their model, the information proved to have little value." (Neely and Whiteman 1996, p. 12). Hence, as long as the correct specification of the model is not ensured and the applied instruments have only low predictive power, interpreting the estimated parameters may be questionable due to biased estimates or estimates with large standard deviations.

Furthermore, interpreting the rejection of a model may be questionable for two reasons. First, simulation studies find that in small samples the chi square statistic may overreject the overidentifying restrictions because the asymptotic standard errors are likely to be understated, e.g., Kocherlakota (1990c) and Ferson and Foerster (1994)[4]. Second, the GMM estimates may be sensitive to aspects of the data that are of little economic importance. Therefore, the failure to obtain a close fit, i.e. the statistical rejection, may not provide strong evidence against the underlying economic theory (cf. Hoover (1995)).

An important issue particularly for the present analysis is that estimation techniques such as GMM, although econometrically rigorous, are limited in their ability to isolate those characteristics of the data not accommodated by the model. Hence, the estimation approach does not indicate how to modify the specification to obtain a better fit (cf. Canova (1994)).

To summarize, investigating equilibrium asset pricing models empirically by applying the estimation approach, particularly the GMM technique, takes account of the statistical properties of both the moments of the observed time series and the estimated parameters. On the other hand, it is plagued by estimation problems and problems concerning the interpretation of the estimated parameters. Most of these "technical problems" may be solved by carefully applying the estimation techniques and by additional analyses. However, depending on the purpose of the analysis some of their characteristics may not be desirable. For example, the estimation may be

[4]Ferson and Foerster (1994) propose to use an "iterated GMM" technique with superior finite sample properties. They argue that in certain cases a simple adjustment of the standard errors may also be appropriate.

influenced by aspects of the data that are of little economic importance so that a rejection of the model may not allow to reject the underlying economic theory and it may fail to indicate how the model could be improved. The findings of studies empirically investigating intertemporal asset pricing models using estimation techniques have been surveyed in section 4.2.2.3 and section 4.2.3.3.

Calibration The calibration approach is widely used in empirical investigations of real business cycle theories. It was introduced to macroeconomics by Kydland and Prescott (1982) and can be interpreted as a realization of a procedure asked for by Lucas (1980):

> "Our task ... is to write a ... program that will accept specific economic policy rules as 'input' and will generate as 'output' statistics describing the operating characteristics of time series we care about, which are predicted to result from these policies." (Lucas 1980, p. 288).

While estimation does not require equilibrium models, calibration is generally based on equilibrium considerations. In a first step a model economy is constructed which, given stochastic processes for the exogenous variables, allows to generate time series of equilibrium realizations of the endogenous variables dependent on particular parameter values of the model in question by simulating realizations of the exogenous variables[5]. The second step is to vary the parameter values for the model economy such that it mimics the reality as closely as possible along a limited number of dimensions. Thus, calibration means finding the model parameters by simulating time series of the equilibrium realizations of the endogenous variables with different parameter values and selecting those values which best match specified properties of the simulated time series with those of observed time series. Often, findings in other applied areas of economics are used to restrict the parameter ranges or to pre-set specific parameters (cf. Kydland and Prescott (1982)). Gregory and Smith (1990) demonstrate that as a consequence of this use of additional information, calibration may be more efficient than GMM estimation. However, the selected parameter values do not necessarily provide the best fit in some statistical sense (cf. Kydland and Prescott (1996)).

[5]Under specific assumptions concerning the stochastic processes of the exogenous variables, closed-form solutions of the endogenous variables can be calculated analytically so that the simulation is not necessary. A central problem of the calibration approach is the correct specification of the exogenous stochastic processes because the calibration results may not be robust to misspecifications of the exogenous processes.

The calibration procedure is illustrated using the example introduced above[6]. (1) Start by specifying the model economy. Assume an equilibrium model with a representative agent with constant relative risk aversion and with exogenous production, $\mathrm{E}_t\{\phi(\tilde{\lambda}_{t+1},\theta)\cdot\tilde{R}^i_{t+1}\} = 1\ \forall i,t$. Extend this model to a general equilibrium model by assuming that production equals dividends and consumption in each period (cf. Lucas (1978)). $\tilde{\lambda}_{t+1}$ denotes the gross growth rate of production, dividends and consumption from date t to date $t+1$. Further assume this endowment process to be i.i.d. and stationary. In this case, the current realization of consumption growth conveys no information about future consumption growth and it follows that the gross return on the market from date t to date $t+1$, $\tilde{R}^M_{t+1}$, is proportional to the gross growth rate of consumption, i.e. $\tilde{R}^M_{t+1} = \Phi\tilde{\lambda}_{t+1}$. The constant proportionality factor Φ depends on the preferences of the representative investor and on the expected growth rate of consumption. Finally choose the distribution of consumption growth to be consistent with the time series of observed growth rates of consumption. The model economy is now specified and allows to analyze which equilibrium rates of return on the market and which equilibrium risk-free rate of return are implied, depending on the preference parameters chosen. (2) Finally, try to match specific properties of the implied equilibrium rates of return with those of observed rates of return by varying the parameters of the preference function, θ.

Using the calibration approach allows to match aspects of the data different from those adopted in estimation procedures such as GMM. It especially allows to focus on economically important aspects. In turn, the calibration results can only be interpreted with respect to the properties of the data that are to be matched. Generally, any property of the simulated and observed data may be matched, but as Gregory and Smith (1990) show, this may yield inconsistent estimators for the preference parameters. Gregory and Smith further show that in the special case of matching population moments of simulated data to sample moments of observed data the estimators are consistent. Therefore, unconditional means, variances, and covariances typically are the focus in calibration. Exercises calibrating asset pricing models typically concentrate on these moments of the risk-free rate and the risk-premium on the market portfolio[7].

As discussed above, rejecting a model using the estimation and testing approach may not provide strong evidence against the model's underlying economic theory, given the measurement problems and the abstract nature of intertemporal asset pricing models. Kydland and Prescott (1982) argue that focusing on properties of the data that are less affected by noise introduced by approximations and measurement errors relative to formal test statistics,

[6]See section 5.2 for a detailed discussion of the example.

[7]In this context Gregory and Smith (1990) show that matching means yields parameter estimates that are robust to certain misspecifications of the endowment process (cf. footnote 5 within this chapter).

the failure of the model to mimic the behavior of observed data with respect to these properties provides stronger evidence against the model and the underlying theory. The focus on economically important aspects of the data strengthens this argument. In addition, the calibration approach indicates the dimensions along which the approximation of the observed data is poor and may therefore indicate how the model can be improved in order to obtain a better fit. It indicates, for example, whether the observations lie in the attainable set of solutions of the model or not.

Besides these aspects, the calibration approach has further advantages. The impact of the preference specification and its parameters on the implied equilibrium rates of return can easily be examined by changing one parameter, holding the others constant (see chapter 8), whereas an analytical comparative statics approach is often impossible for more sophisticated utility specifications. Similarly, given a calibrated model, the impact of different realizations of the exogenous stochastic inputs on the equilibrium rates of return can easily be analyzed. Thus, one can draw conclusions concerning the impact of the macroeconomic activity on the expected rates of return and on the co-movement of the rates of return on risk-free and risky assets (see section 9.2).

On the other hand, calibration does not take account of the statistical properties of both the moments of the observed time series trying to be matched and the values implied by the model to which they are compared. Therefore, statistical inference is impossible without further analyses. In the literature different methods are proposed to formally test calibrated models[8]. Watson (1993) measures the fit of a model calibrated to match first moments by the stochastic error that must be added to the variables in the model in order to exactly match the second moments of the simulated data to those of observed data. The larger the required error, the poorer the fit of the model. An intuitive way of testing calibrated models is conducting Monte Carlo analyses, i.e. generating a large number of realizations of the simulated time series and comparing the distribution of the moments of these simulated time series with the moments of the observed time series to obtain statistical inference (see Gregory and Smith (1991) and the chapters 8 and 9). Canova (1994) suggests a Monte Carlo procedure that randomizes not only over the exogenous stochastic process but also over the parameters.

To summarize, despite the disadvantage that strong assumptions have to be made in a general equilibrium model[9] the calibration approach for empirically investigating equilibrium asset pricing models has several advantages. It allows to focus on economically important aspects of the data and may

[8]See Gregory and Smith (1991), Watson (1993), Canova (1994) and DeJong, Ingram, and Whiteman (1996).

[9]However, Friedman (1953) argues that as long as a model succeeds in reproducing selected features of reality, the assumptions upon which the model is based do not matter.

indicate properties of the data that are not accommodated by the model. This may provide stronger evidence against the underlying theory than the rejection through formal test statistics. On the other hand, the calibration results must be interpreted with respect to the aspects of the data that are focused on and the assumption that the exogenous endowment process is correctly specified. An advantage of the calibration approach is that it allows to analyze how the equilibrium rates of return are affected by changes in the preference parameters, the preference specification, and the exogenous endowment process. Monte Carlo analyses of calibrated models allow to obtain statistical inference.

Literature Discussion The advantages and disadvantages as well as the distinction between the estimation approach and the calibration approach, especially concerning their applicability to real business cycle theory, are broadly discussed in the literature[10]. This discussion is briefly reviewed.

Some authors criticize the distinction drawn between both approaches. Hansen and Heckman (1996) argue that "the distinction drawn between calibrating and estimating the parameters of a model is artificial at best. Moreover, the justification for what is called 'calibration' is vague and confusing. In a profession that is already too segmented, the construction of such artificial distinctions is counterproductive" (p. 91). They further argue that each estimation method can be described by a particular loss function that summarizes the weights put on deviations of a model's predictions from the data and that the calibration approach is "estimation based on other loss functions" (p. 92). They argue that calibration should only be the starting point of an empirical analysis of general equilibrium models. Nevertheless, they admit that "properly used and qualified simulation methods can be an important source of information and an important stimulus to high-quality empirical economic research" (p. 101). Also Quah (1995) criticizes the classification of researchers in those that "have chosen, not to apply standard diagnostic testing, but to use a model's operating characteristics to evaluate its success or failure" (p. 1594) and those that "evaluate models using statistical or aesthetic criteria unmotivated by economic research" (p. 1594). He writes "Isn't every measure of model fit just some re-writing of an econometric test statistic, and vice versa? Isn't every calibrated guess of a numerical value of a parameter just an estimate – no more, no less – a particular realisation of an estimator (which, in turn, is just a random variable) conditioned on observed data?" (p. 1594).

Other authors argue that both approaches should be distinguished and are in favor of one or the other approach. Kydland and Prescott (1996) argue

[10] See Hoover (1995), Quah (1995), Kydland and Prescott (1996), Hansen and Heckman (1996) and especially the November 1995 issue of the Economic Journal and the December 1994 issue of the Journal of Applied Econometrics.

that since a theory is an explicit set of instructions for building a model economy to address certain questions and not a collection of assertions about the behavior of the actual economy "statistical hypothesis testing, which is designed to test assertions about actual systems, is not an appropriate tool for testing economic theory. ... the ultimate test of a theory is whether its predictions are confirmed – that is, did the actual economy behave as predicted by the model economy, given the policy rule selected?" (p. 83). Pagan's (1994) characterization of the calibration approach points in the same direction: "'calibration' involves quantitative research in which a theoretical model is taken very seriously rather than a technique for estimating the parameters of a theoretical model" (p. S2). Although Hoover (1995) strongly argues in favor of the calibration approach, too, he asks "Does the calibration methodology amount to a repudiation of econometric estimation altogether?" and concludes "Clearly not. At some level, econometrics still helps to supply the values of the parameters of the models. ... it would be a mistake to treat calibration as simply an alternative form of estimation." (pp. 40-1).

Conclusion Which conclusion should be drawn and which approach should be applied? Although both approaches allow an empirical investigation of equilibrium asset pricing models they should not be seen as alternatives for reaching the same goal. The necessary assumptions differ substantially, both approaches have different strengths and weaknesses, the application of one or the other approach allows to answer different questions and the results obtained should be interpreted differently. Therefore, the differentiation of both approaches seems justifiable. Whether one or the other approach is applied should depend on the model to be analyzed and the questions posed. The intertemporal asset pricing models that are to be analyzed in the present study are strongly theory based and idealized along many dimensions. Probably the most important dimension is that the models concern aggregates while the economy is composed of individuals. Therefore, these models cannot be expected to be true and to mimic all aspects of observed asset returns. However, the main question is, whether these models can explain economically important aspects such as the observed size of the risk-free rate of return and the equity premium. If not, the empirical investigation should indicate how the models could be improved, for example, by modifying the preference function. Considering these aspects, the unsatisfactory results obtained by authors applying the estimation approach (see the discussion in section 4.2.2.3 and 4.2.3.3) and the fact that until now no calibration exercise has been conducted using German data, the calibration approach is chosen besides the variance bound approach for empirically investigating the parametric models of the market pricing kernel derived in chapter 4. Before this empirical investigation and its results are discussed in the chapters 6 to 9, the remainder of this chapter further illustrates the calibration approach using a simple model economy with i.i.d. production growth and reviews studies that apply the calibration approach in a context related to the present analysis.

5.2 Calibration of a Model Economy with i.i.d. Production Growth: An Illustrative Example

Section 5.2.1 illustrates the calibration approach for a simple model economy. Closed-form solutions for the implied equilibrium rates of return on the market and the equilibrium risk-free rate are derived using time-additive expected utility with constant relative risk aversion (section 5.2.2) and, alternatively, using recursive non-expected utility (section 5.2.3). In chapter 8 this simple model economy will be generalized, e.g., the i.i.d. endowment process will be replaced by a Markov state-switching endowment process.

5.2.1 The Model Economy

The model economy is based on the following assumptions[11]:

1. The economy is a one good, closed economy. In each period aggregate production $\tilde{m}_t$ equals aggregate dividends $\tilde{d}_t$ and also aggregate consumption $\tilde{c}_t$.

2. The production process and hence the dividend and consumption process is exogenously given. The stochastic process of the gross growth rate of production $\tilde{\lambda}_{t+1} = \tilde{m}_{t+1}/m_t$ is i.i.d. and stationary.

3. The preferences of the representative investor are given and exhibit constant relative risk aversion (CRRA).

The assumption of CRRA-preferences ensures that the risk aversion and, therefore, the intertemporal marginal rate of substitution of the representative agent are not affected by the growing level of wealth in this growth economy. Under the assumed i.i.d. stationary process of production growth the current and past realizations of the growth rate contain no additional information about future growth. From these assumptions it follows that (1) if a one-period risk-free asset exists, its return is constant over time, and (2) the ex-dividend market value of the market portfolio at date t, P_t^M,

[11] This model economy can be interpreted as a one-state version of Lucas' (1978) exchange economy with i.i.d. and stationary production growth.

is proportional to the level of production at date t, m_t. The proportionality factor Θ, $P_t^M = \Theta \cdot m_t$, is constant over time. Θ depends on the parameters of the preference function and the expected future production growth. Since aggregate production is paid off as dividends in this model, Θ can be interpreted as a price-dividend-ratio. This ratio is constant in this simple model. The random gross return on the market portfolio follows as

$$\tilde{R}_{t+1}^M = \frac{\tilde{m}_{t+1} + \tilde{P}_{t+1}^M}{P_t^M} = \frac{\tilde{\lambda}_{t+1} \cdot m_t + \Theta \cdot \tilde{\lambda}_{t+1} \cdot m_t}{\Theta \cdot m_t} = \frac{1+\Theta}{\Theta} \tilde{\lambda}_{t+1}. \tag{5.1}$$

The growth rate of production, $\tilde{\lambda}_{t+1}$, and Θ determine the implied market return.

5.2.2 Equilibrium Rates of Return Implied by Time-additive Expected Utility

Now, solve for Θ and derive the implied equilibrium rate of return on the market portfolio and the implied equilibrium risk-free rate of return for time-additive expected utility with constant relative risk aversion (cf. section 4.2.2.1).

The first order condition for the optimization problem of the representative investor implies the Euler equation (see equations (2.2) and (4.26))

$$1 = \mathrm{E}\left\{\beta \left(\frac{\tilde{c}_{t+1}}{c_t}\right)^{-\alpha} \tilde{R}_{t+1}^i\right\} = \mathrm{E}\left\{\beta \cdot \tilde{\lambda}_{t+1}^{-\alpha} \cdot \tilde{R}_{t+1}^i\right\} \qquad \forall i. \tag{5.2}$$

Using equation (5.1) this Euler equation yields in the case of the market portfolio

$$\begin{aligned} 1 &= \mathrm{E}\left\{\beta \cdot \tilde{\lambda}_{t+1}^{-\alpha} \cdot \tilde{\lambda}_{t+1} \frac{1+\Theta}{\Theta}\right\} \\ \Leftrightarrow \frac{1+\Theta}{\Theta} &= \beta^{-1} \left(\mathrm{E}\left\{\tilde{\lambda}_{t+1}^{1-\alpha}\right\}\right)^{-1} \end{aligned}$$

and thus

$$\tilde{R}_{t+1}^M = \beta^{-1} \left(\mathrm{E}\left\{\tilde{\lambda}_{t+1}^{1-\alpha}\right\}\right)^{-1} \tilde{\lambda}_{t+1}. \tag{5.3}$$

The risk-free rate of return, R_{t+1}^f, can be obtained from the Euler equation (5.2) as

$$\begin{aligned} 1 &= \mathrm{E}\left\{\beta \cdot \tilde{\lambda}_{t+1}^{-\alpha} \cdot R_{t+1}^f\right\} \\ \Leftrightarrow R_{t+1}^f &= \beta^{-1} \left(\mathrm{E}\left\{\tilde{\lambda}_{t+1}^{-\alpha}\right\}\right)^{-1}. \end{aligned} \tag{5.4}$$

At this point the idea of calibration can easily be summarized. A given class of preferences, here, time-additive expected utility with constant relative risk aversion, and a given process for the exogenous endowment, here, stationary i.i.d. growth rates of production, together imply the equilibrium rates of return. In this simple example the closed-form solutions for both the implied equilibrium return on the market portfolio (equation (5.3)) and the implied equilibrium risk-free rate of return (equation (5.4)) can be derived analytically[12]. Both depend exclusively on the random growth rate of the exogenous endowment and the preference parameters. For a given exogenous endowment process, e.g., a process estimated from consumption data observed historically, the equilibrium rates of return implied by the model depend exclusively on the preference parameters. Calibration means, that one tries to match specific moments of these implied rates of return with the corresponding moments of historically observed rates of return by varying the preference parameters.

5.2.3 Equilibrium Rates of Return Implied by Recursive Non-expected Utility

Alternatively to the time-additive expected utility the parametrization of recursive non-expected preferences introduced in section 4.2.3.1 is used. In this case the first order condition for the optimization problem of the representative investor implies the Euler equation (see equations (2.2) and (4.46))

$$
\begin{aligned}
1 &= \mathrm{E}\left\{\left[\beta\left(\frac{\tilde{c}_{t+1}}{c_t}\right)^{-\rho}\right]^{\frac{1-\alpha}{1-\rho}}\left(\tilde{R}^M_{t+1}\right)^{\frac{1-\alpha}{1-\rho}-1}\tilde{R}^i_{t+1}\right\} \\
&= \mathrm{E}\left\{\left[\beta\cdot\tilde{\lambda}^{-\rho}_{t+1}\right]^{\frac{1-\alpha}{1-\rho}}\left(\tilde{R}^M_{t+1}\right)^{\frac{1-\alpha}{1-\rho}-1}\tilde{R}^i_{t+1}\right\} \qquad \forall i. \qquad (5.5)
\end{aligned}
$$

Using equation (5.1) the Euler equation (5.5) yields in the case of the market portfolio

$$
\begin{aligned}
1 &= \mathrm{E}\left\{\left[\beta\cdot\tilde{\lambda}^{-\rho}_{t+1}\right]^{\frac{1-\alpha}{1-\rho}}\left(\tilde{\lambda}_{t+1}\frac{1+\Theta}{\Theta}\right)^{\frac{1-\alpha}{1-\rho}}\right\} \\
\Leftrightarrow \left(\frac{1+\Theta}{\Theta}\right)^{\frac{1-\alpha}{1-\rho}} &= \beta^{-\frac{1-\alpha}{1-\rho}}\left(\mathrm{E}\left\{\tilde{\lambda}^{1-\alpha}_{t+1}\right\}\right)^{-1} \\
\Leftrightarrow \frac{1+\Theta}{\Theta} &= \beta^{-1}\left(\mathrm{E}\left\{\tilde{\lambda}^{1-\alpha}_{t+1}\right\}\right)^{-\frac{1-\rho}{1-\alpha}}
\end{aligned}
$$

and thus

$$\tilde{R}^M_{t+1} = \beta^{-1}\left(\mathrm{E}\left\{\tilde{\lambda}^{1-\alpha}_{t+1}\right\}\right)^{-\frac{1-\rho}{1-\alpha}}\tilde{\lambda}_{t+1}. \qquad (5.6)$$

[12]This may not be the case for more complicated stochastic processes for the exogenous endowment or for preferences that do not exhibit CRRA.

Using this implied equilibrium return on the market, the risk-free rate of return, R^f_{t+1}, can be obtained from the Euler equation (5.5) as follows

$$\begin{aligned} 1 &= \mathrm{E}\left\{\left[\beta\cdot\tilde{\lambda}_{t+1}^{-\rho}\right]^{\frac{1-\alpha}{1-\rho}}\left(\beta^{-1}\left(\mathrm{E}\left\{\tilde{\lambda}_{t+1}^{1-\alpha}\right\}\right)^{-\frac{1-\rho}{1-\alpha}}\tilde{\lambda}_{t+1}\right)^{\frac{1-\alpha}{1-\rho}-1}R^f_{t+1}\right\} \\ \Leftrightarrow \frac{1}{R^f_{t+1}} &= \mathrm{E}\left\{\beta\cdot\tilde{\lambda}_{t+1}^{-\alpha}\left(\mathrm{E}\left\{\tilde{\lambda}_{t+1}^{1-\alpha}\right\}\right)^{\frac{1-\rho}{1-\alpha}-1}\right\} \\ \Leftrightarrow R^f_{t+1} &= \beta^{-1}\left(\mathrm{E}\left\{\tilde{\lambda}_{t+1}^{-\alpha}\right\}\right)^{-1}\left(\mathrm{E}\left\{\tilde{\lambda}_{t+1}^{1-\alpha}\right\}\right)^{1-\frac{1-\rho}{1-\alpha}}. \end{aligned} \tag{5.7}$$

Note, that for $\alpha = \rho$ the implied equilibrium returns (equations (5.6) and (5.7)) equal their counterparts for time-additive expected utility with constant relative risk aversion (equations (5.3) and (5.4)). In this case the recursive non-expected utility specification is identical with the time-additive expected utility specification with constant relative risk aversion.

5.2.4 Excursus: Equivalence of Both Utility Specifications under i.i.d. Consumption Growth

As shown in section 4.2.3.1, recursive non-expected utility (RNEU) is identical with time-additive expected utility with constant relative risk aversion (TAEU), if the elasticity of intertemporal substitution σ equals the reciprocal of the parameter of risk aversion α. If $\alpha \neq 1/\sigma$, both utility specifications are different and, thus, may not imply the same equilibrium rates of return, given a discount factor β which exclusively measures time preference. RNEU applies the "true" elasticity of intertemporal substitution σ, whereas TAEU restricts $\sigma = 1/\alpha$. Therefore, RNEU should have more explanatory power than TAEU. It will now be shown that under the assumption of i.i.d. consumption growth this is not true. Instead, both utility specifications are equivalent in the sense that for given levels of time preference, risk aversion and elasticity of intertemporal substitution the preference parameters of both utility functions can be chosen so that both imply the same equilibrium rates of return[13].

Assume that consumption growth is i.i.d. and that the representative investor has fixed values of risk aversion α, discount factor β and elasticity of intertemporal substitution σ. Further assume that $1/\sigma = \rho \neq \alpha$. Then, comparing the risk-free rate implied by TAEU, $R^f_{t+1,TAEU}$ (equation (5.4)),

[13]The irrelevance of separating the parameter of risk aversion and the elasticity of intertemporal substitution under i.i.d. consumption growth was emphasized by Kocherlakota (1990a). However, the illustration given here is less complicated and more intuitive.

with the risk-free rate implied by RNEU, $R^f_{t+1,RNEU}$ (equation (5.7)), yields

$$R^f_{t+1,RNEU} = R^f_{t+1,TAEU} \left(\mathrm{E}\left\{ \tilde{\lambda}^{1-\alpha}_{t+1} \right\} \right)^{1-\frac{1-\rho}{1-\alpha}} . \tag{5.8}$$

The two risk-free rates differ because TAEU does not take into account the difference between α and ρ, whereas RNEU accounts for this difference. Therefore, assume that $R^f_{t+1,RNEU}$ is the "correct" risk-free rate. Notice, that for given α and ρ the term $(\mathrm{E}\{\tilde{\lambda}^{1-\alpha}_{t+1}\})^{1-\frac{1-\rho}{1-\alpha}}$ is a constant. The idea now is to find an adjusted discount factor β' so that, using this discount factor in TAEU instead of the original discount factor, TAEU also implies the "correct" risk-free rate. It is easy to see from equation (5.8) that

$$\beta' = \beta \left(\mathrm{E}\left\{ \tilde{\lambda}^{1-\alpha}_{t+1} \right\} \right)^{\frac{1-\rho}{1-\alpha}-1} . \tag{5.9}$$

The adjusted discount factor β' may be interpreted as the "true" discount factor which exclusively measures the time preference, times a compensation for the implicitly chosen elasticity of intertemporal substitution which differs from the investor's "true" elasticity of intertemporal substitution. Given a positive average consumption growth, it follows for $\alpha > \rho$ that $\beta' > \beta$, and vice versa. β' may therefore exceed unity. Assume, for example, that risk aversion α is 5 which in TAEU implies an elasticity of intertemporal substitution σ of .2. The true σ, however, is assumed to be .8[14]. Assume further that $\beta = .99$ and that $\mathrm{E}\{\tilde{\lambda}^{1-\alpha}_{t+1}\} = .88$ for $\alpha = 5$ which implies that consumption growth is about three percent. For this example the adjusted discount factor β' equals 1.116. If β' is erroneously interpreted as the normal discount factor this would indicate a negative time preference, although the true time preference is positive. This example illustrates the result of Kocherlakota (1990b) that in growth economies with TAEU equilibria may exist for $\beta > 1$ and that $\beta > 1$ may be economically plausible[15].

So far, it has been shown that an adjusted discount factor can be found for TAEU such that TAEU implies the same equilibrium risk-free rate of return as RNEU[16]. However, to be equivalent both utility specifications must further imply the same equilibrium rates of return for any other asset i using the same adjusted discount factor in TAEU. Under i.i.d. consumption

[14] $\alpha = 5$ and $\sigma = .8$ can both be accepted as plausible (see discussion in section 7.2.3).

[15] In the TAEU-framework the intuition of this result is as follows. Consider a highly risk averse investor with a positive time preference and TAEU. As a consequence of the preference restriction this investor is extremely unwilling to substitute intertemporally. If consumption growth is positive on average and the risk-free rate is low, this investor would have a strong desire to borrow from the future. This is not consistent with equilibrium. A low risk-free rate is consistent with equilibrium only if the investor has a negative time preference.

[16] Notice, that up to this point this result is not surprising because it is always possible to equalize two implied risk-free rates by multiplying one of them with a constant factor.

growth, this is indeed true. Consider the return on the market. Using the adjusted discount factor β', TAEU implies the same equilibrium rate of return on the market as RNEU (see equation 5.3 and 5.6):

$$\begin{aligned} \tilde{R}^M_{t+1,TAEU} &= \beta'^{-1} \left(\mathrm{E}\left\{ \tilde{\lambda}^{1-\alpha}_{t+1} \right\} \right)^{-1} \tilde{\lambda}_{t+1} \\ &= \beta^{-1} \left(\mathrm{E}\left\{ \tilde{\lambda}^{1-\alpha}_{t+1} \right\} \right)^{-\frac{1-\rho}{1-\alpha}} \tilde{\lambda}_{t+1} \\ &= \tilde{R}^M_{t+1,RNEU}. \end{aligned} \tag{5.10}$$

If the adjusted discount factor β' is used in TAEU, both utility specifications also imply the same equilibrium rates of return for any other asset i. This can be seen by replacing β with β' in the Euler equation (5.2) of the TAEU and comparing the resulting expression with the Euler equation (5.5) of the RNEU after $\tilde{R}^M_{t+1}$ in equation (5.5) is replaced by expression (5.6). Thus, under i.i.d. consumption growth the parameters α and β can be chosen in TAEU to fully reflect the risk aversion, the elasticity of intertemporal substitution and the time preference, so that TAEU and RNEU imply the same equilibrium rates of return. However, since the adjusted β cannot be interpreted as the usual discount factor the clear separation of the three dimensions of preferences in RNEU is more satisfying from a theoretical point of view also under i.i.d. consumption growth. Moreover, the equivalence of both utility specifications depends on the restrictive assumption of i.i.d. consumption growth. If this assumption is not satisfied, there certainly exists an adjusted discount factor for TAEU such that TAEU and RNEU imply the same equilibrium risk-free rates of return, but using this adjusted discount factor, TAEU and RNEU do not imply the same equilibrium rates of return for other assets. This becomes evident in section 8.2, assuming that consumption growth follows a two-state Markov switching model.

5.3 Review of Studies Applying the Calibration Approach

This section briefly surveys studies applying the calibration approach in order to match the risk-free rate and the equity premium implied by equilibrium models to the corresponding observed values. Existing calibration exercises modify the simple example shown in section 5.2. After discussing the initial study of Mehra and Prescott (1985) the survey is structured along different modifications of the key assumptions of their approach.

Mehra and Prescott (1985) attempt to calibrate the parameters of time-additive expected utility with constant relative risk aversion in a discrete state-space version of the Lucas (1978) economy. The Lucas economy is a closed economy defined in real terms and characterized by one good, a representative agent, complete and frictionless asset markets and the equality of production, dividends and consumption. Mehra and Prescott model the exogenously given stochastic growth rate of production as a symmetric Markov state-switching model with two states with deterministic outcomes, which means that the growth rate of production switches between two deterministic values occurring with the same unconditional probability of .5. The switching model is chosen such that the mean, variance, and autocorrelation of the growth rate equal the corresponding values of historically observed consumption growth. They find that the equity premia on US stocks observed in the 90 years from 1889 to 1978 are too high to be explained by conventionally accepted values of risk aversion below ten[17] and a discount factor below one. This finding is termed the "Equity Premium Puzzle". If, however, risk aversion is chosen sufficiently large to explain the observed equity premia, then the observed risk-free rates are too low to be explained. This finding is termed the "Risk-free Rate Puzzle"[18]. Siegel (1992) extends the time period to the period 1802 to 1990. Over this period the average risk-free rate is larger and the average equity premium is smaller than over the period analyzed by Mehra and Prescott[19]. Nevertheless, the observed values are still inconsistent with the equilibrium model.

Many attempts have been made to solve these puzzles. The remainder of this section discusses attempts along the following modifications of the model:

(1) relaxing the assumed equality of production, dividends and consumption,
(2) modifying the exogenous stochastic process,
(3) modifying the preference specification of the representative agent,
(4) relaxing the assumed completeness and frictionlessness of asset markets, e.g., incorporating additional risks such as uninsurable idiosyncratic income risks or constraints such as transactions costs and liquidity constraints.

These modifications maintain the assumption of rational individuals[20] and with exception of (4) they also maintain the assumption of complete and frictionless asset markets[21]. These assumptions will also be maintained in the calibration exercises in chapter 8.

---

[17]While Kandel and Stambaugh (1991, pp. 65-9) argue that also parameters of risk aversion, α, up to 30 may be plausible "a vast majority of economists believe that values for α above ten ... imply highly implausible behavior" (Kocherlakota 1996, p. 52).

[18]This term was originally introduced by Weil (1989).

[19]For the average rates of return calculated by Mehra and Prescott (1985) and Siegel (1992) see section 1.2.

[20]A stream of research not discussed here attempts to resolve the Equity Premium Puzzle by relaxing the assumption of rational individuals, e.g., Kurz and Beltratti (1997).

[21]The following survey also includes some studies related to the Equity Premium Puzzle- and Risk-free Rate Puzzle-literature that do not apply the calibration approach, in particular in point (4). Considering the numerous studies in this field of research the survey does not claim to enclose all. Also, studies exploring the effects of a demand for money, e.g., Bansal and Coleman II (1996), are not discussed. However, the main

(1) In reality stockholders do not receive total production as dividends since payments on fixed claims such as interest payments and wages are deducted before. The introduction of leverage proposed by Mehra and Prescott (1985) relaxes the equality of dividends and production. Aggregate production equals aggregate consumption in each period, but production is paid off in two parts, a risk-free part which is the payoff on fixed claims and a risky residual, the payoff on equity. Leverage increases the riskiness of the payoff on equity so that models with leverage may yield higher equilibrium risk premia. Benninga and Protopapadakis (1990) show that the introduction of leverage helps to solve the Equity Premium Puzzle[22]. Holding leverage constant, production, respectively consumption, and dividends are strongly correlated. In reality, however, the aggregate payoff on fixed claims need not be constant and the dividend on the stock market represents only a fraction of the payoff on the economy's aggregate equity. Therefore, a strong correlation between aggregate production and dividends may be too restrictive. Cecchetti, Lam, and Mark (1993), Hung (1994) and Bonomo and Garcia (1996) relax this restriction by separately modeling production growth and dividend growth. They substitute the univariate stochastic process for production (= consumption = dividend) growth by a bivariate process for production (= consumption) growth and dividend growth with the correlation between both set to the historically observed value. The estimated bivariate processes usually exhibit a larger variability of dividend growth than of consumption growth, which may increase the equity premia. The smaller correlation between consumption growth and dividend growth may on the other hand decrease the equity premia. Two alternative approaches can be found in Abel (1994) and Mankiw and Zeldes (1991). Abel (1994) explicitly includes labor income to break the equality between dividends and consumption. Mankiw and Zeldes (1991) do not directly relax the equality of production, consumption and dividends but argue that not all individuals hold stocks and that therefore consumption data of stockholders rather than aggregate consumption should be used. They observe that consumption of stockholders and non-stockholders differs considerably and that this difference in consumption explains some, but not all, of the Equity Premium Puzzle.

(2) Modifications of the exogenous stochastic process enclose Markov state-switching models with more than two states as well as models where the outcomes are stochastic instead of deterministic. In models with states with deterministic outcomes the exogenous variable is deterministic within each state, whereas in switching models with states with stochastic outcomes it

streams of research related to the present analysis are discussed. Surveys of the Equity Premium Puzzle- and Risk-free Rate Puzzle-literature are also provided by Abel (1991) and Kocherlakota (1996).

[22]See Kandel and Stambaugh (1990), (1991) for other studies applying the leverage approach.

is stochastic within each state. In the latter case a state-dependent distribution of the variable is assumed and the switching model is estimated directly from the data. In contrast, switching models with states with deterministic outcomes are usually assumed to be symmetric, i.e. the unconditional probabilities of all states are equal, and the states are chosen to fit the first-order autocorrelation and the first and second moments of the observed time series. Rietz (1988) adds a third state with a deterministic outcome to the framework of Mehra and Prescott (1985) representing a strong crash occurring with low probability. This modification raises the variability of consumption growth and, thus, raises the equilibrium equity premium for a given level of risk aversion. Mehra and Prescott (1988), however, argue that the assumed reduction in consumption of more than 25 percent per annum in the crash state is too large to provide a reasonable resolution of the Equity Premium Puzzle. Kandel and Stambaugh (1990), (1991) analyze a symmetric endowment process with four states with stochastic outcomes. Cecchetti and Mark (1990) and Cecchetti, Lam, and Mark (1990) investigate asymmetric endowment processes with two states directly estimated from the time series of the endowment variable. An asymmetric process with three states is explored by Bonomo and Garcia (1994). Modelling consumption growth and dividend growth separately, the switching between univariate distributions is replaced by the switching between bivariate distributions. Models with bivariate state-switching processes are estimated by Cecchetti, Lam, and Mark (1993), Hung (1994) and Bonomo and Garcia (1996). Abel (1994) uses conditional i.i.d. stochastic shocks instead of a state-switching model. However, Kocherlakota (1996) demonstrates that the Equity Premium Puzzle cannot be resolved by solely modifying the exogenous stochastic process.

(3) In modifying the preference specification of the representative investor the first step was to allow the discount factor in the time-additive expected utility specification to increase beyond one. As shown in section 5.2.4, discount factors larger than one may be plausible because with time-additive expected utility the elasticity of intertemporal substitution and the risk aversion are not separately specifiable but one is the reciprocal of the other (see section 4.2.1.2). Benninga and Protopapadakis (1990) find that discount factors larger than one in combination with leverage can resolve the Equity Premium Puzzle and the Risk-free Rate Puzzle. This indicates that the recursive non-expected preferences introduced in section 4.2.3.1 allowing to distinguish between risk aversion and the elasticity of intertemporal substitution may help to resolve these puzzles. Weil (1989), Kandel and Stambaugh (1991) and Hung (1994) use this intertemporally non-separable utility specification in calibration exercises. The last two articles find promising results. Epstein and Zin (1990) suggest to combine the recursive non-expected preferences with "first-order risk aversion", which means that in one-stage lotteries the required risk premium is proportional to the standard deviation

of the outcome rather than to the variance. The implications of first-order risk aversion for the equity premium are also explored by Bekaert, Hodrick, and Marshall (1997). Another approach is to use utility specifications with habit formation which are also not intertemporally separable. The habit level in consumption increases the variability of the utility gained from consumption and, thus, produces higher equity premia. For studies applying this approach see, e.g., Abel (1990), Constantinides (1990), Ferson and Constantinides (1991), Ni (1993) and Campbell and Cochrane (1995).

(4) In an early paper Mankiw (1986) questions the use of aggregate data. He argues that unless aggregate shocks to income affect all investors equally, asset returns in incomplete markets depend on the distribution of the shocks among the population and the concentration of aggregate shocks is an important determinant of the equity premium[23]. This study is related to studies assuming incomplete markets by including uninsurable idiosyncratic risks, e.g., risks due to uncertain individual labor income. Some of these studies, e.g., Weil (1992) and Gollier and Schlesinger (1997), argue theoretically that under specific assumptions, e.g., standard risk aversion[24], individuals facing an additional idiosyncratic risk reduce their investments in risky assets also when the idiosyncratic risk is independent of the assets' risks, which in turn raises the equilibrium equity premium. A shortcoming of these studies is that they analyze two-period models, at best, and, thus, ignore intertemporal self-insurance through savings and dissavings (cf. Kocherlakota (1996)). Quantitatively investigating dynamic incomplete markets models, Telmer (1993) and Lucas (1994) show that individuals can closely approximate the allocation in complete markets by savings and dissavings when the shocks to individual income are not permanent. Thus, idiosyncratic risks and other incompletenesses of markets are not likely to significantly increase the equilibrium equity premium[25]. Moreover, Bonomo and Garcia (1996) demonstrate that the additional risk necessary to make the variability of consumption sufficiently high to be consistent with the observed equity premium is implausibly large. In addition to idiosyncratic risks transactions costs, borrowing constraints and short sale constraints are analyzed by Heaton and Lucas (1992) and Lucas (1994). Fisher (1994) and He and Modest (1995) examine market frictions without idiosyncratic risks. They find that the equity premium is substantially larger only if trading stocks is much more costly than trading risk-free assets.

Overall, many attempts have been made to solve the Equity Premium Puzzle and the Risk-free Rate Puzzle, most of which assume complete and frictionless markets. While evidence is accumulating that more sophisticated

---

[23]See also Danthine, Donaldson, and Mehra (1992).

[24]Standard risk aversion means that the utility function exhibits decreasing absolute risk aversion and decreasing absolute prudence (cf. Kimball (1993)).

[25]Assuming permanent income shocks instead, alters this result as is shown by Constantinides and Duffie (1996).

preference specifications help to resolve the Risk-free Rate Puzzle the evidence is not that clear for the Equity Premium Puzzle. Most of the attempts do not substantially increase the explained part of the equity premium, at least not with reasonable modifications of Mehra and Prescott's original approach. Until now the literature has exclusively focused on the US[26]. The calibration exercises in chapter 8 provide evidence for Germany. Most of the modifications discussed under (1) to (3) are applied to German data. Further modifications are suggested and analyzed.

5.4 Summary of Chapter 5

Parametric models of the market pricing kernel can empirically be investigated either by estimating the parameters so as to maximize the fit of the asset pricing model derived from the parametric model of the market pricing kernel or by specifying a model economy and an exogenous endowment process and comparing the equilibrium rates of returns implied by a reasonably parametrized market pricing kernel with observed rates of return. The latter approach is known as the calibration approach. It attempts to choose reasonable parameters of the market pricing kernel such that the implied rates of return mimic specific aspects of observed rates of return. Although both approaches have strengths and weaknesses the calibration approach is chosen for the present analysis because it provides stronger evidence for or against the underlying theory and may indicate why a specific parametric model of the market pricing kernel might fail to produce equilibrium rates of return that mimic observed data. The illustrative example of calibration reveals that under i.i.d. endowment growth the parametric models of the market pricing kernel derived from time-additive expected utility with constant relative risk aversion respectively derived from recursive non-expected utility do not differ in explanatory power. Furthermore, it is demonstrated that with time-additive expected utility it may be plausible to allow discount factors larger than one. The first calibration exercises that have been conducted unveiled the Equity Premium Puzzle and the Risk-free Rate Puzzle. Many attempts have been made to resolve these puzzles but up to now none of them has led to a widely accepted resolution of the puzzles. The calibration exercises in chapter 8 provide evidence concerning these puzzles for German data and several modifications of the model, alternatively applying time-additive expected utility with constant relative risk aversion and recursive non-expected utility.

[26] At least to my knowledge calibration exercises for other countries do not exist.

Part II

Intertemporal Asset Pricing: Empirical Analysis

Chapter 6

Overview and Description of Data

6.1 Overview

This section provides an overview of the empirical analysis. The analysis proceeds in two steps. The first step (chapter 7) is to apply the variance bound approach in order to analyze the information content of different asset return data sets for intertemporal asset pricing and to compare the potential of the parametric models of the market pricing kernel derived in section 4.2 to be consistent with these returns. Variance bounds are calculated from German stock, bond and money market returns in the period 1968 to 1994[1]. The second step is to apply the calibration approach (chapter 8) and to evaluate calibrated models using simulation techniques (chapter 9) in order to analyze whether the parametric models of the market pricing kernel can explain specific properties of the German one-period risk-free rate and the German equity premium observed in the period 1960 to 1994[2].

The purpose of the first step is to investigate the following aspects: (1) Empirical analyses of parametric models of the market pricing kernel, such as calibration exercises or studies estimating consumption-based intertemporal asset pricing models, are usually restricted to few assets, for example, the equity premium or a risk-free and one risky asset. To analyze the sensitivity of the results of such studies with respect to the assets applied, the

[1]The variance bound approach has been introduced in section 3.3. Besides Meyer (1997), an earlier study of mine that contains some of the results presented in chapter 7, I am aware of only one other study calculating variance bounds for Germany. In this study, Missong (1994) uses quarterly and yearly returns on only two assets, a risk-free asset and the aggregate stock market. In addition, no tax adjustment is made.

[2]The calibration approach has been introduced in chapter 5. Besides Meyer (1996), an earlier study of mine that contains some of the results presented in chapter 8, I am aware of no other calibration exercises for Germany.

variance bounds implied by different sets of assets are compared, because the variance bounds reflect the information content of the different sets of assets. (2) The application of parametric models of the market pricing kernel is often restricted to quarterly or yearly time intervals by the availability of macroeconomic data. Shorter or longer time intervals may change the results. Therefore, the bounds placed on the market pricing kernel by monthly, quarterly and yearly real returns are compared. (3) Results of empirical studies may further be sensitive to the method of return calculation. Returns calculated from end-of-period figures embody noise which cannot be found in macroeconomic data, so that returns calculated from period averages may be better comparable to macroeconomic data (e.g., consumption data) used in parametric models of the market pricing kernel. Therefore, the impact of different return calculation methods on the variance bounds is analyzed. (4) Further, the impact of different tax rates and the different taxation of private and institutional investors is investigated. The analyses (1) to (4) serve as an initial step for the empirical investigation of specific parametric models of the market pricing kernel. They provide information about which data, which time interval, which method of return calculation and which tax adjustment should be used for this investigation. (5) Finally, it is investigated whether the parametric models of the market pricing kernel derived in section 4.2 satisfy the variance bounds. This investigation provides a first impression of the capability of the parametric models of the market pricing kernel to jointly price different assets. Furthermore, it illustrates the Equity Premium Puzzle and the Risk-Free Rate Puzzle for Germany, which indicates that these puzzles, which are more thoroughly discussed in the calibration exercises in chapter 8, are not implied by special empirical tools such as calibration or estimation.

The purpose of the second step is to answer the following questions: (1) Can model economies with the derived parametric models of the market pricing kernel be calibrated to match the first and second moments of the historically observed risk-free rate and equity premium? (2) Do market pricing kernels derived under the assumption of the more general recursive non-expected utility allow to better match the risk-free rate and the equity premium than the market pricing kernel derived under the assumption of time-additive expected utility with constant relative risk aversion? (3) Do modifications of the simple example of calibration given in section 5.2, such as different stochastic processes for the exogenous endowment or the introduction of leverage, allow to improve the capacity of the model economy to match the historically observed rates of return? (4) Do calibrated equilibrium models also account for properties of the observed rates of return other than the first and second moments?

Which data are used? The first part covers the period 1968 to 1994 and necessitates the calculation of time series of real[3] rates of return on one-period risk-free assets, on bonds with maturities from 1 to 10 years, on the aggregate bond market and on the aggregate stock market. For the different analyses rates of return must be calculated for monthly, quarterly and yearly time intervals, for different taxation scenarios, applying different methods of return calculation. The calibration exercises in the second part cover the period 1960 to 1994. They require the calculation of "benchmark returns" on one-period risk-free assets and on the aggregate stock market that are to be matched by the equilibrium rates of return implied by the calibrated model economies. In addition, both parts of the analysis employ data on consumption. The first part uses consumption data to calculate first and second moments of the parametric models of the market pricing kernel. The second part uses consumption data for estimating the exogenous endowment process.

6.2 Description of Data

First, the selection of the time series of asset prices is discussed. Then the calculation of the return time series, especially of the tax adjusted real returns, is described and stylized facts of these rates of return are presented. Finally, consumption data are introduced.

6.2.1 Asset Price Data

Subsection 6.2.1.1 discusses alternative time series of asset prices for the empirical investigation. In subsection 6.2.1.2 the investment horizon, i.e. the length of the time intervals over which returns are measured, and the different methods of return calculation are clarified, before the derived nominal rates of return are analyzed. A method to estimate dividend returns is developed in subsection 6.2.1.3.

6.2.1.1 Original Time Series

Stock Prices As a proxy for stock prices the following stock market performance index is taken:

[3]Since the parametric models of the market pricing kernel that are to be investigated have been derived from utility functions in terms of real consumption the empirical analysis is conducted in real terms.

DAFOX — Capital-weighted German stock market performance index, constructed especially for research needs (*Deutscher Aktien-Forschungsindex*). It includes all German stocks traded in the first segment (Amtlicher Markt) at the Frankfurt Stock Exchange and is adjusted for cash dividends and proceeds from the sale of rights. Daily observations for the period 1/1960 to 12/1994 are provided by the Deutsche Finanzdatenbank[4] [5].

Alternatively, the DAX (*Deutscher Aktien Index*), a widely used capital-weighted German stock market performance index, could have been used. The DAX includes the 30 most actively traded German stocks and is adjusted for dividends and proceeds from the sale of rights[6]. The DAX was introduced in 1987. To gain a historical time series it was linked with the former calculated Index of the Börsen-Zeitung (BZ-Index), which itself was linked to the Index of the Hardy-Bank before April 1981 (cf. Mella (1988)). The Index of the Hardy-Bank was not adjusted for dividends and the BZ-Index was equally weighted. Therefore, the DAX before 1987 misestimates the performance of the German stock market.

The upper part of table 6.1 lists means and standard deviations of nominal returns calculated from these indices for the period 1960 to 1994 and for various subperiods. The returns are derived from monthly observations and expressed in percent per annum[7]. The returns calculated from the DAX before 1986 are almost equal to the returns on the DAFOX-PR which is not adjusted for dividends. Stehle, Huber, and Maier (1996) recalculate the history of the DAX using today's calculation method of the DAX and selecting the stocks that would have been selected for the index applying the criteria used today. End-of-month observations of the recalculated DAX (DAX-SHM) in the period 1/1960 to 12/1987 are taken from table 3 in Stehle, Huber, and Maier (1996). Table 6.1 shows that the returns on the DAX-SHM are close to those on the DAFOX. Nevertheless, the DAFOX is

[4]For a detailed description of the DAFOX and the subindices available see Göppl and Schütz (1993) and Göppl, Lüdecke, Schlag, and Schütz (1995).

[5]For the calculation of tax adjusted returns capital gains must be separated from dividend payments to account for differences in taxation (see section 6.2.1.3). For this purpose the capital-weighted German stock market *price* index, DAFOX_PR, is used in addition. It contains the same stocks as the DAFOX but reflects only capital gains and proceeds from the sale of rights; in contrast to the DAFOX dividends are not included. Such indices are usually called capital appreciation indices. A capital appreciation index measures the increase in wealth assuming that just the capital gain, and not any income generated, is reinvested (cf. Siegel (1992, p. 29)).

[6]Daily observations in the period 1/1960 to 12/1994 are provided by the Deutsche Finanzdatenbank.

[7]Annualized means are calculated by compounding monthly means. The standard deviations are annualized by multiplying the standard deviations of monthly returns by the square root of 12.

used instead of the DAX-SHM for three reasons. First, the DAFOX includes all stocks traded in the first segment of the Frankfurt Stock Exchange and therefore represents a broader fraction of the stock market than the DAX-SHM. The use of a broader stock market index has two effects, (1) the standard deviation of the returns decreases due to the diversification effect, and, (2) the mean return increases resulting from the inclusion of smaller firms and the well-known firm-size-effect[8]. A comparison of the returns on the DAFOX and the returns on the DAX-SHM in table 6.1 reveals both effects, a higher mean and a lower volatility of the returns on the DAFOX as compared to the returns on the DAX-SHM. Second, for the DAFOX daily observations are available, and, third, the combination of the DAFOX and the DAFOX-PR allows to separate dividend payments and capital gains (see section 6.2.1.3). This separation is necessary for the calculation of tax adjusted returns.

Money Market Rates Money market rates are used to represent the return on one-period risk-free assets for monthly and quarterly time intervals[9].

MMR_x x-month money market rate ($x \in \{1, 3\}$). Weekly averages from 1/1960 to 2/1967 and daily observations from 3/1967 to 12/1994 are provided by the Deutsche Bundesbank.

The money market rates are provided in percent per annum. Monthly average money market rates and the standard deviation of the monthly averages are listed in the second panel of table 6.1. For comparison, means and standard deviations are also given for the one-day money market rate (MMR_1D)[10].

Bond Prices The empirical investigation necessitates a measure of holding period returns on bonds, i.e. the returns on bonds realized over a specific period. These returns consist of coupon payments and capital gains. The following bond performance index is used:

---

[8]The firm-size-effect describes the empirically observed higher returns on stocks of small firms (see Banz (1981)). For the German stock market this effect is documented in Oertmann (1994a), (1994b) and Stehle (1997).

[9]One could argue that since the models are based on consumption and investment decisions of individuals and since private investors have no access to interbank deposits the interest rate on risk-free investments that are accessible to private investors should be used as risk-free rate. However, in the period 1975 to 1996 the difference between this interest rate and the three-month money market rate was less than one percentage point (cf. Deutsche Bundesbank (1996, p. 14)) so that the use of money market rates seems justifiable.

[10]Daily observations of the one-day money market rate for the period 1/1960 to 12/1994 are provided by the Deutsche Bundesbank.

Table 6.1: Annualized means and standard deviations (in parentheses) of monthly percentage changes in the stock and bond indices and in the consumer price index for the period 1960 to 1994 and different subperiods. Means and standard deviations are also given for the monthly averages of money market rates and average yields to maturity of bonds. All values are expressed in percent per annum.

Time series	1960-1967	1968-1976	1977-1985	1986-1994	1968-1994	1960-1994
Nobs[a]	96	108	108	108	324	420
Percentage Change in German Stock Market Indices						
DAFOX	6.87 (19.64)	5.51 (15.51)	17.19 (12.70)	6.28 (19.66)	9.54 (16.22)	8.93 (17.04)
DAFOX-PR	3.57 (19.64)	1.78 (15.52)	13.61 (12.78)	4.01 (19.57)	6.36 (16.21)	5.71 (17.03)
DAX	4.03 (18.34)	1.63 (17.37)	12.63 (13.81)	7.29 (20.98)	7.09 (17.62)	6.39 (17.77)
DAX-SHM	5.42 (20.27)	3.71 (16.78)	17.90 (14.22)	6.08 (21.45)	9.07 (17.76)	8.23 (18.34)
Average German Money Market Rates						
MMR_1D	3.65 (1.05)	5.98 (3.01)	6.50 (2.59)	6.43 (2.07)	6.30 (2.59)	5.70 (2.58)
MMR_1	4.07 (1.20)	6.59 (2.91)	6.76 (2.71)	6.54 (2.08)	6.63 (2.59)	6.05 (2.58)
MMR_3	4.53 (1.22)	7.00 (2.99)	6.94 (2.73)	6.62 (2.09)	6.85 (2.63)	6.32 (2.57)
Percentage Change in German Bond Performance Indices						
REX_1	. .	7.38 (1.42)	7.50 (1.41)	6.90 (.96)	7.26 (1.28)	. .
REX_2	. .	7.45 (2.08)	7.65 (2.45)	6.74 (1.73)	7.28 (2.11)	. .
REX_3	. .	7.40 (2.61)	7.74 (3.36)	6.65 (2.45)	7.26 (2.83)	. .
REX_4	. .	7.25 (3.07)	7.80 (4.06)	6.62 (3.10)	7.22 (3.43)	. .
REX_5	. .	7.23 (3.50)	7.92 (4.81)	6.64 (3.66)	7.26 (4.02)	. .
REX_6	. .	7.25 (3.88)	8.12 (5.38)	6.71 (4.14)	7.36 (4.50)	. .
REX_7	. .	7.23 (4.23)	8.32 (5.89)	6.71 (4.58)	7.42 (4.94)	. .
REX_8	. .	7.05 (4.53)	8.49 (6.36)	6.61 (4.99)	7.38 (5.34)	. .
REX_9	. .	6.72 (5.90)	8.94 (7.88)	6.28 (5.88)	7.31 (6.61)	. .
REX_10	. .	6.11 (5.60)	8.94 (7.88)	6.10 (6.24)	7.04 (6.63)	. .
REX	. .	7.12 (3.43)	8.06 (4.38)	6.61 (3.62)	7.26 (3.82)	. .
Average Yield to Maturity of German Public Bonds						
AYTM_1	. .	7.16 (1.38)	7.21 (2.13)	6.55 (1.81)	6.97 (1.82)	. .
AYTM_4	. .	8.05 (1.27)	7.59 (1.48)	6.85 (1.36)	7.50 (1.46)	. .
AYTM_7	. .	8.23 (1.20)	7.83 (1.32)	7.16 (1.05)	7.74 (1.27)	. .
AYTM_9	. .	8.25 (1.22)	7.89 (1.23)	7.15 (.97)	7.75 (1.23)	. .
AYTM	6.58 (.77)	8.16 (1.18)	7.73 (1.45)	7.02 (1.18)	7.64 (1.32)	7.40 (1.32)
Percentage Change in the German Consumer Price Index						
CPI	2.49 (1.01)	4.79 (.94)	3.78 (.83)	2.36 (.73)	3.64 (.88)	3.37 (.92)

[a]Number of observations.

REX_x German bond performance index measuring a weighted average of interest rates and capital gains of 3 synthetic bonds with a constant maturity of x years (x ∈ {1, 2, ..., 10}) and coupons of 6, 7.5 and 9 percent. REX denotes the weighted average of REX_x with $x \in \{1, 2, \ldots, 10\}$. End-of-month observations of these indices for the period 1/1968 to 12/1994 are provided by the Deutsche Börse AG[11].

The third panel of table 6.1 exhibits the annualized means and standard deviations of monthly holding period returns calculated from the REX_x. The returns on long bonds have been much more volatile than those on short bonds, but investors have not been compensated for this extra variation over the period 1968 to 1994. Only over the subperiod 1977 to 1985 long bonds have yielded higher returns than short bonds. For comparison, means and standard deviations of monthly averages of the yield to maturity of German public bonds issued by the government, the Federal Railway and the Federal Postal Administration are given in the next panel. AYTM_x denotes the average yield to maturity of bonds with a maturity between x and x+1 years. AYTM is the average yield to maturity of German public bonds for all maturities[12]. The results obtained for the average yield to maturity differ from those for the returns on the REX_x. The standard deviation of the average yield to maturity decreases with the maturity of the bonds and over all periods long bonds exhibit higher yields to maturity than short bonds.

Consumer Price Index Inflation is measured by the following consumer price index:

CPI Seasonally adjusted consumer price index for all West German private households. For the period 1/1960 to 12/1994 monthly and yearly values are provided by the Statistisches Bundesamt. Quarterly values come from the Deutsches Institut für Wirtschaftsforschung (DIW). The index values are average prices over the time intervals (month, quarter, year) with the basis 1991=100.

Mean and standard deviation of the annualized percentage change in the monthly consumer price index are listed in the last row of table 6.1.

---

[11] For the calculation of tax adjusted returns capital gains must be separated from coupon payments to account for differences in taxation. For this purpose the German bond price indices, REX_PR_x, measuring exclusively capital gains, are used in addition.

[12] For AYTM, monthly averages in the period 1/1960 to 2/1977 and daily observations in the period 3/1977 to 12/1994 are provided by the Deutsche Bundesbank. For x between 1 and 6 monthly observations from 1/1967 to 1/1986 and daily observations from 2/1986 to 12/1994 are available. For x between 7 and 9 monthly observations from 1/1967 to 8/1986 and daily observations from 9/1986 to 12/1994 are available. For x between 1 and 9 the monthly observations in the period 1/1967 to 3/1973 exclusively cover German government bonds.

Figure 6.1: Annualized means and standard deviations of nominal monthly holding period returns on German one-month interbank deposits, on German bonds with maturities from 1 to 10 years and on German stocks.

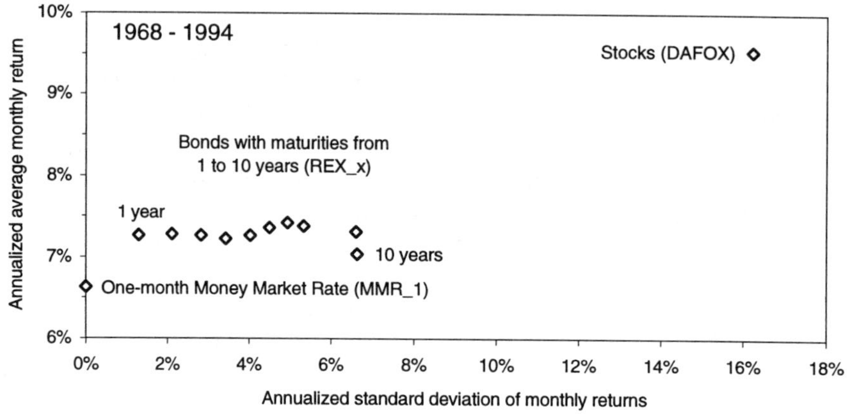

Summary In the empirical investigation the stock returns on the DAFOX, the bond returns on the REX and REX_x ($x \in \{1, \ldots, 10\}$) and the money market rates MMR_1 and MMR_3 will be used. The consumer price index CPI measures inflation and will be applied to deflate nominal returns. To get a first impression of the cross section of these returns, the annualized means and standard deviations of monthly nominal returns are plotted in figure 6.1. The standard deviation of the one-month money market rate is set equal to zero, because for monthly time intervals the one-month money market rate represents the one-period risk-free rate. For efficient portfolios the theory predicts a linear relation, the so-called "Capital Market Line" (see, for example, Franke and Hax (1994, pp. 345-6)), between expected returns and standard deviations. However, the analyzed indices cannot be assumed to be efficient. They are less than perfectly correlated, so that portfolios of these indices lead to dominant risk-return tradeoffs.

6.2.1.2 Length of Time Interval and Method of Return Calculation

One goal is to investigate the differences in the information content of monthly, quarterly and yearly real returns. Therefore, at least three sets of return time series are calculated. However, asset price data are observed with a high frequency so that one has to clarify how asset returns should be calculated to be comparable to quarterly or yearly macroeconomic data used in parametric models of the market pricing kernel. Consider, for example, a stock market index with daily observations. For this index quarterly returns can be calculated from end-of-quarter observations. This is the usual method for calculating returns over a reporting period and it was applied to monthly time intervals in the previous section. However, these returns may not be comparable to growth rates of macroeconomic data. Breeden, Gibbons, and Litzenberger (1989) and Grossman, Melino, and Shiller (1987) point out that reported consumption, for example, is the integral of the consumption flow over the reporting period and they argue that this may result in lower estimates for the variance of consumption and of consumption growth as compared to end-of-period observations of asset prices. Using the average of the daily observed asset prices over the reporting period instead of the last observation may increase the comparability with reported consumption data. Therefore, for each quarter the average of daily asset prices and then the returns from these averages are calculated, alternatively[13]. Similarly, yearly returns are calculated from end-of-year observations and from yearly averages. Finally, five sets of return time series are obtained: End-of-month returns (M), end-of-quarter returns (Q1), quarterly average returns (Q2), end-of-year returns (Y1) and yearly average returns (Y2).

For the money market rates the last observation of the time interval is used in the sets M, Q1 and Y1 and the average money market rates over the time interval are used in the sets Q2 and Y2. The one-period risk-free rate is given by the one-month money market rate in the set M, the three-month money market rate in the sets Q1 and Q2, and the return on bonds with a maturity of one year in the sets Y1 and Y2. For the monthly time interval the three-month money market rate represents no feasible one-period investment opportunity and is thus not considered. The same argument applies for the one-month money market rate for quarterly time intervals and for the one-month and three-month money market rates in the case of yearly time intervals. Because the money market rates are provided in percent per annum the one-month money market rates are divided by 12 and the three-month money market rates are divided by four to make them comparable to the growth rates of the other time series.

CPI-M is set equal to the monthly consumer price index provided by the Statistisches Bundesamt. Similarly, CPI-Q2 and CPI-Y2 are set to the original quarterly and yearly time series. CPI-Q1 consists of the March, June, September and December observations of CPI-M and CPI-Y1 consists of the December observations of CPI-M.

Table 6.2 lists the means and standard deviations of the calculated time series. All figures are percentage changes. Each column represents one of the sets of time series described above. To save space the means and standard deviations of the returns on the REX_x are given for four of the ten maturities, only. Over the periods 1968 to 1994 and 1960 to 1994 all assets yielded positive mean returns. Except for the returns on the DAFOX-Y1 and on

---

[13] Working (1960) theoretically compares first differences between averages over successive segments of a random chain and first differences between correspondingly positioned terms in the chain.

Table 6.2: Means and standard deviations (in parentheses) in percent of nominal returns on different stock and bond indices and interbank deposits, of the estimated dividend return and of the change in the consumer price index for monthly, quarterly and yearly time intervals for the period 1968 to 1994 and for the period 1960 to 1994.

Time Series	Set of Return Time Series[a] M	Q1	Q2	Y1	Y2
	Period 1968-1994[b]				
Nobs	324	108	108	27	27
DAFOX	$.76^{***}_{\circ\circ\circ}$ (4.68)	$2.34^{***}_{\circ\circ\circ}$ (8.54)	$2.25^{***}_{\circ\circ\circ}$ (7.04)	$10.43^{**}_{\circ\circ}$ (23.28)	$10.24^{***}_{\circ\circ\circ}$ (16.96)
DAFOX_PR	$.51^{**}_{\circ\circ}$ (4.68)	$1.60^{*}_{\circ\circ}$ (8.60)	$1.50^{**}_{\circ\circ}$ (7.06)	7.25 (22.85)	$7.04^{**}_{\circ}$ (16.69)
DIV[c]	$.32^{***}_{\circ\circ\circ}$ (.43)	$.97^{***}_{\circ\circ\circ}$ (.78)	$.97^{***}_{\circ\circ\circ}$ (.93)	$3.95^{***}_{\circ\circ\circ}$ (.93)	$3.99^{***}_{\circ\circ\circ}$ (.92)
MMR[d]	$.55^{***}_{\circ\circ\circ}$ (.21)	$1.72^{***}_{\circ\circ\circ}$ (.66)	$1.71^{***}_{\circ\circ\circ}$ (.66)	. .	. .
REX_1	$.59^{***}_{\circ\circ\circ}$ (.37)	$1.77^{***}_{\circ\circ\circ}$ (.74)	$1.77^{***}_{\circ\circ\circ}$ (.69)	$7.26^{***}_{\circ\circ\circ}$ (1.75)	$7.32^{***}_{\circ\circ\circ}$ (1.71)
REX_4	$.58^{***}_{\circ\circ\circ}$ (.99)	$1.76^{***}_{\circ\circ\circ}$ (2.05)	$1.77^{***}_{\circ\circ\circ}$ (1.84)	$7.26^{***}_{\circ\circ\circ}$ (4.67)	$7.41^{***}_{\circ\circ\circ}$ (4.56)
REX_7	$.60^{***}_{\circ\circ\circ}$ (1.43)	$1.82^{***}_{\circ\circ\circ}$ (2.90)	$1.81^{***}_{\circ\circ\circ}$ (2.48)	$7.47^{***}_{\circ\circ\circ}$ (6.37)	$7.67^{***}_{\circ\circ\circ}$ (6.21)
REX_10	$.57^{***}_{\circ\circ\circ}$ (1.91)	$1.72^{***}_{\circ\circ\circ}$ (3.53)	$1.71^{***}_{\circ\circ\circ}$ (2.87)	$7.06^{***}_{\circ\circ\circ}$ (7.51)	$7.24^{***}_{\circ\circ\circ}$ (6.76)
REX	$.59^{***}_{\circ\circ\circ}$ (1.10)	$1.77^{***}_{\circ\circ\circ}$ (2.17)	$1.77^{***}_{\circ\circ\circ}$ (1.99)	$7.30^{***}_{\circ\circ\circ}$ (5.20)	$7.46^{***}_{\circ\circ\circ}$ (5.04)
CPI	$.30^{***}_{\circ\circ\circ}$ (.25)	$.90^{***}_{\circ\circ\circ}$ (.58)	$.88^{***}_{\circ\circ\circ}$ (.50)	$3.65^{***}_{\circ\circ\circ}$ (1.98)	$3.62^{***}_{\circ\circ\circ}$ (1.89)
	Period 1960-1994[b]				
Nobs	420	140	140	35	35
DAFOX	$.72^{***}_{\circ\circ\circ}$ (4.92)	$2.23^{***}_{\circ\circ\circ}$ (9.36)	$2.06^{***}_{\circ\circ\circ}$ (7.52)	$9.78^{**}_{\circ\circ}$ (23.65)	$8.63^{***}_{\circ\circ\circ}$ (16.59)
DAFOX_PR	$.46^{*}_{\circ}$ (4.92)	$1.48^{*}_{\circ\circ}$ (9.38)	$1.30^{**}_{\circ}$ (7.52)	6.56 (23.13)	5.48^{*} (16.36)
DIV[c]	$.31^{***}_{\circ\circ\circ}$ (.41)	$.92^{***}_{\circ\circ\circ}$ (.75)	$.93^{***}_{\circ\circ\circ}$ (.87)	$3.77^{***}_{\circ\circ\circ}$ (.97)	$3.75^{***}_{\circ\circ\circ}$ (1.02)
MMR[d]	$.50^{***}_{\circ\circ\circ}$ (.21)	$1.59^{***}_{\circ\circ\circ}$ (.65)	$1.58^{***}_{\circ\circ\circ}$ (.64)	. .	. .
CPI	$.28^{***}_{\circ\circ\circ}$ (.27)	$.83^{***}_{\circ\circ\circ}$ (.58)	$.83^{***}_{\circ\circ\circ}$ (.48)	$3.42^{***}_{\circ\circ\circ}$ (1.84)	$3.35^{***}_{\circ\circ\circ}$ (1.80)

[a]M denotes the end-of-month returns, Q1 the end-of-quarter returns, Q2 the quarterly average returns, Y1 the end-of-year returns and Y2 the yearly average returns.

[b]*** (**, *) [∘ ∘ ∘ (∘∘, ∘)] indicates that the hypothesis of zero mean can be rejected at a 1 (5, 10) percent level of significance on the basis of the parametric t-statistic [the nonparametric Wilcoxon-Sign-Rank statistic].

[c]Nominal dividend return on stocks including the tax credit. The calculation of DIV is described in section 6.2.1.3.

[d]One-month money market rate (MMR_1) for the set M and three-month money market rate (MMR_3) for the sets Q1 and Q2.

the DAFOX_PR-M (-Q1, -Q2, -Y1, -Y2) the hypothesis of zero mean can be rejected at a one percent level of significance. The means of the DAFOX-Y1 and the DAFOX_PR-M (-Q1, -Q2, -Y2) are different from zero at a level of significance around five percent. Only the mean of the DAFOX_PR-Y1 is not significantly different from zero. The growth rate of the consumer price index is highly significantly positive, so that real asset returns may not be significantly positive.

Comparing the standard deviations of the returns over the different time intervals reveals that the time series do not meet the usual assumptions of stochastic independence over time and stationary conditional distribution. Otherwise, neglecting the effects of compound interest, the standard deviation should be proportional to the square root of the length of the time interval over which the return is measured, i.e. the standard deviation of quarterly returns should equal $\sqrt{3}$ times the standard deviation of monthly returns and the standard deviation of yearly returns should equal $\sqrt{4}$ times the standard deviation of quarterly returns. However, the standard deviation of the return on the DAFOX-Y2 (16.96) in the period 1968-1994, for example, is larger than $\sqrt{4}$ times the standard deviation of the return on the DAFOX-Q2 (14.08). Autocorrelation in the quarterly returns may account for this difference[14].

To test for autocorrelations, models with autocorrelated error terms are estimated for the returns on the DAFOX. For monthly returns up to 13 lags are included, for quarterly and yearly returns up to 5 lags. Nonsignificant autoregressive parameters are removed in the order of least significance. Table 6.3 lists the Maximum Likelihood estimates. σ_ε^2 is the variance of the conditional distributions of the returns[15]. The estimation results show no clear structure of autocorrelation. For the sets Q2 and Y2 a positive first-order autocorrelation of .25 should be observed resulting from averaging of the prices within the time intervals (see Working (1960, p. 918)). For Y2 the number of observations may be too low to find this autocorrelation. Applying the formula given in footnote 14 to the returns on the DAFOX-Q2 in the period 1968-1994 ($\sqrt{4 + 6 \cdot .31 + 4 \cdot .31^2 + 2 \cdot .31^3} \cdot 6.72 = 16.87$) almost yields the standard deviation of the return on the DAFOX-Y2 (16.96).

---

[14]For example, with first-order autocorrelation ρ_1 in one-period returns $\tilde{r}^1$, neglecting the effects of compound interest, the variance of N-period returns $\tilde{r}^N$ can be calculated as

$$\begin{aligned} \text{var}(\tilde{r}^N) &= \text{var}(\sum_{i=1}^{N} \tilde{r}_i^1) = N \cdot \text{var}(\tilde{r}^1) + 2 \sum_{i=1}^{N} \sum_{j=i+1}^{N} \text{cov}(\tilde{r}_i^1, \tilde{r}_j^1) \\ &= \text{var}(\tilde{r}^1) \cdot \left(N + 2 \sum_{i=1}^{N} \sum_{j=i+1}^{N} \text{corr}(\tilde{r}_i^1, \tilde{r}_j^1) \right) \\ &= \text{var}(\tilde{r}^1) \cdot \left(N + 2 \sum_{i=1}^{N} \sum_{j=i+1}^{N} \rho_1^{(j-i)} \right) . \end{aligned}$$

Therefore, with first-order autocorrelation in quarterly returns and neglecting the effects of compound interest, the standard deviation of yearly returns should equal $\sqrt{4 + 6\rho_1 + 4\rho_1^2 + 2\rho_1^3}$ times the standard deviation of quarterly returns. Similarly, the standard deviation of quarterly returns should equal $\sqrt{3 + 4\rho_1 + 2\rho_1^2}$ times the standard deviation of monthly returns.

[15]For first-order autocorrelation the unconditional variance can be obtained as $\sigma_\varepsilon^2/(1-\rho_1^2)$.

Table 6.3: Maximum Likelihood estimates of models with autocorrelated error terms for monthly, quarterly and yearly nominal returns on the DAFOX.

Autoregressive Model for the Return on the DAFOX $X_t = \bar{X} + \varepsilon_t - \sum_{\tau=1}^{N} \rho_\tau \varepsilon_{t-\tau}$, ε_t is i.i.d. with $E\{\varepsilon_t\} = 0$ and $std(\varepsilon_t) = \sigma_\varepsilon$					
	M[a]	Q1	Q2	Y1	Y2
			Period 1968-1994[b]		
$\bar{X}$ (σ_ε)	.76 (4.65)	2.34 (8.54)	2.26 (6.72)	10.43 (23.28)	10.24 (16.96)
	$\rho_{10} = .120$		$\rho_1 = .310$		
			Period 1960-1994[b]		
$\bar{X}$ (σ_ε)	.72 (4.89)	2.20 (9.17)	2.03 (7.13)	9.78 (23.65)	8.41 (15.09)
	$\rho_1 = .115$	$\rho_5 = -.223$	$\rho_1 = .326$		$\rho_2 = -.428$

[a]M denotes the end-of-month returns, Q1 the end-of-quarter returns, Q2 the quarterly average returns, Y1 the end-of-year returns and Y2 the yearly average returns.

[b]ρ_τ denotes the remaining significant autoregressive parameter after nonsignificant autoregressive parameters are removed in order of least significance starting with $N = 13$ [5, 5, 5, 5] for the set M [Q1, Q2, Y1, Y2].

However, in other cases the results are not that clear and even if no autocorrelation is observable the conditional distributions may not be stationary. Thus, even when autocorrelation is considered the length of the chosen time interval may affect the results of empirical investigations. Figuring this argument and the unclear structure of autocorrelation, autocorrelations are not considered subsequently. Instead, the impact of the length of the time interval chosen is explicitly investigated using the original sample moments.

Comparing the nominal end-of-period returns, Q1 and Y1, with the period average returns, Q2 and Y2, shows that the mean returns are almost the same for both methods of return calculation. Only for the longer period (1960-1994) the differences between the means increase beyond three percent. The differences between the standard deviations are larger. The standard deviations of period average returns are always lower than those of end-of-period returns. For example, the standard deviation of the yearly average returns on the DAFOX is 30 percent smaller than that of the end-of-year returns. This finding corresponds to the theoretical results of Working (1960, p. 917) showing that the variance of first differences between average prices over time intervals approximates 2/3 of the variance of first differences between correspondingly positioned prices.

6.2.1.3 Estimating Dividend Returns

For the calculation of tax adjusted returns on the stock market capital gains and dividend payments must be separated because of differences in the taxation of capital gains and dividend payments (see section 6.2.2). The capital gains of stocks are represented by the stock market price index DAFOX_PR. Unfortunately, no time series of dividend payments with a higher than yearly frequency is available for Germany. The Statistisches Bundesamt publishes a monthly time series of annual dividend returns. This time series includes the dividends paid within the last 12 month or, if available, the announcements of future dividend payments. From this time series one cannot calculate monthly or quarterly dividend returns and even for measuring yearly dividend returns the inclusion of the dividend announcements may bias the results. Therefore, the difference in the return on the DAFOX, i.e. the stock market performance index, and the return on the DAFOX_PR, i.e. the stock market price index, is used to estimate a time series of dividend returns. This method can be applied for any length of the time interval.

The DAFOX is adjusted for dividend payments applying the method of "opération blanche" (see Göppl and Schütz (1993, p. 511)) which assumes that the dividends are directly reinvested in the dividend paying stock (see Sauer (1991)). Since 1977 German investors receive dividends with a 36 percent tax credit (since 1994: 30 percent) which is deductible from the personal income tax. This tax credit is not included in the DAFOX, thus assuming implicitly a personal tax rate of 36 (30) percent. To obtain the dividend return before taxation this tax credit must be added.

The estimation is based on the following assumptions: (1) the dividends are continuously paid over the period and (2) reinvesting in the dividend paying stock equals reinvesting in the whole index portfolio. From these assumptions it follows that on average half of the dividend payments are reinvested in the stock market index over the same period implying the adjustment factor $(1 - .5 \cdot \mathrm{DAFOX}_t)$ in the following equation. The nominal dividend return including the tax credit from date $t-1$ to t, DIV_t, can then be estimated as

$$\mathrm{DIV}_t = \frac{\mathrm{DAFOX}_t - \mathrm{DAFOX\_PR}_t}{(1 - .5 \cdot \mathrm{DAFOX}_t)(1 - \tau_{credit})}, \tag{6.1}$$

where DAFOX_t and $\mathrm{DAFOX\_PR}_t$ are the nominal returns on the stock performance and the stock price index and τ_{credit} is the tax credit the investors receive on dividends. For the period 1960 to 1994 the level of τ_{credit} is given in the last column of table 6.4.

Means and standard deviations of the estimated dividend returns including the tax credit are listed in table 6.2. Over the period 1960 to 1994 the

Figure 6.2: Comparison of the time series of estimated yearly dividend returns including the tax credit (DIV-Y1) with the monthly time series of annual dividend returns published by the Statistisches Bundesamt and a time series taken from Bay (1990, table 14).

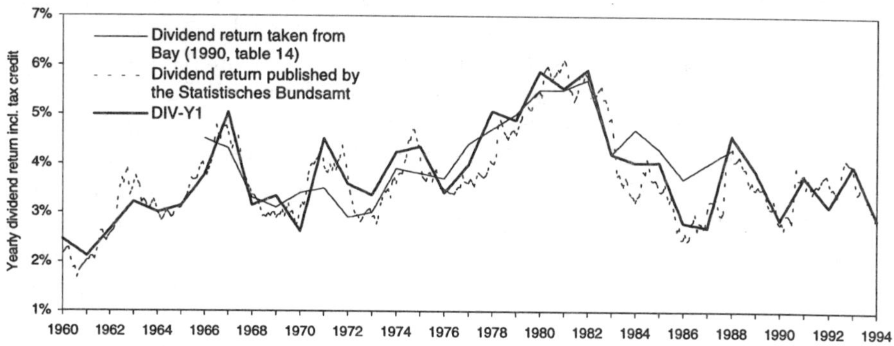

3.77 percent mean annual dividend return including the tax credit (DIV-Y1) is almost equal to the 3.68 percent mean published by the Statistisches Bundesamt. To further assess the quality of the estimated time series of dividend returns the estimated yearly time series DIV-Y1 is compared with the monthly time series of annual dividend returns including the tax credit published by the Statistisches Bundesamt and a time series taken from Bay (1990, table 14). Figure 6.2 graphs these time series and shows that the three time series are quite similar[16]. The yearly values of DIV-Y1 and of the time series taken from Bay are depicted at January instead of at June or July because the time series of the Statistisches Bundesamt leads dividend payments about six month due to the inclusion of dividend announcements.

Employing the estimation method to monthly returns, i.e. DAFOX-M and DAFOX-PR-M, and analyzing the seasonal structure of the estimated dividend returns DIV-M (see figure 6.3) yields the same plausible structure found by other authors for the German stock market. Bay (1990, p. 76, figure 1), for example, illustrates the number of dividend payments in each month of the year over the period 1966 to 1988 and Morawietz (1994, p. 123, figure 2.6) analyzes the aggregate dividend payment of 519 German corporations in each month of the year 1991.

Overall, the developed method to estimate dividend returns from the DAFOX and the DAFOX_PR allows to estimate the dividend return for any time interval and yields a good approximation of dividend returns.

---

[16]The correlation of the time series DIV-Y1 and the time series taken from Bay is .80. The correlation of the first differences is .66.

Figure 6.3: Seasonal structure of dividend payments illustrated by the average estimated monthly dividend return including the tax credit (DIV-M) for each month of the year for the period 1960 to 1994.

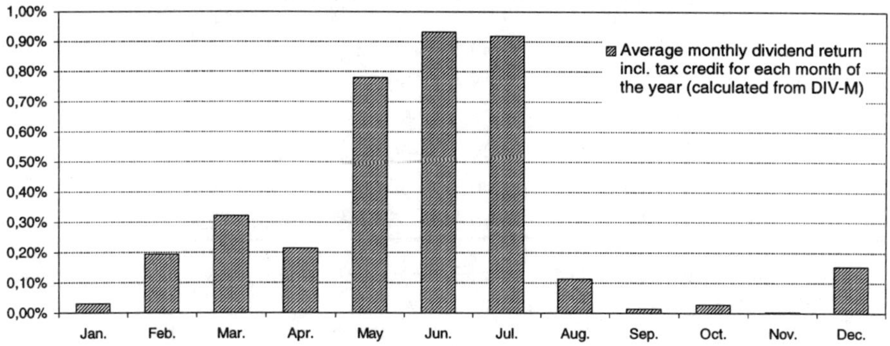

6.2.2 Real Tax Adjusted Returns

In the previous section the time series of asset prices have been described, different time series of nominal returns have been calculated and time series of dividend returns have been estimated. Now, time series of real tax adjusted returns for different taxation scenarios are calculated from these time series. Subsection 6.2.2.1 clarifies the differences in the taxation of private and corporate investors and defines the taxation scenarios. A description of the tax adjustment procedure follows in subsection 6.2.2.2. Stylized facts of the tax adjusted real returns are presented in section 6.2.2.3.

6.2.2.1 German Tax Rates, Taxation Scenarios and the Marginal Investor

The taxation of capital gains and of dividend and interest payments at the level of the investors differs across investor types. For private investors capital gains of securities are not taxed if the security is held for at least six months. In the case of stocks further requirements are that the private investor holds less than 25 percent of the shares of a company and that less than one percent of the shares of a company are sold[17]. Dividend and interest payments are taxed at the personal progressive income tax rate[18]. The maximum marginal tax rate for private investors, τ_{Priv}^{max}, has changed over time. The second row of table 6.4 lists τ_{Priv}^{max} for the period 1960 to 1994. Church tax is not included because church membership is voluntary. For

[17]See §2.1 no. 5 and §22 no. 2 EStG.
[18]See §2.1 no. 5 and §20.1 no. 1 EStG.

Table 6.4: Private and corporate tax rates and tax credit on dividend payments.

Time period	τ_{Priv}^{max} [a]	τ_{Corp} [b]	τ_{credit} [c]
1960-1967	53.00 %	51.00 %	.00 %
1968-1974	54.59 %	52.53 %	.00 %
1975-1976	57.68 %	52.53 %	.00 %
1977-1978	57.68 %	57.68 %	37.08 %
1979-1989	56.00 %	56.00 %	36.00 %
1990-1993	53.00 %	50.00 %	36.00 %
1994	53.00 %	45.00 %	30.00 %

[a] Maximum private marginal tax rate including supplementary tax but not church tax.
[b] Corporate tax rate including supplementary tax.
[c] Tax credit on dividend payments including supplementary tax. The change in the taxation of dividends from 1976 to 1977 became effective for dividends paid for fiscal years ending after 1976. Therefore, dividend payments for fiscal years ending with 1976 are not affected by the new regulation. The analysis of Bay (1990, p. 98) reveals that the new regulation became effective for dividends paid in the fourth quarter of 1977. In the empirical analysis $\tau_{credit} = 0$ is therefore used until October 1977.

corporate investors capital gains as well as dividend and interest payments are subject to the corporate tax rate, τ_{Corp}. τ_{Corp} is given in the third row of table 6.4[19].

Taking into account these differences in the taxation of different types of investors one cannot make generalizing statements about the after tax returns received by the investors. In addition, the returns received may also depend on the investment horizon, i.e. the length of the time interval, the amount of money an investor invests and the transactions costs the investor faces. Nevertheless, with the exception of the different time intervals, these additional aspects are ignored throughout the analysis, i.e. it is assumed that there are no transactions costs and that there exist no differences in the investment opportunities for different types of investors.

[19] At this point, two further types of investors, namely, institutional investors such as charitable institutions and investment funds and foreign investors, could have been discussed. But since no taxation scenario for these investors will be analyzed this discussion is omitted. For a discussion of the taxation of these types of investors see Amihud and Murgia (1997, pp. 400-1).

Table 6.5: Taxation of interest and dividend payments and of capital gains in the different taxation scenarios.

Scenario		S1	S2	S3	S4
τ_{ID}	Tax rate on interest and dividend payments	0	36%	τ_{Priv}^{max} [a]	τ_{Corp} [b]
τ_{CG}	Tax rate on capital gains	0	0	0	τ_{Corp}

[a] Maximum private marginal tax rate including supplementary tax but not church tax.
[b] Corporate tax rate including supplementary tax.

To account for the differences in the taxation of different types of investors, four taxation scenarios are investigated:

- Scenario S1 is the zero tax scenario.
- The scenarios S2 and S3 refer to private investors. These scenarios are based on the following assumptions: (1) the private investors hold the securities for at least six months, (2) no private investor holds 25 percent or more of the shares of a company, (3) no private investor sells one percent or more of the shares of a company or, if these assumptions do not hold, (4) private investors conceal capital gains at their tax declaration. These assumptions imply that capital gains are not taxed. Scenario S2 refers to a private investor with a marginal tax rate of 36 percent; S3 refers to a private investor with the maximum marginal tax rate.
- Scenario S4 refers to a corporate investor, for whom capital gains as well as dividend and interest payments are subject to the corporate tax rate. It is assumed that capital gains are taxed in the period in which the market value of the security rises[20] and that capital losses are reduced by the corporate tax rate because they can be deducted from other income[21].

The taxation of capital gains and of interest and dividend payments in the four taxation scenarios is summarized in table 6.5[22].

[20] In reality the capital gain is not taxed until the gain is realized by selling the security. Hence, on average capital gains are subject to the corporate income tax in later periods than assumed here.

[21] This assumption implies that the corporate investor's income is positive in each period or that losses in one period can be deducted from positive income in other periods.

[22] Stehle and Hartmond (1991) calculate average returns on German stocks and one-month interbank deposits for private investors with a tax rate of 0 percent, 36 percent and 56 percent in the period 1954 to 1988.

Recall that the main question of this study is whether the historically observed rates of return are consistent with intertemporal equilibrium asset pricing. Therefore, an aspect different from the sole description of rates of return earned by different types of investors becomes central, namely, the question which investor determines the stock market, i.e. which investor is the marginal investor. The returns received by this investor should then be used as the benchmark for the analysis of different equilibrium models. Searching the marginal investor means asking which investor is willing to pay the highest price for stocks.

One possibility to reveal information about the marginal investor is to analyze the decline in stock prices subsequent to dividend payments. Let D denote the dividend per share and ΔP the ex-dividend decline in the stock price. ΔP can be interpreted as the capital gain if earnings would have been retained. This capital gain is the alternative to the dividend payment D. An investor is indifferent between the dividend payment and the capital gain from retained earnings if, considering taxation, both yield the same increase in wealth. Hence, the investor is indifferent between both if $(1 - \tau_{CG})\Delta P = (1 - \tau_{ID})D$, where τ_{CG} is the investor's tax rate on capital gains and τ_{ID} is the investor's tax rate on interest and dividend payments. For corporate investors both tax rates are equal so that they are indifferent if $\Delta P = D$. Hence, if corporate investors determine the ex-dividend decline in the stock price, ΔP should be equal or close to D[23]. For private investors, however, τ_{CG} is equal to zero, so that private investors are indifferent if $\Delta P = (1 - \tau_{ID})D$. Hence, if private investors with marginal tax rates τ_{ID} larger than zero determine ΔP, ΔP should be smaller than D and should decrease for increasing τ_{ID}. The marginal tax rate of the marginal investor can therefore be estimated from observed dividend payments and the corresponding subsequent decline in the stock prices.

Bay (1990) applies this approach. He estimates the implied marginal tax rate of the marginal investor in Germany from dividend payments and the ex-dividend decline in stock prices. For the years 1966 to 1988 he finds an average marginal tax rate of 29.9 percent (Bay 1990, p. 153, table 6). The estimated marginal tax rate increases over time and especially exhibits a strong increase after the elimination of the double taxation of dividends in 1977. Bay calculates an average marginal tax rate of 17.7 percent before and of 44.7 percent after the elimination of the double taxation of dividends. These results suggest that the decline in stock prices subsequent to dividend payments are determined by private investors with a marginal tax rate that is lower than the maximum marginal income tax rate.

Another argument in favor of the hypothesis that private investors determine the stock prices is as follows. An investor is willing to pay a higher price for a stock as long as the risk premium contained in the after tax

[23] With transactions costs corporate arbitrageurs do not exactly equalize ΔP and D.

return received by the investor is higher than the risk premium required by the investor. Now, assume that all investors are equally risk averse. Then, those investors are willing to pay the highest stock prices that receive the highest after tax risk premia on stocks. These investors are the private investors with the maximum marginal tax rate[24]. Combining this argument and the results of Bay (1990) provides evidence that scenario S2, i.e. the scenario of a private investor with a marginal tax rate of 36 percent, may be considered as the scenario representing the marginal investor.

6.2.2.2 Calculating Real Tax Adjusted Returns

Using the tax rate on interest and dividend payments, τ_{ID}, and the tax rate on capital gains, τ_{CG}, given in table 6.5 and the percentage growth in the consumer price index, CPI_t, in each period the real tax adjusted return on each asset is calculated for each scenario as

$$\frac{1 + (1 - \tau_{ID})\mathrm{ID}_t + (1 - \tau_{CG})\mathrm{CG}_t}{1 + \mathrm{CPI}_t} - 1.$$

ID_t denotes the asset's nominal return from interest and dividend payments and CG_t is the asset's nominal return from capital gains.

For interbank deposits the return from interest payments equals the nominal money market rate, ID_t=MMR_x$_t$. The return from capital gains is zero ($\mathrm{CG}_t=0$) because interbank deposits exhibit no capital gains or losses. In the case of bonds the return from capital gains is measured by the nominal return on the bond price indices, CG_t=REX_PR_x$_t$. The return from interest payments is calculated from the difference in the nominal returns on the bond performance and the bond price index, ID_t=REX_x$_t$-REX_PR_x$_t$. The return from capital gains of stocks is represented by the nominal return on the stock price index (CG_t=DAFOX_PR$_t$) and the return from dividend payments is measured by the nominal dividend return including the tax credit (ID_t=DIV$_t$) estimated in section 6.2.1.3.

6.2.2.3 Stylized Facts about Real Tax Adjusted Returns

First, the results for the zero tax scenario in the period 1968 to 1994 are analyzed. The upper part of table 6.6 contains the means and standard deviations for these returns. On average all assets yielded positive real returns before taxation (S1). Except for the end-of-year returns on stocks (DAFOX-Y1) all means are significantly different from zero on a level of significance of at least five percent. Means and standard deviations are also given for

[24]This is a consequence of the difference in the taxation of capital gains and of dividend payments for private investors. This is illustrated in the tables 6.6 and 6.7 in section 6.2.2.3.

the equity premium (EPR), i.e. the risk premium on stocks over the one-period risk-free rate. For monthly time intervals EPR is calculated as the return on the DAFOX minus the one-month money market rate (MMR_1). For quarterly time intervals the one-period risk-free rate is the three-month money market rate (MMR_3) and for yearly time intervals it is the return on bonds with a maturity of one year (REX_1). The average real equity premium is positive but not significantly different from zero. In general, the mean returns are only slightly affected by the method of return calculation, whereas the standard deviations, especially those of the stock returns, are up to 25 percent smaller for period average returns (Q2 and Y2) than for end-of-period returns (Q1 and Y1). Comparing the standard deviations over the different time intervals reveals that the time series do not meet the usual assumptions of stochastic independence over time and stationary conditional distributions[25].

The lower part of table 6.6 lists the correlation coefficients for the real end-of-quarter returns before taxation (Q1, S1). Except for the correlation of the three-month money market rates (MMR_3) and the stock returns (DAFOX) and the correlation of MMR_3 and the equity premium (EPR), all returns are significantly positively correlated. The stock returns are positively correlated with the holding period returns on bonds because an increase [decrease] of interest rates decreases [increases] the prices of bonds and simultaneously decreases [increases] the prices of stocks because it makes the investment in stocks less [more] attractive. Not surprisingly, the correlation between the returns on bonds increases for decreasing differences in maturities. A comparable result is obtained for the correlation between the three-month money market rates and the holding period returns on bonds. For maturities of bonds increasing from 1 to 10 years the correlation of the bond returns and the money market rates decreases from .43 to .21.

Analyzing the tax adjusted returns for the scenarios S2 to S4 listed in table 6.7 yields the following results. Private investors with a marginal tax rate of 36 percent (S2) received positive average real after tax returns on all assets under consideration. However, only the money market rates and the returns on the one-year bond are significantly different from zero on a level of significance of at least five percent. Except for the set Y1 the tax adjusted real returns on stocks are different from zero on a ten percent level of significance. Private investors with the highest marginal tax rate (S3) received negative real tax adjusted returns on interbank deposits and bonds. The tax adjusted returns on stocks are still positive. The difference between the returns on stocks and the one-period risk-free rate increases from scenario S1 to S3 because stock returns are only partially affected by the taxation of private investors. For example, for yearly average returns (Y2) the equity premium rises from 3.55 percent in scenario S1 to 4.75 per-

[25] These findings are similar to those for the nominal returns discussed in section 6.2.1.2.

Table 6.6: The upper panel lists means and standard deviations (in parentheses) in percent of monthly, quarterly and yearly real before tax returns on the stock market, the bond market and interbank deposits and of the equity premia in the period 1968 to 1994. The lower panel exhibits correlation coefficients for the end-of-quarter returns (Q1).

Time Series	Set of Return Time Series[a] M	Q1	Q2	Y1	Y2
	S1: Zero tax scenario[b]				
DAFOX	$.54^{**}_{\circ\circ}$ (4.71)	$1.67^{**}_{\circ\circ\circ}$ (8.60)	$1.58^{**}_{\circ\circ}$ (7.13)	7.39 (22.95)	$7.30^{**}_{\circ}$ (17.25)
MMR[c]	$.25^{***}_{\circ\circ\circ}$ (.26)	$.82^{***}_{\circ\circ\circ}$ (.62)	$.82^{***}_{\circ\circ\circ}$ (.53)	. .	. .
REX_1	$.29^{***}_{\circ\circ\circ}$ (.46)	$.87^{***}_{\circ\circ\circ}$ (.97)	$.88^{***}_{\circ\circ\circ}$ (.87)	$3.52^{***}_{\circ\circ\circ}$ (2.62)	$3.60^{***}_{\circ\circ\circ}$ (2.54)
REX_4	$.28^{***}_{\circ\circ\circ}$ (1.04)	$.86^{***}_{\circ\circ\circ}$ (2.16)	$.88^{***}_{\circ\circ\circ}$ (1.92)	$3.52^{***}_{\circ\circ\circ}$ (4.98)	$3.70^{***}_{\circ\circ\circ}$ (4.91)
REX_7	$.30^{***}_{\circ\circ\circ}$ (1.47)	$.91^{***}_{\circ\circ\circ}$ (2.97)	$.92^{***}_{\circ\circ\circ}$ (2.54)	$3.72^{***}_{\circ\circ}$ (6.52)	$3.95^{***}_{\circ\circ\circ}$ (6.42)
REX_10	$.27^{**}_{\circ\circ\circ}$ (1.95)	$.82^{**}_{\circ\circ}$ (3.60)	$.82^{***}_{\circ\circ\circ}$ (2.92)	$3.34^{**}_{\circ\circ}$ (7.56)	$3.54^{**}_{\circ\circ}$ (6.97)
REX	$.29^{***}_{\circ\circ\circ}$ (1.15)	$.87^{***}_{\circ\circ\circ}$ (2.27)	$.89^{***}_{\circ\circ\circ}$ (2.07)	$3.56^{***}_{\circ\circ\circ}$ (5.46)	$3.75^{***}_{\circ\circ\circ}$ (5.35)
EPR[d]	.29 (4.70)	.85 (8.59)	.76 (7.16)	3.83 (21.74)	3.55 (15.58)

Correlation coefficients for end-of-quarter returns before taxation (Q1, S1)[e]

	MMR_3	EPR	REX	REX_1	REX_4	REX_7	REX_10
DAFOX	.0439	.9974***	.2912***	.2931***	.2796***	.2870***	.3007***
MMR_3		-.0287	.2720***	.4301***	.2942***	.2355**	.2126**
EPR			.2716***	.2621***	.2584***	.2701***	.2854***
REX				.8337***	.9829***	.9736***	.9329***
REX_1					.8560***	.7637***	.7168***
REX_4						.9720***	.9174***
REX_7							.9422***

[a] M denotes the end-of-month returns, Q1 the end-of-quarter returns, Q2 the quarterly average returns, Y1 the end-of-year returns and Y2 the yearly average returns.

[b] *** (**, *) [∘∘∘ (∘∘, ∘)] indicates that the hypothesis of zero mean can be rejected at a 1 (5, 10) percent level of significance on the basis of the parametric t-statistic [the nonparametric Wilcoxon-Sign-Rank statistic].

[c] One-month money market rate (MMR_1) for the set M and three-month money market rate (MMR_3) for the sets Q1 and Q2.

[d] Equity premium: EPR-M = DAFOX-M − MMR_1, EPR-Q1 = DAFOX-Q1 − MMR_3-Q1, EPR-Q2 = DAFOX-Q2 − MMR_3-Q2, EPR-Y1 = DAFOX-Y1 − REX_1-Y1 and EPR-Y2 = DAFOX-Y2 − REX_1-Y2

[e] *** (**, *) indicates that the hypothesis of zero correlation can be rejected at a 1 (5, 10) percent level of significance.

cent in scenario S2 and 5.37 percent in scenario S3. In scenario S3 the equity premium becomes significantly positive. On the other hand, the riskiness of stocks measured by the standard deviation of stock returns is nearly unaffected by the taxation of private investors. Hence, as a result of the German tax system, the highest risk premia are earned by private investors with the

Table 6.7: Means and standard deviations (in parentheses) in percent of monthly, quarterly and yearly real tax adjusted returns on the stock market, the bond market and interbank deposits and of the equity premia in the period 1968 to 1994.

Time Series	Set of Return Time Series[a]									
	M		Q1		Q2		Y1		Y2	
S2: Private investor with marginal tax rate of 36%										
DAFOX	$.42_{\circ}$	(4.71)	$1.32_{\circ\circ}$	(8.61)	$1.24^{*}_{\circ}$	(7.12)	6.02	(22.86)	5.91^{*}	(17.21)
MMR	$.06^{***}_{\circ\circ\circ}$	(.24)	$.21^{***}_{\circ\circ\circ}$	(.52)	$.21^{***}_{\circ\circ\circ}$	(.41)	.	.	.	.
REX_1	$.08^{***}_{\circ\circ\circ}$	(.46)	$.24^{**}_{\circ\circ\circ}$	(.97)	$.25^{***}_{\circ\circ\circ}$	(.87)	$1.00^{*}_{\circ\circ}$	(2.58)	$1.07^{**}_{\circ\circ}$	(2.50)
REX_4	$.07_{\circ}$	(1.04)	.23	(2.15)	.24	(1.92)	.98	(4.90)	1.15	(4.82)
REX_7	$.08_{\circ}$	(1.46)	.26	(2.96)	.26	(2.53)	1.09	(6.38)	1.31	(6.28)
REX_10	.06	(1.95)	.19	(3.58)	.19	(2.91)	.80	(7.48)	1.01	(6.81)
REX	$.07_{\circ}$	(1.16)	.23	(2.31)	.24	(2.07)	.99	(5.35)	1.18	(5.24)
EPR	$.37_{\circ}$	(4.68)	$1.12_{\circ}$	(8.56)	1.03	(7.09)	5.03	(21.61)	4.75	(15.51)
S3: Private investor with maximum marginal tax rate										
DAFOX	.36	(4.70)	$1.14_{\circ}$	(8.62)	1.05	(7.12)	5.28	(22.79)	5.16	(17.18)
MMR	$-.05^{***}_{\circ\circ\circ}$	(.24)	$-.12^{**}_{\circ\circ}$	(.50)	$-.11^{***}_{\circ\circ\circ}$	(.40)	.	.	.	.
REX_1	-.04	(.46)	-.10	(.97)	-.09	(.87)	-.35	(2.56)	-.28	(2.48)
REX_4	-.04	(1.04)	-.11	(2.14)	-.10	(1.91)	-.38	(4.84)	-.21	(4.75)
REX_7	-.04	(1.46)	-.10	(2.95)	-.09	(2.52)	-.32	(6.30)	-.10	(6.19)
REX_10	-.05	(1.95)	-.15	(3.56)	-.15	(2.90)	-.55	(7.38)	-.35	(6.70)
REX	-.04	(1.16)	-.11	(2.33)	-.10	(2.07)	-.39	(5.28)	-.20	(5.16)
EPR	$.41_{\circ}$	(4.68)	$1.26_{\circ\circ}$	(8.54)	$1.16^{*}_{\circ}$	(7.06)	5.66	(21.55)	5.37^{*}	(15.48)
S4: Corporate investor										
DAFOX	.08	(2.22)	.26	(4.07)	.24	(3.37)	1.40	(10.73)	1.48	(8.36)
MMR	$-.04^{***}_{\circ}$	(.24)	$-.09^{*}$	(.50)	$-.08^{**}_{\circ\circ}$	(.40)	.	.	.	.
REX_1	-.03	(.31)	-.08	(.68)	-.06	(.59)	-.24	(2.04)	-.19	(1.97)
REX_4	-.03	(.54)	-.08	(1.13)	-.06	(1.00)	-.26	(2.85)	-.15	(2.81)
REX_7	-.02	(.72)	-.06	(1.47)	-.05	(1.27)	-.17	(3.48)	-.04	(3.42)
REX_10	-.04	(.94)	-.11	(1.75)	-.10	(1.44)	-.37	(3.98)	-.24	(3.67)
REX	-.03	(.59)	-.08	(1.18)	-.06	(1.06)	-.24	(3.04)	-.13	(2.98)
EPR	.12	(2.18)	.36	(3.99)	.32	(3.31)	1.65	(9.87)	1.61	(7.14)

[a]For descriptions see footnotes of table 6.6.

maximum marginal tax rate.

In the scenario of a corporate investor (S4), the average real stock returns after taxation are still positive but they are smaller than without taxation of capital gains. As for the scenario S3 the tax adjusted real returns on interbank deposits and bonds are negative. The equity premium calculated from yearly average returns equals only 1.61 percent. However, the riskiness of stocks and also of bonds measured by the standard deviation of their

Table 6.8: Means and standard deviations (in parentheses) in percent of monthly, quarterly and yearly real returns before taxation and real tax adjusted returns on the stock market and interbank deposits and of the equity premia in the period 1960 to 1994.

Time Series	Set of Return Time Series[a]									
	M		Q1		Q2		Y1		Y2	
	S1: Zero tax scenario									
DAFOX	$.50^{**}_{\circ\circ}$	(4.95)	$1.57^{*}_{\circ\circ}$	(9.40)	$1.39^{**}_{\circ\circ}$	(7.58)	6.81^{*}	(23.34)	$5.81^{**}_{\circ}$	(16.84)
MMR	$.23^{***}_{\circ\circ\circ}$	(.28)	$.75^{***}_{\circ\circ\circ}$	(.63)	$.74^{***}_{\circ\circ\circ}$	(.54)	.	.	.	.
EPR	.27	(4.93)	.82	(9.35)	.65	(7.58)	.	.	.	.
	S2: Private investor with marginal tax rate of 36%									
DAFOX	.39	(4.95)	$1.24_{\circ}$	(9.41)	1.06^{*}	(7.57)	5.50	(23.23)	4.51	(16.79)
MMR	$.05^{***}_{\circ\circ\circ}$	(.25)	$.19^{***}_{\circ\circ\circ}$	(.53)	$.18^{***}_{\circ\circ\circ}$	(.42)	.	.	.	.
EPR	.34	(4.92)	$1.05_{\circ}$	(9.33)	.88	(7.52)	.	.	.	.
	S3: Private investor with maximum marginal tax rate									
DAFOX	.33	(4.95)	1.07	(9.42)	.89	(7.56)	4.81	(23.16)	3.82	(16.76)
MMR	$-.05^{***}_{\circ\circ\circ}$	(.25)	$-.11^{**}_{\circ\circ}$	(.51)	$-.12^{***}_{\circ\circ\circ}$	(.40)	.	.	.	.
EPR	.38	(4.91)	$1.18_{\circ}$	(9.32)	1.00	(7.50)	.	.	.	.
	S4: Corporate investor									
DAFOX	.08	(2.37)	.27	(4.53)	.19	(3.63)	1.33	(11.13)	.96	(8.21)
MMR	$-.04^{***}_{\circ\circ}$	(.25)	$-.08^{*}$	(.51)	$-.09^{***}_{\circ\circ}$	(.40)	.	.	.	.
EPR	.12	(2.33)	.35	(4.43)	.28	(3.57)	.	.	.	.

[a]For descriptions see footnotes of table 6.6.

returns is significantly smaller and equals about 50 percent of that in the scenarios S1 to S3. This risk reduction results from the fact that capital losses can be deducted from other income so that the taxation of capital gains not only reduces the gains but also the losses. The risk-return tradeoff measured by the Sharpe-ratio, the ratio of an asset's average risk premium and the standard deviation of its returns, in scenario S4 is worse than in the scenarios S2 and S3. It is almost the same for the scenarios S1 and S4. The results obtained for the period 1960 to 1994 presented in table 6.8 are similar.

Overall, the average equity premia are positive for all types of investors. Increasing the tax rate of the private investor reduces the risk-free rate and increases the equity premium. Private investors with the highest marginal tax rate receive a negative real risk-free rate and a significantly positive equity premium. Comparing the left and the right graph in figure 6.4 illustrates that these results are not affected by the method of return calculation. However, the standard deviation of period average returns is lower than that of end-of-period returns.

Figure 6.4: Annualized average real quarterly one-period risk-free rate and equity premium for the four taxation scenarios. Results for end-of-quarter returns (Q1) are given in the left graph, those for quarterly average returns (Q2) in the right graph.

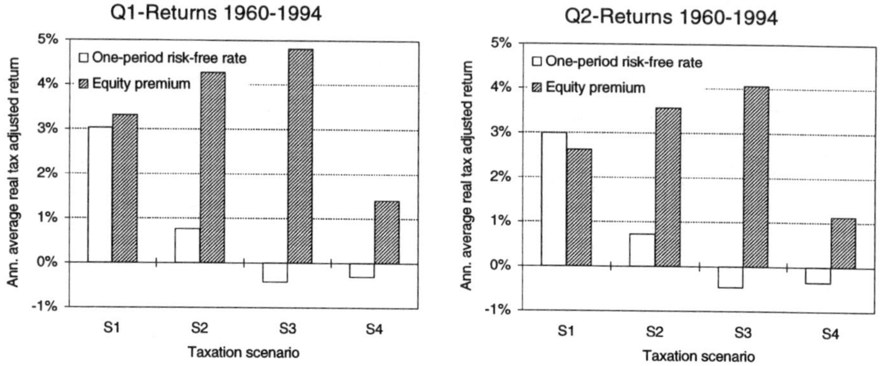

6.2.3 Consumption Data

The empirical investigation of the parametric models of the market pricing kernel derived in chapter 4 requires consumption data. Therefore, two issues concerning the measurement of consumption need to be clarified. First, which time series should be chosen given a model economy with a single consumption good? Second, how can the distribution of benefits over time be measured, given that consumption data always measure consumption expenditures instead of consumption from which utility is gained in the same period?

Independently answering the first question would result in the use of the time series of aggregate consumption. However, the second question implies that only those consumption expenditures which exclusively generate utility in the same period should be used. This is only true for nondurable consumption. Therefore, the time series of consumption expenditures on nondurables and services is usually applied in this context, although some services, for example medical expenditures, may also generate utility in subsequent periods[26]. But how can the use of only one component of consumption be justified with respect to the first question? The justification is usually based on the assumption that in the time-additive expected utility framework the utility function of current consumption, $u(\cdot)$, is addi-

[26] In their study analyzing the nature of nondurable consumption Eichenbaum and Hansen (1990) find that consumption goods that are classified as nondurable generate positive utility in subsequent time periods. However, this result may be sensitive to the temporal aggregation bias, i.e. that the reporting periods of consumption are too long.

tively separable across different components of consumption, at least across durable and nondurable goods (see, for example, Deaton (1992, pp. 10-2)). Under this assumption, in the optimum, the intertemporal marginal rates of substitution are equal for all components so that the use of any component of consumption, e.g., nondurable consumption, can be justified theoretically[27]. As pointed out by Epstein and Zin (1991, p. 272), a similar argument does not apply for recursive non-expected utility. However, since a reliable measure of consumption is necessary the present analysis follows other studies and employs per capita consumption expenditures on nondurables and services.

Unfortunately, no long time series of consumption expenditures on nondurables and services is available for Germany. On a yearly basis such a time series can be constructed from detailed data on consumption expenditures published by the Statistisches Bundesamt but on a quarterly basis no detailed data are available. Therefore, up to this point, studies applying German consumption data have worked on a yearly basis or have used total consumption expenditures on a quarterly basis (e.g., Funke (1992), Wolters (1992) and Missong (1994)). In contrast, this study approximates a time series of quarterly per capita consumption expenditures on nondurables and services from available data. Before the approximation procedure is presented, the available data are described.

Consumption Expenditures The West German data on aggregate consumption expenditures of all private households are from two different sources[28]. First, seasonally adjusted quarterly consumption expenditures on food (C1-Q)[29], clothing and shoes (C2-Q), housing and energy (C3-Q), and other goods and services (C4-Q) from January 1960 to December 1994 have been provided by the Deutsches Institut für Wirtschaftsforschung (DIW). Adding the four components yields the time series of quarterly total aggregate consumption expenditures (C-Q). By aggregating the quarterly expenditures over the quarters of each year the time series of yearly aggregate consumption expenditures, C1-Y to C4-Y and C-Y, are obtained. Second, on a yearly basis more detailed data on aggregate consumption expenditures are available from the Statistisches Bundesamt. The following subgroups of the yearly time series C3-Y and C4-Y are used: housing (C3_1-Y), energy (not including fuel) (C3_2-Y), furniture and other goods for housekeeping (C4_1-Y), cosmetics and medical expenditures (C4_2-Y), transportation (including vehicles) (C4_3-Y), information transmission (C4_4-Y), education,

[27]However, Eichenbaum and Hansen (1990) find mixed results when examining whether preferences are completely separable between durable and nondurable consumption goods.

[28]East German data are never included.

[29]"Q" ("Y") denotes time series consisting of quarterly (yearly) data.

Figure 6.5: Comparison of the constructed time series of yearly aggregate consumption expenditures on nondurables and services with a published time series for the period 1980 to 1994. In addition, the growth rates of both time series are graphed.

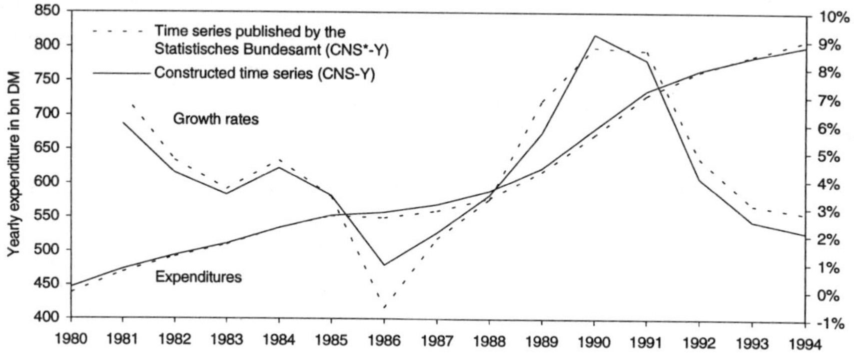

entertainment, and leisure (C4_5-Y), and personal equipment (C4_6-Y)[30]. An explicit time series of annual aggregate expenditures on nondurables and services (CNS*-Y) is published by the Statistisches Bundesamt since 1980[31].

The construction of the quarterly per capita consumption expenditures on nondurables and services proceeds in three steps. In the first step the time series of yearly aggregate consumption expenditures on nondurables and services (CNS-Y) is constructed. Taking into account the nature of the components of consumption (durable or nondurable), a subset of the time series C1-Y, C2-Y, C3_1-Y, C3_2-Y, and C4_1-Y to C4_6-Y is chosen that mimics the time series CNS*-Y as closely as possible in the period 1980 to 1994. It is found, that the time series of the following combination of components of consumption comes closest to the time series CNS*-Y: CNS-Y = C1-Y + .2 · C2-Y + C3_2-Y + C4_x-Y, where C4_x-Y = C4_2-Y + C4_4-Y + C4_5-Y + C4_6-Y. Figure 6.5 compares the constructed time series CNS-Y with the published time series CNS*-Y.

[30] These time series are taken from table 3.2.2.2, Fachserie 18, Reihe 1.2 (1960-1979) respectively Reihe 1.3 (1980-1994), "Käufe der privaten Haushalte im Inland nach Verwendungszwecken" of the Statistisches Bundesamt. C3_1-Y equals row 12, C3_2-Y equals row 15, C4_1-Y equals row 20, C4_2-Y equals row 22, C4_3-Y equals the sum of rows 28 to 31, C4_4-Y equals row 32, C4_5-Y equals row 33, and C4_6-Y equals row 34. In contrast to the data provided by the DIW these time series only include the national population's expenditures in Germany.

[31] Sum of the rows 33 (nondurables) and 38 (other services) of table 3.2.2.1, Fachserie 18, Reihe 1.3 "Privater Verbrauch nach Lieferbereichen und nach Dauerhaftigkeit und Wert der Güter".

The second step is to construct the time series of quarterly aggregate consumption expenditures on nondurables and services (CNS-Q). C1-Q and C2-Q are available. To obtain C3_2-Q and C4_x-Q the dynamic structure of C3-Q is imposed on C3_2-Y and the dynamic structure of C4-Q is imposed on C4_x-Y. This approach seems reliable because on an annual basis C3-Y and C3_2-Y exhibit a high correlation of .944. The correlation of the growth rates of C3-Y and C3_2-Y is .874. C4-Y and C4_x-Y exhibit a correlation of .999 and the correlation of the growth rates equals .917. To impose the dynamic structure, first, the relative proportion of C3-Y spent on C3_2-Y, C3'-Y = C3_2-Y / C3-Y, and the relative proportion of C4-Y spent on C4_x-Y, C4'-Y = C4_x-Y / C4-Y, are calculated on a yearly basis. Then the quarterly relative proportions C3'-Q and C4'-Q are interpolated from the yearly values. This is done as follows: A continuous curve is fitted to the data by connecting successive straight line segments. Interval midpoints are used as the break points and ordinates are chosen so that the average of the estimates for the four quarters of each year equals the yearly input value. Using the interpolated quarterly proportions of C3-Q and C4-Q, quarterly aggregate consumption expenditures on nondurables and services are obtained as CNS-Q = C1-Q + .2 · C2-Q + C3'-Q · C3-Q + C4'-Q · C4-Q.

In the third step per capita consumption expenditures are derived by dividing C-Q, C-Y, CNS-Q and CNS-Y by the quarterly, respectively yearly average West German population computed from the average monthly population published by the Statistisches Bundesamt.

Growth Rates of Consumption From the four time series of per capita consumption expenditures discrete growth rates are calculated and deflated by the growth rates of the quarterly average consumer price index (CPI-Q2), respectively the yearly average consumer price index (CPI-Y2).

Table 6.9 lists the means and the standard deviations of the real growth rates of quarterly and yearly per capita total consumption expenditures (C-Q and C-Y) and of quarterly and yearly per capita consumption expenditures on nondurables and services (CNS-Q and CNS-Y) for the period 1960 to 1994. All mean growth rates are significantly positive. The 2.30 percent average growth of per capita consumption of nondurables and services is almost equal to the 2.39 percent mean observed for the US in the period 1959 to 1978 (cf. Mehra and Prescott (1985, p. 147)). Total consumption has grown stronger than the consumption of nondurables and services. This seems plausible since with growing wealth the population can spend a larger fraction of wealth on durables and luxury goods.

The variability of real per capita consumption growth is low as compared to the variability of real asset returns (see table 6.8). This is a common finding

Table 6.9: Means and standard deviations of quarterly and yearly real percentage growth of per capita total consumption expenditures (C-Q and C-Y) and per capita consumption expenditures on nondurables and services (CNS-Q and CNS-Y) for the period 1960-1994.

C-Q		CNS-Q		C-Y		CNS-Y	
Mean[a]	Std	Mean	Std	Mean	Std	Mean	Std
$.695^{***}_{\circ\circ\circ}$	(.846)	$.562^{***}_{\circ\circ\circ}$	(.835)	$2.831^{***}_{\circ\circ\circ}$	(2.077)	$2.303^{***}_{\circ\circ\circ}$	(1.836)

[a] *** [∘ ∘ ∘] indicates that the hypothesis of zero mean can be rejected at a 1 percent level of significance on the basis of the parametric t-statistic [the nonparametric Wilcoxon-Sign-Rank statistic].

of empirical analyses but in this study it may be stronger because, first, the chosen time period does not include wars or great depressions and, second, the German social system may additionally smooth consumption by increasing national debt during recessions. Campbell (1996), for example, reports standard deviations of yearly real growth rates of consumption for time periods including wars and depressions. For the US a standard deviation of 3.32 percent is reported for the period 1890 to 1992 and for the United Kingdom a standard deviation of 2.92 percent is given for the period 1919 to 1993. Authors investigating other periods report even higher standard deviations of real consumption growth for the US, e.g., Mehra and Prescott (1985): 3.57 percent (1889-1978) or Cecchetti, Lam, and Mark (1990): 3.79 percent (1889-1985). The observed standard deviation of German consumption growth in the time period 1960 to 1994 is low as compared to these figures. This finding should be kept in mind when comparing the results of this study with those of other studies.

As for the asset returns the comparison of the quarterly standard deviations with the yearly standard deviations shows that these time series are not stochastically independent over time and/or exhibit nonstationary conditional distributions. Since the yearly standard deviations are again larger than $\sqrt{4}$ times the quarterly standard deviations, the moments are biased in the same direction as those of asset returns.

Correlation of Consumption Growth and Stock Returns Table 6.10 exhibits coefficients for the correlation between real consumption growth and real stock returns. For quarterly and yearly time intervals the correlation between end-of-period returns and consumption growth and between period average returns and consumption growth are listed. The figures for lag=0 represent the correlation between stock returns and consumption growth in the same period. Those for lag=−1 and lag=−2 give the corre-

Table 6.10: Correlation of quarterly and yearly growth rates of per capita consumption expenditures (C-Q and C-Y) and per capita consumption expenditures on nondurables and services (CNS-Q and CNS-Y) with quarterly and yearly end-of-period returns on the stock market (DAFOX-Q1 and DAFOX-Y1) and, alternatively, with quarterly and yearly period average returns on the stock market (DAFOX-Q2 and DAFOX-Y2), both before taxation (S1) for the period 1960 to 1994.

	DAFOX-Q1					DAFOX-Q2				
Lag[a]	−2	−1	0	1	2	−2	−1	0	1	2
C-Q	-.134	-.059	.016	.029	.085	-.191**	-.018	-.025	.094	.161*
CNS-Q	-.122	-.106	-.038	-.027	.045	-.202**	-.095	-.111	.052	.103
	DAFOX-Y1					DAFOX-Y2				
Lag	−2	−1	0	1	2	−2	−1	0	1	2
C-Y	-.057	-.364**	-.244	.225	.310*	-.358**	-.554***	-.046	.412**	.169
CNS-Y	-.148	-.316*	-.323*	.145	.316*	-.351**	-.555***	-.178	.363**	.134

[a]Lag denotes the lag of consumption growth relative to the stock returns, i.e. lag=−2 denotes the correlation of stock returns with consumption growth two periods ago and lag=2 denotes the correlation of stock returns with consumption growth two periods ahead. *** (**, *) indicates that the hypothesis of zero correlation can be rejected at a 1 (5, 10) percent level of significance.

lation between the current stock returns and consumption growth in the previous period, respectively consumption growth two periods ago. Similarly, the figures for lag=1 and lag=2 give the correlation between current stock returns and consumption growth in the next period, respectively consumption growth two periods ahead.

Stock returns and consumption growth in the same period (lag=0) are negatively correlated in seven out of eight cases but only for the yearly time intervals and the growth rates of CNS the correlation is significantly negative. Stock returns and lagged consumption growth (lag=−1 and lag=−2) are significantly negatively correlated, especially for yearly time intervals. In contrast, consumption growth and lagged stock returns (lag=1 and lag=2) are positively correlated. The significance of the correlation is again larger for yearly time intervals[32]. The positive correlation between consumption growth and stock returns in the previous periods may be due to the fact

[32]Lynch (1996, table I) reports comparable figures for the US in the period 1959 to 1991. Per capita real consumption of nondurables and services has grown on average by .49 percent per quarter with a standard deviation of .49. The correlation between consumption growth and stock returns in the previous quarter equals .1605. Campbell (1996, p. 2) reports a quarterly correlation between real stock returns and real consumption growth of .21 for the US. This correlation increases to .34 at a one-year horizon.

that investors decrease [increase] consumption after periods with low [high] stock returns because these stock returns have decreased [increased] their wealth.

The period average returns on stocks (Q2) are stronger correlated with consumption growth than the end-of-period returns (Q1). This finding supports the hypothesis that period average returns should better coincide with consumption data than end-of-period returns. A comparison of the correlation coefficients for total consumption expenditures (C) with those for consumption expenditures on nondurables and services (CNS) shows no clear results. This suggests that investments in durables are not generally more affected by changes in stock returns than expenditures on nondurables and services.

Chapter 7

Analyzing Variance Bounds of the Market Pricing Kernel

The empirical analysis of the variance bounds for the market pricing kernel, i.e. of the bounds for the relationship between the first and the second unconditional moment of the market pricing kernel, covers the period 1968 to 1994 and proceeds in two steps. First, in section 7.1 the variance bounds are estimated employing different time series of asset returns. The sensitivity of the bounds with respect to changes in (1) the underlying set of assets, (2) the method of return calculation, (3) the taxation scenario applied, and (4) the length of the time intervals is analyzed. These analyses provide insights into the information content of different asset return data sets for testing parametric models of the market pricing kernel. Second, in section 7.2 different parametric models of the market pricing kernel are evaluated using the estimated variance bounds, i.e. the implications of the estimated variance bounds for the parameters of the parametric models of the market pricing kernel are analyzed. The investigated parametric models are those derived from the consumption-based asset pricing model alternatively assuming time-additive expected utility with constant relative risk aversion, time-additive expected logarithmic utility and recursive non-expected utility. Notice that the purpose of this analysis is to visualize the models' capability to jointly price different assets and to illustrate how this capability is affected by changes in the models' parameters but not to formally test the parametric models of the market pricing kernel. For formal tests sampling errors in the variance bounds and the means and standard deviations of the parametrized market pricing kernel must be considered in addition. But, as discussed in section 3.3.3, even then the variance bound approach is an unsatisfactory tool for formally testing parametrized market pricing kernels.

7.1 Implications of Different Return Time Series for the Market Pricing Kernel

7.1.1 Bounds Implied by Different Sets of Assets

First, the variance bounds implied by different sets of assets are analyzed to compare the information content of these sets of assets. The comparison is based on the end-of-quarter returns for the zero tax scenario (Q1, S1). Figure 7.1 depicts the variance bounds placed on the market pricing kernel by different sets of assets. The differences between these regions allow us to determine whether a specific market pricing kernel could lie in the region generated by one set of assets and outside the region generated by another set. If this is the case, a particular model may correctly price the assets of one set but not the assets of the other. As discussed in section 3.3.2, the bound imposed by the equity premium (EPR), i.e. the risk premium on stocks (DAFOX) over the one-period risk-free rate (MMR_3), turns out to be a straight line. The volatility necessary for a market pricing kernel to satisfy this restriction depends only slightly on the expected value of the market pricing kernel. Hence, the variance bound does not contain information about the expected value of the market pricing kernel. The restriction placed on the standard deviation of the market pricing kernel by stock returns (DAFOX) is weaker only for market pricing kernels with an expected value in a small range around the reciprocal of the expected gross return on stocks, which equals .9836 (cf. table 6.6). Market pricing kernels with expected values strongly differing from this value must exhibit a larger standard deviation to be consistent with stock returns than to be consistent with equity premia.

The bound becomes more restrictive, when the holding period returns on the aggregate bond market (REX) are used in combination with stock returns. The intuition is as follows: With stocks and bonds together portfolios can be constructed with superior risk-return tradeoffs resulting in an upward shift of the variance bound. Since the average holding period return on the aggregate bond market (REX-Q1: .87, cf. table 6.6) and the average one-period risk-free rate (MMR_3-Q1: .82) are almost equal, the minimum of the parabola obtained from the aggregate stock returns (DAFOX) in combination with the one-period risk-free rates (MMR_3) is nearly equal to that of the parabola obtained from the aggregate stock returns in combination with the aggregate bond returns (REX). For market pricing kernels with expected values smaller or larger than the minimum of both parabolas the bound calculated from the one-period risk-free rates and the aggregate stock returns is more restrictive. This is a consequence of the low volatility of the money market rates and the low correlation between the money market rates and the returns on the aggregate stock market. The low positive

Figure 7.1: Bounds imposed on the standard deviation of the market pricing kernel by real end-of-quarter returns before taxation (Q1, S1) of different sets of assets.

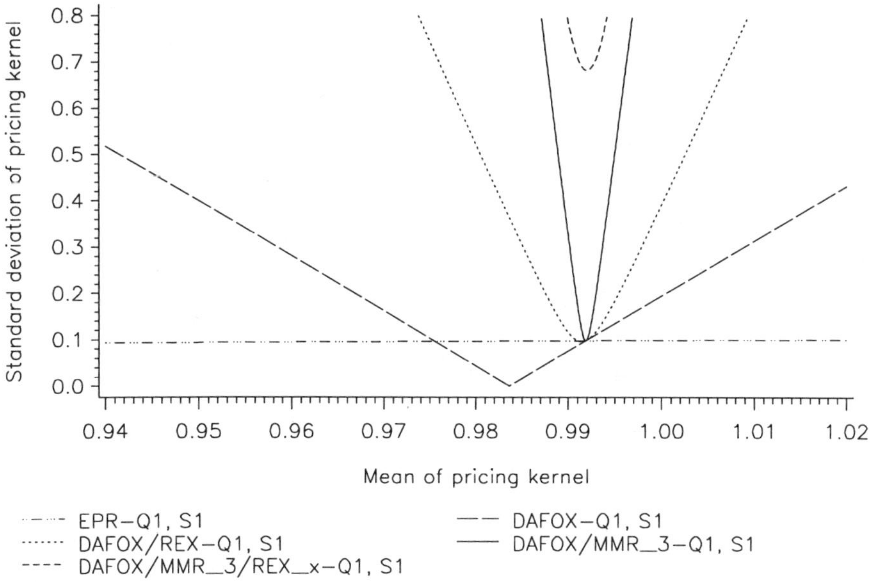

correlation of both returns yields a better diversification and therefore a superior risk-return tradeoff.

Adding aggregate bond returns to the set containing the one-period risk-free rates and the aggregate stock returns does not sharpen the bound. Therefore, this bound is not shown in figure 7.1. Hence, the holding period returns on the aggregate bond market do not contain additional information about the distribution of the market pricing kernel. Thus, including aggregate bond returns in addition to one-period risk-free rates and aggregate stock returns in tests of intertemporal asset pricing models should not change the results. This, however, is not true for disaggregate bond market data as can be seen from the bound calculated from one-period risk-free rates, stock returns and holding period returns on bonds with maturities from 1 to 10 years. A market pricing kernel must exhibit a quarterly standard deviation of at least 70 percent to be consistent with these returns.

7.1.2 Sensitivity to the Method of Return Calculation

The empirical results discussed above are obtained using end-of-quarter returns (hereafter denoted as Q1-returns). Now, it is analyzed to what extent the determined variance bounds are affected by alternatively using period average returns (Q2-returns).

Figure 7.2: Bounds imposed on the standard deviation of the market pricing kernel by real end-of-quarter returns before taxation (Q1, S1) and, alternatively, by real quarterly average returns before taxation (Q2, S1).

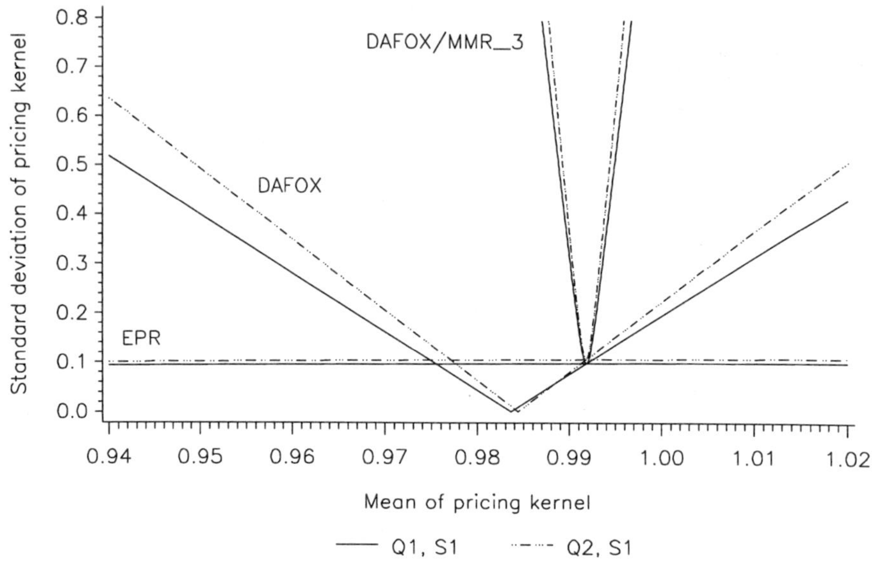

It is evident from figure 7.2 that the bounds imposed by Q2-returns are more restrictive than those imposed by Q1-returns, i.e. for a given mean the market pricing kernel must exhibit a higher standard deviation to be consistent with the variance bounds calculated from Q2-returns than to be consistent with the variance bounds calculated from Q1-returns. This observation is not intuitively clear because Q2-returns were expected to be better comparable with consumption data than Q1-returns. However, as demonstrated in section 3.3.2 the assets' Sharpe-ratios determine the variance bounds. In addition, the discussion of the stylized facts of end-of-period returns and period average returns in section 6.2.1.2 has shown that both returns exhibit a comparably large equity premium but that period average returns on stocks are less risky. Hence, period average returns exhibit a larger Sharpe-ratio than end-of-period returns, and, therefore, are more challenging for asset pricing models in the sense that they require the market pricing kernel to be more volatile. Thus, the moments of the Q2-returns contain information about the distribution of the market pricing kernel that is not contained in the moments of the Q1-returns. However, the difference of the bounds calculated from the one-period risk-free rates in combination with the stock returns is quite small so that it seems unlikely that this difference will have a substantial effect on the investigation of parametric models of the market pricing kernel.

7.1.3 Effects of Taxation

In order to examine whether the restrictions placed on the market pricing kernel by tax adjusted returns differ from those implied by before tax returns the estimated regions for the four taxation scenarios graphed in figure 7.3 are compared.

The discussion starts with the restrictions placed on the standard deviation of the market pricing kernel by the different equity premia. As compared to the zero tax scenario (S1), these restrictions are stronger for the taxation scenarios referring to private investors (S2 and S3) and weaker for the taxation scenario referring to the corporate investor (S4). The distance between the bounds for the scenarios S1 and S4 is small, whereas the distance between the bounds for S1 and S2 or S3 is larger. Recalling the results of the previous chapter, this observation is not surprising. The bound calculated from equity premia is determined by the Sharpe-ratio of the stock returns. The discussion of the stylized facts of tax adjusted returns in section 6.2.2.3 has shown that the Sharpe-ratio for scenario S1 is almost equal to that for scenario S4, although the average equity premium is much smaller for S4 than for S1, because the riskiness of the stocks is reduced in scenario S4 through the deductibility of capital losses from taxable income. The Sharpe-ratios for the scenarios S2 and S3 exceed that for scenario S1. Although the riskiness of stocks is almost unaffected by the taxation of private investors they earn a higher equity premium when taxes are taken into account. Hence, the market pricing kernel must exhibit a higher volatility to be consistent with the equity premia in the scenarios S2 and S3 than to be consistent with those in the scenarios S1 or S4. Thus, with respect to the equity premium, using tax adjusted returns earned by private investors provides a stronger test for parametric models of the market pricing kernel than using returns before taxation.

The restrictions placed on the standard deviation of the market pricing kernel by the combination of one-period risk-free rates and aggregate stock returns differ strongly across the four taxation scenarios. The left-most restriction represents the zero tax scenario. The regions of the other scenarios restrict the expected value of the market pricing kernel to higher values.

Overall, the results indicate that the chosen taxation scenario may affect the results of empirical investigation of parametric models of the market pricing kernel. Furthermore, different market pricing kernels may exist for the different types of investors in the economy. Therefore, the tax adjusted returns received by the marginal investor who determines the asset prices should be used for further analyses. As discussed in section 6.2.2.1, the scenario S2 can be considered to represent the taxation of the marginal investor at the German stock market.

Figure 7.3: Bounds imposed on the standard deviation of the market pricing kernel by equity premia and by the combination of one-period risk-free rates and aggregate stock returns calculated from end-of-quarter figures (Q1) for the taxation scenarios S1 to S4.

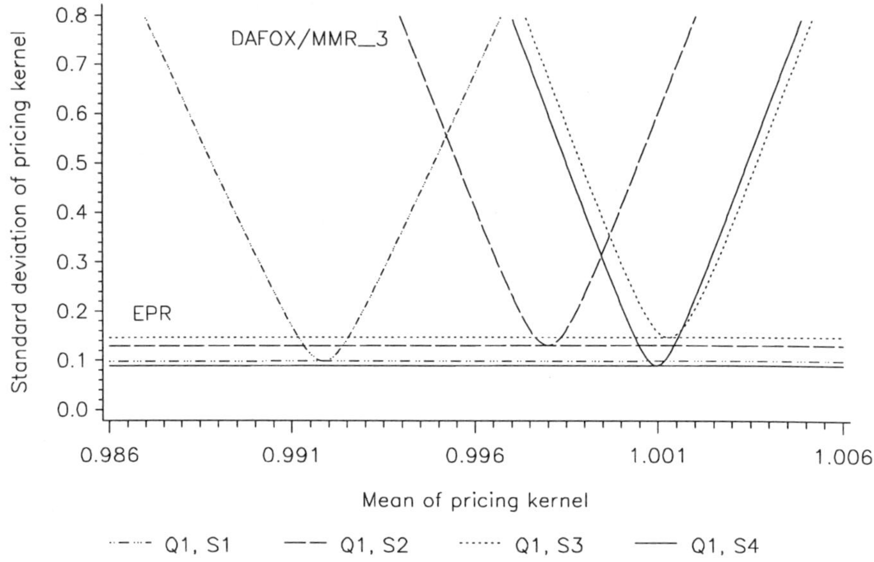

7.1.4 Sensitivity to the Length of the Time Intervals

The sensitivity of the variance bounds to the chosen length of the time intervals over which the returns are measured is analyzed by comparing the restrictions placed on quarterly market pricing kernels by returns calculated for monthly, quarterly and yearly time intervals. The quarterly basis is chosen for two reasons. First, all the results discussed until now are obtained using quarterly returns. Second, German macroeconomic variables are available on a quarterly and a yearly basis; therefore, it is likely that studies evaluating specific parametric models of the market pricing kernel for Germany use quarterly data to comprise a higher number of observations. Differences in the restrictions on quarterly market pricing kernels can only be expected if the asset returns are stochastically dependent over time and/or exhibit a nonstationary conditional distribution, and the monthly [yearly] returns are converted to quarterly returns by multiplying the standard deviation by $\sqrt{3}$ [$\sqrt{1/4}$] and taking the mean of the gross returns to the power of 3 [1/4], implicitly assuming stochastic independence over time and a stationary conditional distribution. However, the discussion in section 6.2.1.2 has shown that the asset returns do not meet these assumptions. Therefore, differences in the variance bounds for quarterly market pricing kernels calculated from

Figure 7.4: Bounds imposed on the quarterly standard deviation of the market pricing kernel by monthly, quarterly and yearly equity premia, stock returns and the combination of one-period risk-free rates and stock returns. The returns are before taxation (S1) and calculated from end-of-period figures (M, Q1, Y1).

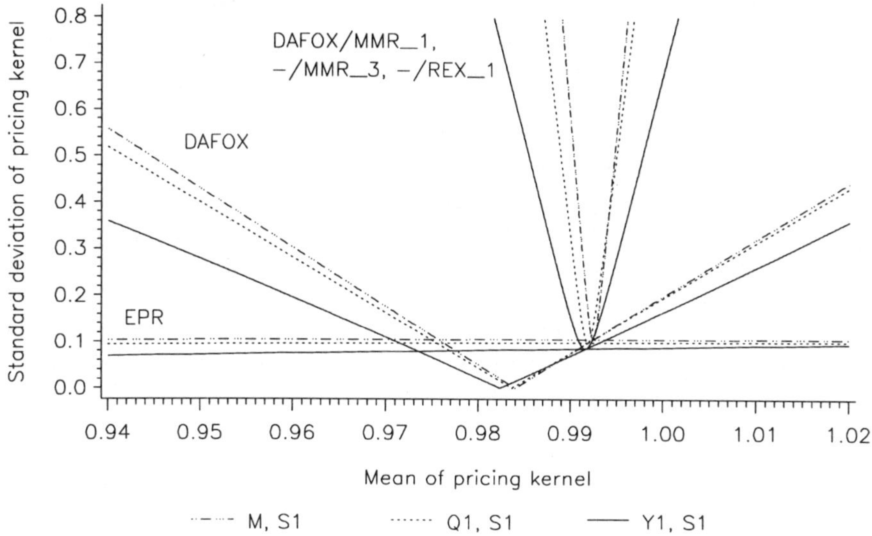

monthly, quarterly and yearly returns should be observed. Subsequently, it is analyzed to which extent the variance bounds are affected by the chosen length of the time intervals.

Figure 7.4 presents the restrictions placed on a quarterly market pricing kernel by monthly, quarterly and yearly returns. The regions estimated from returns on shorter time intervals lie above the regions estimated from returns on longer time intervals. Thus, specific quarterly market pricing kernels may exist that correctly price quarterly returns obtained from yearly time intervals but that do not correctly price quarterly returns obtained from quarterly or monthly time intervals. On the other hand, market pricing kernels that correctly price returns calculated from shorter time intervals should also correctly price returns calculated from longer time intervals. These findings indicate that the length of the time intervals over which the asset returns are measured affects the variance bounds imposed on quarterly market pricing kernels and that the bounds imposed by returns calculated from short time intervals are more restrictive than those imposed by returns calculated from long time intervals. This result is consistent with the findings of section 6.2.1.2. There it was observed that quarterly stock returns exhibit a standard deviation larger than $\sqrt{3}$ times the standard deviation of monthly stock returns and that yearly stock returns exhibit a standard deviation larger than $\sqrt{4}$ times the standard deviation of quarterly stock returns. Therefore, the Sharpe-ratio decreases for longer time intervals, lowering the variance bounds imposed by returns calculated for these time intervals.

However, as demonstrated in section 6.2.3, not only the asset returns violate the assumption of stochastic independence over time and/or stationarity of the conditional distributions but also the growth rates of consumption do violate these assumptions. Moreover, it was found that the moments of the asset returns are biased in the same direction as the moments of the growth rates of consumption. Therefore, the question emerges whether both effects have the same direction so that the above results vanish, when the variance bounds calculated from returns measured over yearly time intervals are compared with a parametrized yearly market pricing kernel and, alternatively, the variance bounds calculated from returns measured over quarterly time intervals are compared with a parametrized quarterly market pricing kernel. This question will be answered in section 7.2.1 using the parametric model of the market pricing kernel derived from the consumption-based asset pricing model assuming time-additive expected utility with constant relative risk aversion.

7.1.5 Summary

The restrictions placed on the mean and the standard deviation of the market pricing kernel by German stock, bond and money market returns in the period 1968 to 1994 have been analyzed. Especially the effects of taxation, different methods of return calculation and different time intervals have been investigated.

It is observed that one-period risk-free rates together with aggregate stock returns are more restrictive to parametric models of the market pricing kernel than equity premia or aggregate stock returns taken separately. This is consistent with the existing empirical evidence that jointly explaining risk-free and risky returns poses a problem to intertemporal asset pricing models. The restrictions placed on the mean and the standard deviation of the market pricing kernel by the combination of one-period risk-free rates and aggregate stock returns is stronger than that of aggregate bond and aggregate stock returns. Aggregate bond returns do not add information about the distribution of the market pricing kernel to the information contained in one-period risk-free rates and aggregate stock returns. Furthermore, period average returns which should be better comparable to growth rates of macroeconomic variables are more restrictive to parametric models of the market pricing kernel than end-of-period returns. This is a consequence of the higher risk premium relative to the standard deviation of period average returns as compared to end-of-period returns. Tax adjusted returns earned by private investors in contrast to returns before taxation necessitate a higher standard deviation of the market pricing kernel. In addition, the use of tax adjusted returns shifts the variance bounds horizontally. Hence, tax adjustment may have an effect on the results of empirical investigations of parametrized market pricing kernels. It is proposed to use scenario S2 to represent the taxation scenario of the marginal investor.

Comparing the variance bounds placed on the quarterly market pricing kernel by asset returns measured over time intervals with different lengths reveals that the bounds imposed by returns measured over short time intervals are more restrictive than those imposed by returns measured over long time intervals. This result indicates that the asset returns are stochastically dependent over time and/or exhibit nonstationary conditional distributions. Since specific parametrizations of the market pricing kernel may also exhibit stochastic dependences over time, returns and the market pricing kernel should be calculated for time intervals with the same length. If the effects of stochastic dependences or nonstationary conditional distributions are different for asset returns and the parametrized market pricing kernel, the chosen length of the time intervals affects the comparison of the variance bound with the parametrized market pricing kernel and, therefore, affects the evaluation of specific parametric models of the market pricing kernel.

7.2 Implications for the Parameters of Parametric Models of the Market Pricing Kernel

This section examines to which extent the unconditional first and second moments of parametric models of the market pricing kernel derived from the consumption-based asset pricing model assuming time-additive expected utility with constant relative risk aversion, assuming time-additive expected logarithmic utility and assuming recursive non-expected utility violate or comply with the estimated variance bounds. The sample moments of the parametrized market pricing kernels are calculated for quarterly data intervals, so that, based on the results of the previous section, variance bounds are employed that are calculated from quarterly asset returns. It has been argued that the taxation scenario S2 best represents the marginal investor. Therefore, mainly the bounds calculated from returns for the taxation scenario S2 are used. The sensitivity of the results to the use of other asset returns is analyzed.

7.2.1 Parametric Model Derived from Time-additive Expected Utility with Constant Relative Risk Aversion

The graphs in figure 7.5 exhibit the variance bounds calculated from end-of-period returns (Q1) for the taxation scenario referring to a private investor with a marginal tax rate of 36 percent (S2). In addition, the sample means and standard deviations of the parametrized market pricing kernel derived from the consumption-based asset pricing model using time-additive expected utility with constant relative risk aversion, $\tilde{\phi}_{t+1} = \beta(\tilde{c}_{t+1}/c_t)^{-\alpha}$ (equation (4.26)), are depicted for different levels of risk aversion α[1] and a discount factor β equal to one. In the upper graph α varies from 0 in increments of 1, in the lower graph in increments of 10. Both graphs show the same variance bounds for different scales. The diamond symbols are used to denote the mean-standard deviation pairs of the market pricing kernel calculated using the time series of consumption expenditures on nondurables and services (CNS), the triangles symbols denote the mean-standard deviation pairs using the time series of total consumption expenditures (C). To be economically feasible, β should be less than one. Only the market pricing kernels for β equal to one are depicted because the effect of a lower discount factor can easily be imagined. For $\tilde{\phi}_{t+1} = \beta(\tilde{c}_{t+1}/c_t)^{-\alpha}$ it follows that $\mathrm{E}\{\tilde{\phi}_{t+1}\} = \beta\mathrm{E}\{(\tilde{c}_{t+1}/c_t)^{-\alpha}\}$ and $\mathrm{std}\{\tilde{\phi}_{t+1}\} = \beta\mathrm{std}\{(\tilde{c}_{t+1}/c_t)^{-\alpha}\}$. Hence, a decrease in β simultaneously reduces the mean and the standard deviation of the market pricing kernel and therefore shifts the diamond and triangle symbols down and to the left.

As can be seen from the upper graph in figure 7.5 the parameter of risk aversion must equal about two or three to generate a market pricing kernel that lies in the area above the variance bound calculated from the aggregate stock returns. However, the parameter of risk aversion must exceed at least 17, using the time series of consumption expenditures on nondurables and services (CNS), respectively 18, using the time series of total consumption expenditures (C), for the market pricing kernel to enter the region estimated from the equity premia. This level of risk aversion seems to be unreasonably high[2]. This finding illustrates the Equity Premium Puzzle, which states that the historically observed equity premium is too high to be explained with plausible levels of risk aversion in a framework with time-additive

[1]Notice that in all subsequent figures Greek characters are replaced by Arabic characters.

[2]For a discussion of the plausible level of risk aversion see, for example, Mehra and Prescott (1985, p. 154). However, a necessary parameter of risk aversion of about 20 is low as compared to the results of Cochrane and Hansen (1992, p. 125) for the US. In a comparable analysis covering the years 1947 to 1990 they find that a parameter of risk aversion of at least 40 is required to generate the variance of market pricing kernels implied by the equity premia.

Figure 7.5: Bounds imposed on the standard deviation of the market pricing kernel by real **tax adjusted end-of-quarter returns** (Q1, S2) of different sets of assets. Mean-standard deviation pairs of a market pricing kernel derived from **time-additive expected utility** with constant relative risk aversion are shown for different levels of risk aversion α (a) and a discount factor $\beta = 1$. To illustrate the effect of large values of risk aversion, the lower graph shows the same variance bounds as the upper graph using a different scale. The mean-standard deviation pairs using the time series of consumption expenditures on nondurables and services (CNS) are represented by diamond symbols, those using the time series of total consumption expenditures (C) are represented by triangle symbols.

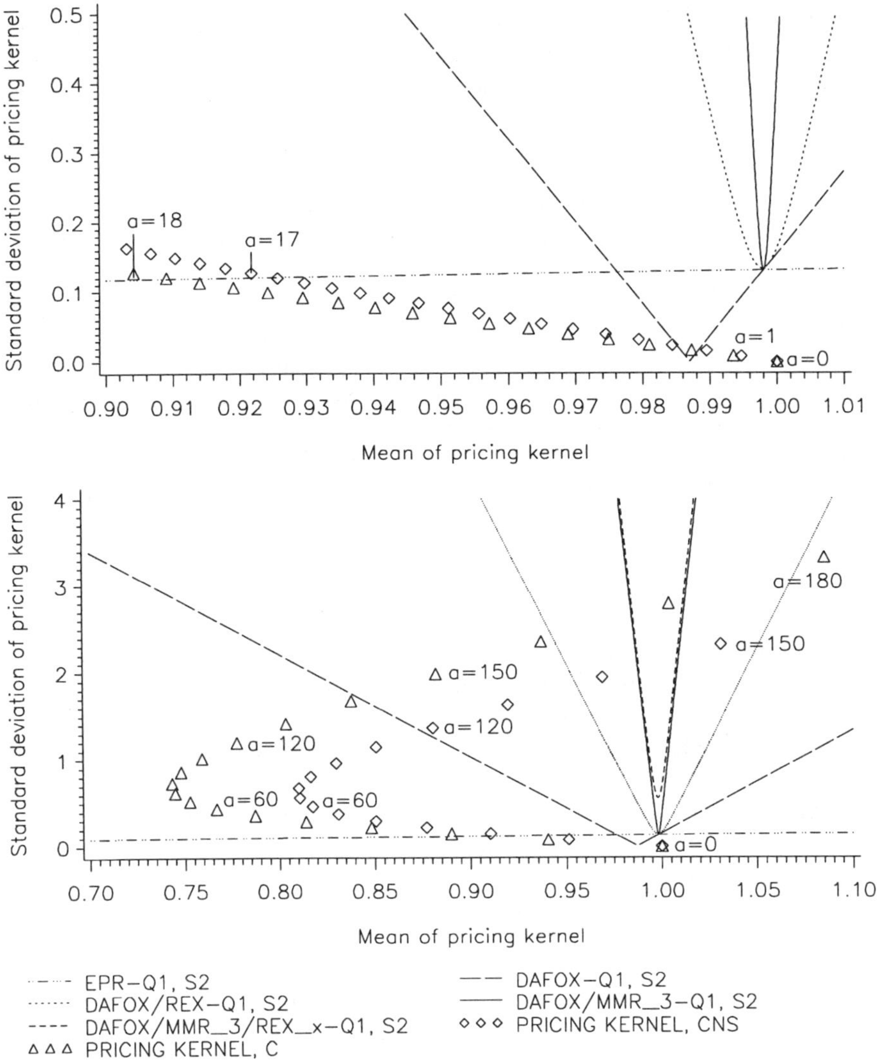

expected utility and constant relative risk aversion. Even if a parameter of risk aversion as large as 17 would be plausible the market pricing kernel would exhibit an expected value of about .922, implying an unreasonably high risk-free rate of 8.5 percent per quarter[3]. The finding that the observed risk-free rate is much smaller than this theoretical value illustrates the Risk-free Rate Puzzle. The lower graph in figure 7.5 shows that the level of risk aversion must rise to about 145 (using CNS) respectively 168 (using C) for the market pricing kernel to enter the region determined by the risky assets and the one-period risk-free rates. The increasing mean of the market pricing kernel for high values of risk aversion was also observed by Hansen and Jagannathan (1991). The initial decline in the mean of the market pricing kernel reflects the dominant role of positive growth rates of consumption, whereas for high levels of risk aversion observations with negative growth rates of consumption dominate the sample mean (Hansen and Jagannathan 1991, p. 250)[4].

Now, assume that a parameter of risk aversion of 17 is plausible. In this case no Equity Premium Puzzle would exist. Then, the diamond and triangle symbols must be shifted to the right to decrease the implied risk-free rate and to enter the region above the variance bound calculated from the one-period risk-free rates and the stock returns. An increase in the discount factor β would have the desired effect. For $\beta = 1$ and $\alpha = 17$ the mean of the market pricing kernel is about .922. To be consistent with the variance bound calculated from one-period risk-free rates in combination with stock returns the expected value of the market pricing kernel must rise to .998. Hence, β must be raised to $.998/.922 = 1.082$. As demonstrated in section 5.2.4, in growth economies with time-additive expected utility equilibria can exist for $\beta > 1$. The assumption of a discount factor smaller than one can therefore be relaxed from this perspective. Moreover, it was argued that as a consequence of the interrelation between risk aversion and elasticity of intertemporal substitution implicit in time-additive expected utility, β larger than one may also be economically plausible. β may not only measure the time preference but may also contain a compensation for the implicitly chosen elasticity of intertemporal substitution, if this value differs from the true value. In a growth economy β beyond one indicates that the true elasticity of intertemporal substitution is higher than the value implied by the chosen level of risk aversion. Hence, the recursive non-expected utility specification introduced in section 4.2.3.1 may at least help to solve the Risk-Free Rate

[3]The gross return on a risk-free asset implied by the market pricing kernel equals the reciprocal of the expected value of the market pricing kernel (see section 2.1.4), i.e. $1/.922 - 1 = .085$. In the analysis of Cochrane and Hansen (1992, p. 125) for the US a risk-free rate in excess of 17 percent per quarter is implied by the market pricing kernel that complies with the variance bound calculated from equity premia.

[4]Notice that in the case of high risk aversion the market pricing kernel is extremely sensitive to consumption and that therefore small errors in aggregate consumption can have large effects on the mean-standard deviation pair.

Puzzle because it allows a separate parametrization of risk aversion and the elasticity of intertemporal substitution. This will be analyzed in section 7.2.3.

Alternatively using the variance bounds calculated from returns for the zero tax scenario (S1) does not considerably change the results discussed above. This is illustrated in figure A.1 in appendix A.2. The parameter of risk aversion necessary for the market pricing kernel to enter the region above the variance bound calculated from the equity premia is slightly smaller (13 respectively 14) as compared to the results obtained for the taxation scenario S2. Also using the variance bounds imposed by period average returns (Q2) does not substantially alter the results (see figure A.2 in appendix A.2). In this case a parameter of risk aversion in excess of 19 respectively 20 is necessary for the market pricing kernel to be consistent with the equity premia. Overall, the results, i.e. the observed Equity Premium Puzzle and the Risk-free Rate Puzzle, are only slightly affected by the chosen time series of consumption growth (C or CNS), by the taxation scenario (S1 or S2) or by the method of return calculation (Q1-returns or Q2-returns). Subsequently, the sensitivity of the results with respect to the length of the time intervals is examined.

The results of section 7.1.4 have indicated that the variance bounds for the quarterly market pricing kernel calculated from returns measured over short time intervals are more restrictive than those calculated from returns measured over long time intervals. However, it was argued that this result may change, when the variance bounds calculated from quarterly returns are compared with a parametrized quarterly market pricing kernel and the variance bounds calculated from yearly returns are compared with a parametrized yearly market pricing kernel. This situation is illustrated in figure 7.6 for tax adjusted (S2) end-of-period returns (Q1, Y1) using the time series of consumption expenditures on nondurables and services (CNS). The mean-standard deviation pairs of the quarterly market pricing kernel are represented by small diamond symbols, those of the yearly market pricing kernel by small square symbols. β is again set equal to one and the parameter of risk aversion varies from 0 in increments of 1 in the upper graph, respectively 10 in the lower graph. The upper graph confirms the previous finding that the bounds calculated from returns measured over short time intervals are more restrictive than those from returns measured over long time intervals. The parameter of risk aversion must exceed 17 for the quarterly market pricing kernel to be consistent with quarterly equity premia, whereas the parameter of risk aversion required for the yearly market pricing kernel to be consistent with the yearly equity premia must only exceed 13. On the other hand, the lower graph shows that considering the variance bound calculated from one-period risk-free rates and stock returns, the bound calculated from yearly returns is more restrictive to the yearly market pricing kernel than the bound calculated from quarterly returns is

Figure 7.6: Bounds imposed on the standard deviation of the market pricing kernel by **quarterly and yearly** equity premia and by the combination of one-period risk-free rates and stock returns. The returns are **tax adjusted** (S2) and calculated from **end-of-period figures** (Q1, Y1). Mean-standard deviation pairs of a quarterly [yearly] market pricing kernel derived from **time-additive expected utility** with constant relative risk aversion are represented by diamond [square] symbols for different levels of risk aversion α (a) and a discount factor $\beta = 1$. To illustrate the effect of large values of risk aversion, the lower graph shows the same variance bounds as the upper graph using a different scale. The time series of consumption expenditures on nondurables and services (CNS) is used.

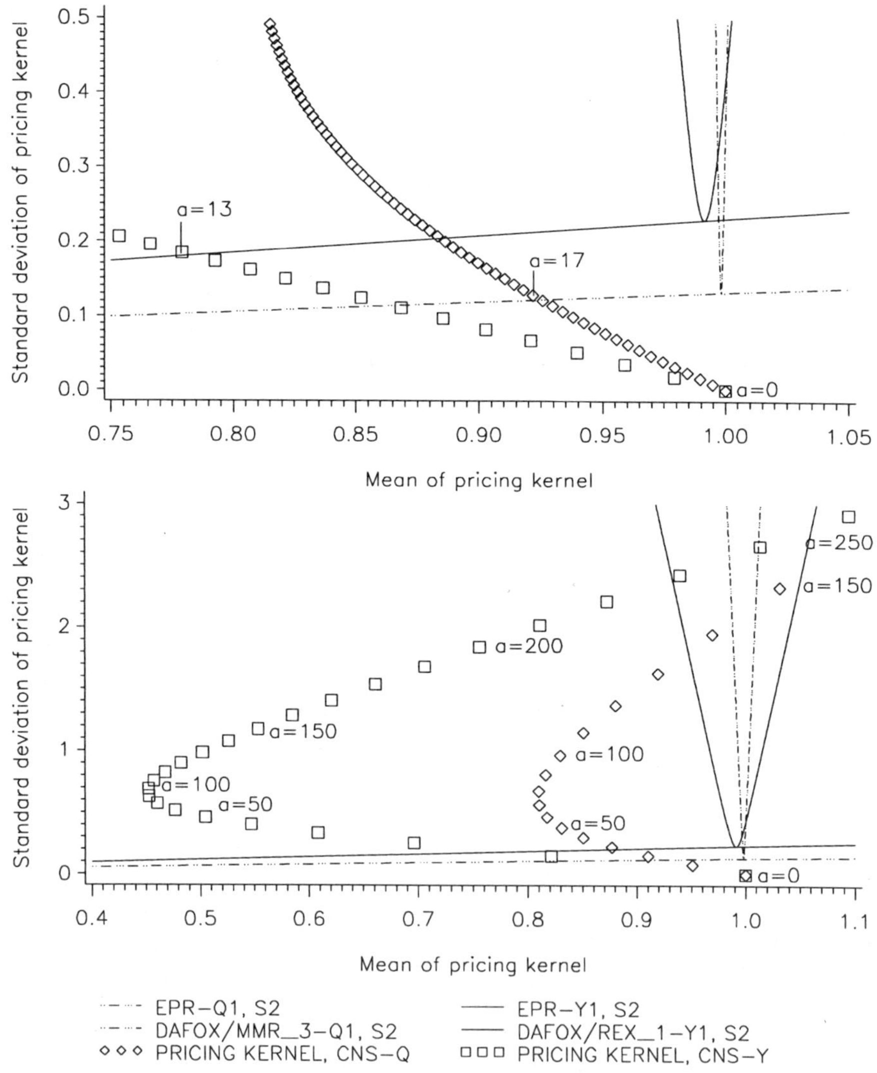

to the quarterly market pricing kernel[5]. This finding contradicts the result above. The explanation appears to be that for longer time intervals fewer observations with a negative growth rate of consumption exit that, as discussed above, dominate the sample means for large values of risk aversion.

7.2.2 Parametric Model Derived from Time-additive Expected Logarithmic Utility

As shown in section 4.2.2.4, for time-additive expected logarithmic utility, i.e. for time-additive expected utility with constant relative risk aversion equal to one, the market pricing kernel equals the reciprocal of the gross return on the aggregate wealth portfolio, $\tilde{\phi}_{t+1} = (\tilde{R}_{t+1}^{M})^{-1}$ (equation (4.35)). Hence, in this case the market pricing kernel can be measured without utilizing consumption data that is burdened with measurement problems. Instead, the problem of measuring the return on the aggregate wealth portfolio arises. The unobservability of this return gives reason for the well-known critique that models that incorporate the return on the aggregate wealth portfolio, such as the static CAPM, are not testable (cf. Roll (1977)). Nevertheless, the return on a broad stock market index is usually used as a proxy for the return on the aggregate wealth portfolio. Therefore, the return on the aggregate wealth portfolio is here measured by the return on the DAFOX; this means that the market pricing kernel assuming time-additive expected logarithmic utility equals the reciprocal of the gross return on the DAFOX.

Figure 7.7 depicts the variance bounds for the market pricing kernel calculated from end-of-quarter (Q1) returns for the taxation scenario S2. The dot denotes the sample mean-standard deviation pair for the reciprocal of the tax adjusted end-of-quarter returns (Q1, S2) on the DAFOX. Not surprisingly, this candidate of the market pricing kernel satisfies the variance bound calculated from the aggregate stock returns, i.e. from the DAFOX. However, this model of the market pricing kernel cannot generate sufficient variability to be consistent with the equity premia, the combination of aggregate stock and bond returns or the combination of one-period risk-free rates and aggregate stock returns. Alternatively using returns before taxation (S1) does not change these results. Thus, although time-additive expected utility with constant relative risk aversion equal to one allows us to calculate the sample mean-standard deviation pair of the parametric model of the market pricing kernel without using consumption data, the model fails to generate enough variability of the market pricing kernel.

[5] In an analysis not presented here, the same result is obtained using the time series of total consumption expenditures (C).

Figure 7.7: Bounds imposed on the standard deviation of the market pricing kernel by real **tax adjusted end-of-quarter returns** (Q1, S2) of different sets of assets and the mean-standard deviation pair of the market pricing kernel derived from **time-additive expected logarithmic utility** represented by the dot symbol.

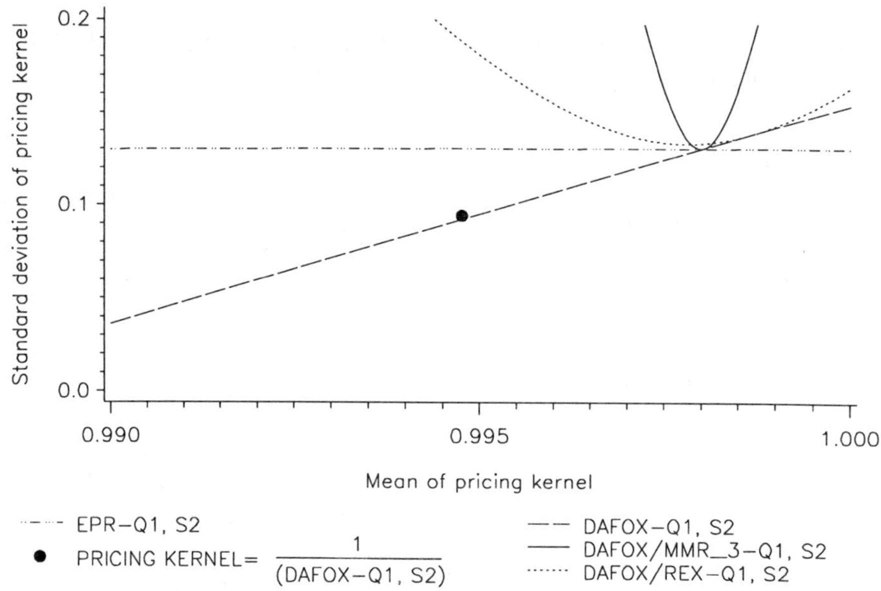

7.2.3 Parametric Model Derived from Recursive Non-expected Utility

The previous two sections have shown that parametric models of the market pricing kernel derived from the consumption-based asset pricing model using time-additive expected utility with constant relative risk aversion violate the variance bounds unless an implausibly high level of risk aversion and a discount factor larger than one are chosen. Furthermore, the findings of section 7.2.1 suggest that parametric models of the market pricing kernel derived from recursive non-expected utility may help to resolve these puzzling results because these models separately parametrize risk aversion and elasticity of intertemporal substitution. Therefore, the parametric model of the market pricing kernel derived from the consumption-based asset pricing model using recursive non-expected utility, $\tilde{\phi}_{t+1} = [\beta(\tilde{c}_{t+1}/c_t)^{-\rho}]^{\frac{1-\alpha}{1-\rho}}(\tilde{R}^M_{t+1})^{\frac{1-\alpha}{1-\rho}-1}$ (equation (4.46)), is now examined. Recall that ρ is the reciprocal of the elasticity of intertemporal substitution σ. As in the previous section, the return on the DAFOX is used as a proxy for the return on the aggregate wealth portfolio.

Figure 7.8 presents the variance bounds calculated from tax adjusted end-of-quarter returns (Q1, S2) and the sample means and standard deviations of the parametric model of the market pricing kernel for different levels of risk aversion α, different levels of the elasticity of intertemporal substitution σ and a discount factor equal to one. A decrease in β shifts the symbols down and to the left as long as $\alpha > 1$ and $\sigma < 1$. α varies from 2 in increments of 2 and σ varies from .01 in increments of .01. In the upper graph the time series of consumption expenditures on nondurables and services (CNS) is used for calculating the sample moments of the market pricing kernel; in the lower graph the time series of total consumption expenditures (C) is used. The results for both time series do not differ substantially. As can be seen, increasing the elasticity of intertemporal substitution beyond the value $1/\alpha$ implied by time-additive expected utility shifts the symbols up and to the right and, thus, raises the variability and the mean of the market pricing kernel. For certain values of the elasticity of intertemporal substitution market pricing kernels with parameters of risk aversion larger than or equal to four satisfy the variance bound calculated from the one-period risk-free rates and the aggregate stock returns.

Consider, for example, the parameter pair $\alpha = 6$ and $\sigma = .39$ in the upper graph. Parameters of risk aversion lower than ten are usually accepted as plausible. For the elasticity of intertemporal substitution estimated from aggregate data Atkeson and Ogaki (1996, table 7) report an average value of .40 in the years 1960 to 1987 for the US and an average value of .27 for India. Elasticities of intertemporal substitution estimated from Indian household data range from .44 to .82 with rich [poor] households exhibiting elasticities of intertemporal substitution at the upper [lower] level indicating a wealth-increasing elasticity of intertemporal substitution (Atkeson and Ogaki 1996, table 6). Considering the values estimated from aggregate data $\sigma = .39$ can be accepted as plausible for Germany. This value is larger than $1/6{=}.17$, the value implied by the same level of risk aversion of 6 in time-additive expected utility, indicating a preference for an early resolution of uncertainty (see section 4.2.3.1). This result is consistent with the findings of experimental research in decision making discussed in section 4.2.1.2. The same argument applies for the other parameter pairs satisfying the variance bounds, e.g., $\alpha = 8$ and $\sigma = .34$.

In the figures A.3 and A.4 in appendix A.2 the figure 7.8 is reproduced for taxation scenario S1 and for period average returns (Q2), respectively. These changes do not affect the results. There still exist plausible parameter pairs satisfying the variance bound calculated from one-period risk-free rates and aggregate stock returns, e.g., $\alpha = 6$ and $\sigma = .34$ in the Q1-S1-case or $\alpha = 6$ and $\sigma = .415$ in the Q2-S2-case. Thus, this analysis indicates that using the parametric model of the market pricing kernel derived from recursive non-expected utility rather than that derived from time-additive expected utility with constant relative risk aversion may help to solve the Equity Premium

Figure 7.8: Bounds imposed on the standard deviation of the market pricing kernel by real **tax adjusted end-of-quarter returns** (Q1, S2) of different sets of assets. Mean-standard deviation pairs of a market pricing kernel derived from **recursive non-expected utility** are shown for different levels of risk aversion α (a), different levels of elasticity of intertemporal substitution σ (s) and a discount factor $\beta = 1$. In the upper graph the time series of consumption expenditures on nondurables and services (CNS) is used; in the lower graph the time series of total consumption expenditures (C) is used.

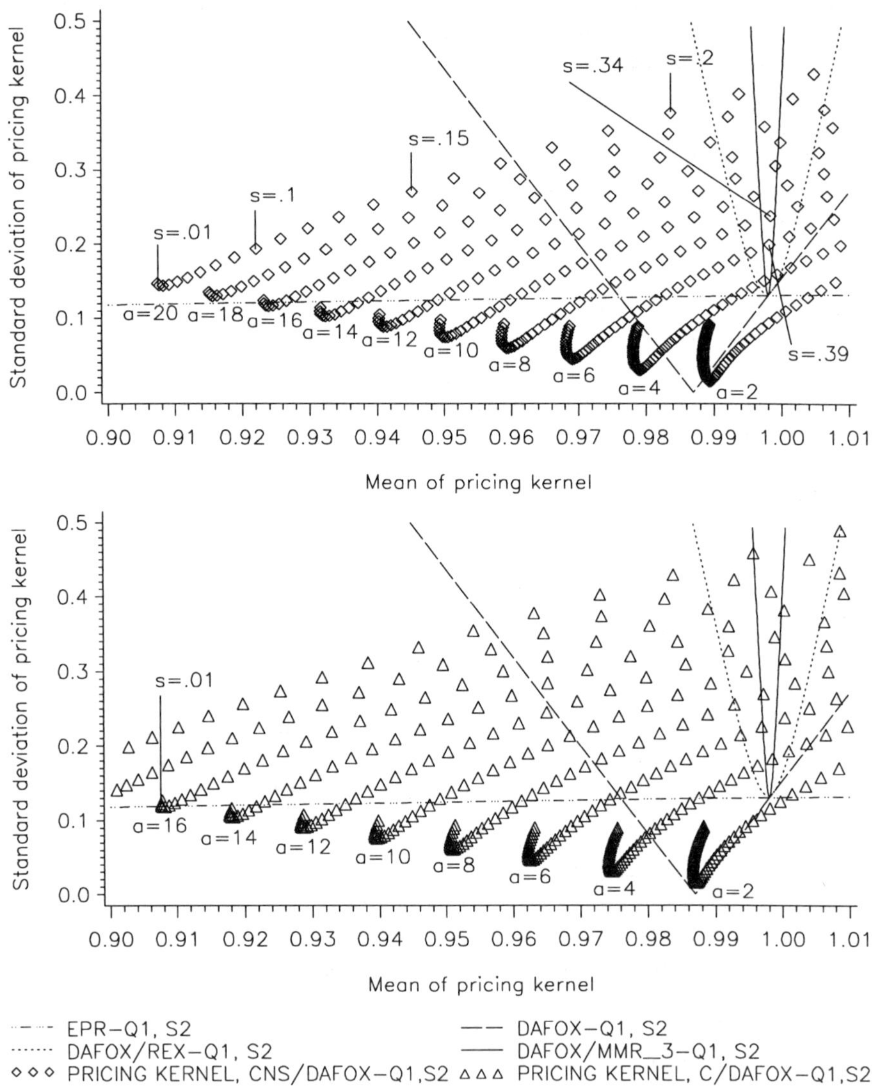

Puzzle and the Risk-free Rate Puzzle. Notice, however, that this analysis uses the aggregate stock return as a proxy for the return on the aggregate wealth portfolio. Since it is unlikely that the aggregate wealth portfolio is as volatile as the stock market the use of the proxy may be conducive to the favorable results.

7.2.4 Summary

This section analyzed the implications of the variance bounds of the market pricing kernel calculated from asset returns for the parameters of different parametric models of the market pricing kernel. The results are compared for parametric models of the market pricing kernel derived using the consumption-based equilibrium asset pricing approach assuming different utility specifications.

Similar to the results for the US, parametric models of the market pricing kernel derived from time-additive expected utility with constant relative risk aversion do not exhibit enough variability to satisfy the German variance bounds for plausible parameters of risk aversion and time preference. This result is consistent with the empirical evidence that pricing risky and risk-free returns jointly poses a problem to intertemporal asset pricing models and illustrates the Equity Premium Puzzle and the Risk-free Rate Puzzle for Germany. The analysis further reveals that these puzzles cannot be explained by changing the length of the time intervals over which the returns are measured, by using different methods of return calculation or by using tax adjusted returns. Also, the use of consumption data that is burdened with measurement problems cannot directly account for these puzzles. This was shown for the special case of unit risk aversion where the market pricing kernel equals the reciprocal of the gross return on the aggregate wealth portfolio that was measured by the gross return on the aggregate stock market.

The results obtained for the parametric model of the market pricing kernel derived from recursive non-expected utility are more favorable. With plausible values of risk aversion and elasticity of intertemporal substitution the market pricing kernel not only satisfies the variance bound calculated from equity premia but also lies above the variance bound imposed by one-period risk-free rates and aggregate stock returns. The findings reveal that the separation of risk aversion and intertemporal substitution may help to solve the puzzles mentioned above. For parametrized market pricing kernels satisfying the variance bounds the elasticity of intertemporal substitution is larger than the reciprocal of the parameter of risk aversion. This indicates that German investors are less averse to intertemporal fluctuations in expected consumption than implied by time-additive expected utility. Moreover, it indicates a preference for an early resolution of uncertainty. This is consistent with findings of empirical research in decision making. These results are not affected by applying a different method of return calculation or using returns before taxation. However, these favorable results are obtained using the return on the aggregate stock market as a proxy for the return on the aggregate wealth portfolio.

Chapter 8

Applying the Calibration Approach

The application of the variance bound approach in the previous chapter has given a first impression of the capability of the different parametric models of the market pricing kernel to jointly price different assets. Now, the calibration approach described in chapter 5 is applied to quarterly German data for the period 1960 to 1994. The goal is to analyze, whether the parametric models of the market pricing kernel derived from time-additive expected utility and, alternatively, derived from recursive non-expected utility can be calibrated to match the first and second moments of the empirically observed risk-free rate and equity premium. The "benchmark returns" that are to be matched are those calculated for the four taxation scenarios (see table 6.8). Different modifications of the original approach of Mehra and Prescott (1985) discussed in section 5.3 are investigated.

A criticism of Mehra and Prescott's approach on the basis of Lucas (1978) is that consumption data are strongly smoothed in comparison with the payoff on the stock market, respectively the aggregate payoff on equity, and, thus, that the assumption of both being equal or strongly correlated is not justifiable. Introducing leverage allows to relax the assumption of the equality of consumption and the payoff on equity. Using a bivariate model with separate exogenous time series measuring consumption (= production) and the payoff on equity additionally allows to relax the assumption of the strong correlation between both. The capability of this bivariate process approach to imply rates of return comparable to those empirically observed is compared with the capability of the univariate process approach with and without leverage.

For the univariate as well as for the bivariate approach the exogenous endowment process is modeled by a Markov state-switching model with states with stochastic outcomes, i.e. the exogenous endowment is assumed to be stochastic in each state. The state-switching model is allowed to be asymmetric, i.e. the conditional probabilities of staying in the states are not restricted to be equal across the states[1]. Furthermore, by allowing the variances and correlations to differ across the states, the exogenous process as well as the derived equilibrium rates of return generalize those of Cecchetti, Lam, and Mark (1993) and Hung (1994)[2]. This generalization has the potential not only to generate time-varying expected returns, but also to generate heteroscedasticity. This may capture the empirically observed increase in the volatility of aggregate stock returns during recessions (cf. Schwert (1989a), (1989b) and Bittlingmayer (1995)). For the evaluation of the estimated state-switching models the Monte Carlo approach proposed by Cecchetti, Lam, and Mark (1990) is generalized[3].

The calibration exercises proceed as follows:

1. The exogenous endowment process is theoretically specified and closed-form solutions for the implied first and second moments of the one-period risk-free rate and the equity premium are derived for the preference specifications.
2. The specified endowment process is estimated from observed data of the endowment variable.
3. If possible, the model is calibrated by choosing reasonable parameters of the preference function so that the implied first moments of the one-period risk-free rate and the equity premium match the corresponding observed values.
4. For calibrated models the implied second moments and the correlation of the risk-free rate and the equity premium are compared with the corresponding observed values. Statistical inference is obtained by simulating the models.

[1] In contrast, the original approach by Mehra and Prescott (1985) was to use a symmetric Markov state-switching model with states with deterministic outcomes. Such a state-switching model, however, does not allow to model excess kurtosis, skewness or asymmetric behavior of the endowment variable. In the study Meyer (1996), I investigated the differences between both approaches. I found that the symmetric bivariate switching model with states with deterministic outcomes implies higher equity premia than the bivariate switching model with states with stochastic outcomes, whereas the equity premia implied by the univariate switching model are not affected. The explanation appears to be that the symmetric bivariate switching model with states with deterministic outcomes implies a perfect correlation between both variables, whereas the bivariate switching model with states with stochastic outcomes does not.

[2] This generalization has already been applied in an earlier study of mine (cf. Meyer (1996)). In an independent study Bonomo and Garcia (1996) proposed a similar generalization.

[3] As discussed in section 5.3, there are other potential ways of modifying the original calibration approach, e.g., applying utility functions with intertemporal dependencies such as habit formation or introducing market frictions or incompleteness, such as borrowing constraints or idiosyncratic background risks. These modifications, however, are not analyzed here.

The chapter is organized as follows. The ensuing section introduces the univariate state-switching model economy. Closed-form solutions for the implied equilibrium rates of return are derived and the switching model for the endowment process is estimated, evaluated and simulated. Section 8.2 presents the calibration results for the basic model. In section 8.3 leverage is introduced and the calibration results for a levered market portfolio are presented. The bivariate approach with separate time series for consumption and the payoff on equity is analyzed in section 8.4. A summary ends the chapter.

8.1 A Model Economy with State-switching Production Growth

8.1.1 Reasons for a State-switching Economy

In the illustrative example in section 5.2 an i.i.d. process, e.g., independent draws from a single normal distribution, was assumed for the exogenous endowment process. Under CRRA preferences an i.i.d. endowment process implies a constant one-period risk-free rate. However, the empirically observed one-period risk-free rate is not a constant (see discussion in section 3.3.1). Furthermore, observed data are often autocorrelated or skewed and exhibit excess kurtosis or nonstationarities like volatility clusters or boom and depression states. Therefore, the assumption of the i.i.d. endowment process is relaxed and a Markov state-switching process for the growth rate of production (= consumption) is introduced instead. The process is stationary but the distribution of the growth rate of production is state dependent. It is assumed that the growth rate of production is normally distributed in each state of the switching model. The resulting unconditional distribution of production growth is a mixture of normal distributions.

The Markov state-switching model with state-dependent normal distributions has several advantages over the i.i.d. normally distributed process. First, the mixture of normal distributions allows to model skewness and excess kurtosis. A mixture of two normal distributions with different variances yields a leptokurtic distribution, whereas a mixture of two normal distributions with different means and different stationary probabilities yields a skewed distribution. In contrast, an ARIMA process with normal error terms, for example, does not allow to model skewness. Second, the Markov state-switching process allows to model first-order autocorrelation as well as nonstationarities such as volatility clusters. Third, conditional expectations can be modeled since at any date t the individuals' expectations about the future growth may depend on the state of nature at date t. However, the main advantage of the Markov state-switching model over other non-i.i.d.

processes such as ARIMA or ARCH is, fourth, that it admits a closed-form solution for the intertemporal asset pricing model and the equilibrium rates of return.

The analysis is restricted to state-switching models with two states for two reasons. First, it is unlikely to get good estimates of the state-switching models with more than two states, given the limited number of observations. Second, two states can be imagined as a boom and a recession state. Nevertheless, the analysis can easily be generalized to a larger number of states.

At each date t the economy is in one of both states. The state variable s_t follows an ergodic Markov switching process with a 2×2 transition matrix and the transition probabilities

$$\text{prob}_{ij} = \text{prob}(s_{t+1} = j | s_t = i) \qquad \text{for} \qquad i, j \in \{1, 2\}.$$

The transition probability prob_{ij} is the probability that state j occurs at date $t+1$, given that the economy is in state i at date t. The sum of all transition probabilities starting in the same state must equal one, i.e. $\text{prob}_{i1} + \text{prob}_{i2} = 1$ for $i \in \{1, 2\}$. If $\text{prob}_{ij} \in (0, 1)$ for $i, j \in \{1, 2\}$, the state variable neither is constant nor does it deterministically alternate between one and two, i.e. the model switches randomly between the two states. In both states the logarithmic growth rate of consumption, $\ln \tilde{\lambda}_t$, is assumed to be normally distributed and serially uncorrelated. The normal distribution in state s is given by $N_s = N(\mu_{cs}, \sigma_{cs}^2)$, $s \in \{1, 2\}$, where $\mu_{cs} = \mathrm{E}\{\ln \tilde{\lambda}_t | s_t = s\}$ and $\sigma_{cs}^2 = \mathrm{var}(\ln \tilde{\lambda}_t | s_t = s)$.

In contrast to the i.i.d. endowment process (see section 5.2.1) the last realization of the growth rate, i.e. the current state of the economy, now contains information about future growth. But as a consequence of the assumed Markov structure the current state contains all information for the subsequent state; past realizations of the growth rate contain no additional information. The ex-dividend market value of the market portfolio at date t, P_t^M, now equals $\Theta_s \cdot m_t$, $s \in \{1, 2\}$, where Θ_s is state-dependent. Recall that m_t is the aggregate production at date t and the proportionality factor Θ_s can be interpreted as a price-dividend-ratio. This ratio can take on different values in each state. In the present model it takes on one of two values. Notice that in reality the current state may not be directly observable. If the individuals must estimate the probability of being in state one or two, the price-dividend-ratio can take on all values between Θ_1 and Θ_2 (cf. Cecchetti, Lam, and Mark (1990, footnote 15)). The gross return on the market portfolio in a period from state i at date t to state j at date $t+1$ can be expressed as

$$\tilde{R}_{ij}^M = \frac{\tilde{\lambda}_j m_t + \tilde{P}_{j,t+1}^M}{P_{i,t}^M} = \frac{\tilde{\lambda}_j m_t + \Theta_j \tilde{\lambda}_j m_t}{\Theta_i m_t} = \tilde{\lambda}_j \frac{1 + \Theta_j}{\Theta_i} \text{ for } i, j \in \{1, 2\}. \quad (8.1)$$

8.1.2 Implied Equilibrium Rates of Return

Since the time-additive expected utility specification with constant relative risk aversion is the special case of the recursive non-expected utility specification where $\alpha = \rho$, it is sufficient to derive the implied equilibrium rates of return for the recursive non-expected utility specification. This derivation follows that of Kandel and Stambaugh (1991) and Hung (1994) but is generalized to allow the variance of the exogenous growth rate of production (= consumption) to differ across the states.

8.1.2.1 Return on the Market Portfolio

It has been noted that for recursive non-expected utility the market pricing kernel depends on the return on the market portfolio. For that reason, the implied equilibrium return on the market has to be determined first.

Using equation (8.1) together with the transition probabilities, prob_{ii} and prob_{ij}, and $\tilde{c}_{t+1}/c_t = \tilde{\lambda}_{t+1}$ the Euler equation (5.5) for an economy in state i yields in the case of the market portfolio

$$
\begin{aligned}
1 &= \sum_{j=1}^{2} \text{prob}_{ij} \mathrm{E}\left\{ \left[\beta \cdot \tilde{\lambda}_j^{-\rho}\right]^{\frac{1-\alpha}{1-\rho}} \left(\tilde{\lambda}_j \frac{1+\Theta_j}{\Theta_i}\right)^{\frac{1-\alpha}{1-\rho}} \right\} \\
\Leftrightarrow \Theta_i^{\frac{1-\alpha}{1-\rho}} &= \sum_{j=1}^{2} \text{prob}_{ij} \beta^{\frac{1-\alpha}{1-\rho}} \mathrm{E}\left\{\tilde{\lambda}_j^{1-\alpha}\right\} (1+\Theta_j)^{\frac{1-\alpha}{1-\rho}} \text{ for } i \in \{1,2\}. \qquad (8.2)
\end{aligned}
$$

Equation (8.2) is a system of two non-linear equations with the unknowns Θ_1 and Θ_2 which in general cannot be solved analytically[4].

For the time-additive expected utility specification, where $\alpha = \rho$, equation (8.2) becomes a system of two linear equations

$$
\Theta_i = \sum_{j=1}^{2} \text{prob}_{ij} \beta \mathrm{E}\{\tilde{\lambda}_j^{1-\alpha}\}(1+\Theta_j) \qquad \text{for} \qquad i \in \{1,2\}, \qquad (8.3)
$$

which can be solved analytically. Solving this system yields

$$
\Theta_i = \frac{q_{ii}(1-q_{jj}) + q_{ij}(1+q_{ji})}{(1-q_{ii})(1-q_{jj}) - q_{ij}q_{ji}} \quad \text{for} \quad i,j \in \{1,2\} \quad \text{and} \quad i \neq j, \qquad (8.4)
$$

where $q_{ii} = \text{prob}_{ii} \beta \mathrm{E}\{\tilde{\lambda}_i^{1-\alpha}\}$ and $q_{ij} = \text{prob}_{ij} \beta \mathrm{E}\{\tilde{\lambda}_j^{1-\alpha}\}$.

[4] For the calibration exercise this system is solved numerically using the SAS procedure "model".

The expected gross return on the market portfolio from state i to state j follows from equation (8.1) as

$$\mathrm{E}\{\tilde{R}_{ij}^M\} = \frac{1+\Theta_j}{\Theta_i}\mathrm{E}\{\tilde{\lambda}_j\} \quad \text{for} \quad i,j \in \{1,2\}. \tag{8.5}$$

Substituting Θ_1 and Θ_2 from equation (8.4) in (8.5) and using $\mathrm{E}\{\tilde{\lambda}_j\} = \exp[\mu_{cj} + 0.5\sigma_{cj}^2]$ and $\mathrm{E}\{\tilde{\lambda}_j^{1-\alpha}\} = \exp[(1-\alpha)\mu_{cj} + 0.5(1-\alpha)^2\sigma_{cj}^2]$, $\mathrm{E}\{\tilde{R}_{ij}^M\}$ can be calculated.

Using the unconditional probabilities of state 1 and 2, prob_1 and prob_2, the unconditional expected gross return on the market portfolio is

$$\mathrm{E}\{\tilde{R}^M\} = \sum_{i=1}^{2}\sum_{j=1}^{2} \mathrm{prob}_i \mathrm{prob}_{ij} \mathrm{E}\{\tilde{R}_{ij}^M\}. \tag{8.6}$$

The variance of the return on the market portfolio from state i to state j follows from equation (8.1) as

$$\mathrm{var}(\tilde{R}_{ij}^M) = \left(\frac{1+\Theta_j}{\Theta_i}\right)^2 \mathrm{var}(\tilde{\lambda}_j) \quad \text{for} \quad i,j \in \{1,2\}, \tag{8.7}$$

where $\mathrm{var}(\tilde{\lambda}_j) = \exp[2\mu_{cj} + \sigma_{cj}^2](\exp[\sigma_{cj}^2] - 1)$.

The unconditional variance of the return on the market portfolio is given by[5]

$$\mathrm{var}(\tilde{R}^M) = \sum_{i=1}^{2}\sum_{j=1}^{2} \mathrm{prob}_i \mathrm{prob}_{ij} \left(\mathrm{var}(\tilde{R}_{ij}^M) + \left(\mathrm{E}\{\tilde{R}_{ij}^M\} - \mathrm{E}\{\tilde{R}^M\}\right)^2\right). \tag{8.8}$$

8.1.2.2 Risk-free Rate of Return

Given the implied equilibrium return on the market portfolio from state i to state j, the implied equilibrium one-period risk-free rate can be calculated. Using the transition probabilities, prob_{ii} and prob_{ij}, and equation (8.1), the

[5] Notice that the unconditional variance consists of the expected conditional variance (first term within the brackets) plus the variance of the conditional expectation (second term within the brackets). A similar formula is applied by Kandel and Stambaugh (1991, equation (22)). Hung (1994, equation (22)) and Cecchetti, Lam, and Mark (1993, equation (22)) do not consider the conditional variance. Calculating only the variance of the conditional expectation they obtain a smaller implied variance. Comparing this variance with the sample variance of historical rates of return is incorrect and produces misleading calibration results.

Euler equation (5.5) can be rewritten for the gross return on a one-period risk-free asset in state i as

$$\begin{aligned}\frac{1}{R_i^f} &= \sum_{j=1}^{2} \text{prob}_{ij} \text{E}\left\{ [\beta\tilde{\lambda}_j^{-\rho}]^{\frac{1-\alpha}{1-\rho}} \left(\tilde{\lambda}_j \frac{1+\Theta_j}{\Theta_i}\right)^{\frac{1-\alpha}{1-\rho}-1} \right\} \\ \Leftrightarrow R_i^f &= \left(\sum_{j=1}^{2} \text{prob}_{ij} \beta^{\frac{1-\alpha}{1-\rho}} \text{E}\{\tilde{\lambda}_j^{-\alpha}\} \left(\frac{1+\Theta_j}{\Theta_i}\right)^{\frac{1-\alpha}{1-\rho}-1} \right)^{-1}, i \in \{1,2\} \quad (8.9)\end{aligned}$$

where $\text{E}\{\tilde{\lambda}_i^{-\alpha}\} = \exp[-\alpha\mu_{ci} + 0.5\alpha^2\sigma_{ci}^2]$.

For time-additive expected utility, where $\alpha = \rho$, equation (8.9) becomes

$$R_i^f = \left(\sum_{j=1}^{2} \text{prob}_{ij} \beta \text{E}\{\tilde{\lambda}_j^{-\alpha}\} \right)^{-1} \quad \text{for} \quad i \in \{1,2\}. \quad (8.10)$$

Using the unconditional probabilities of state 1 and 2, prob_1 and prob_2, the unconditional expected gross return on a one-period risk-free asset follows from equation (8.9) or (8.10) as

$$\text{E}\{\tilde{R}^f\} = \sum_{i=1}^{2} \text{prob}_i R_i^f. \quad (8.11)$$

The unconditional variance of the one-period risk-free rate of return is calculated as

$$\text{var}(\tilde{R}^f) = \sum_{i=1}^{2} \text{prob}_i \left(R_i^f - \text{E}\{\tilde{R}^f\} \right)^2. \quad (8.12)$$

8.1.2.3 Risk Premium on the Market Portfolio

The expected risk premium on the market portfolio over the one-period risk-free rate from state i to state j is

$$\text{E}\{\tilde{p}_{ij}^M\} = \text{E}\{\tilde{R}_{ij}^M\} - R_i^f \quad \text{for} \quad i,j \in \{1,2\},$$

and the unconditional expected risk premium is

$$\text{E}\{\tilde{p}^M\} = \sum_{i=1}^{2}\sum_{j=1}^{2} \text{prob}_i \text{prob}_{ij} \text{E}\{\tilde{p}_{ij}^M\}. \quad (8.13)$$

The conditional variance of the risk premium on the market, $\text{var}(\tilde{p}_{ij}^M)$, equals the conditional variance of the return on the market, $\text{var}(\tilde{R}_{ij}^M)$. The unconditional variance of the risk premium follows as

$$\text{var}(\tilde{p}^M) = \sum_{i=1}^{2}\sum_{j=1}^{2} \text{prob}_i \text{prob}_{ij} \left(\text{var}(\tilde{R}_{ij}^M) + \left(\text{E}\{\tilde{p}_{ij}^M\} - \text{E}\{\tilde{p}^M\}\right)^2 \right). \quad (8.14)$$

The unconditional covariance of the risk premium and the one-period risk-free rate is given by

$$\operatorname{cov}(\tilde{p}^M, \tilde{R}^f) = \sum_{i=1}^{2} \operatorname{prob}_i \left(\mathrm{E}\{\tilde{p}_i^M\} - \mathrm{E}\{\tilde{p}^M\}\right) \left(R_i^f - \mathrm{E}\{\tilde{R}^f\}\right). \qquad (8.15)$$

In contrast to the variance equations (8.8) and (8.14), the covariance equation (8.15) contains no term reflecting the conditional covariance because the conditional covariance is equal to zero as the conditional risk-free rate is deterministic.

8.1.3 The Markov State-switching Model

8.1.3.1 Estimation of the Model

It was argued that the Markov state-switching model with state-dependent normal distributions allows to model excess kurtosis, skewness and first-order autocorrelation. Often, negative skewness and leptokurtosis of high frequency financial time series data are reasons for the hypothesis of a single normal distribution to be rejected. For lower frequency data the assumption of a single normal distribution often cannot be rejected. This is also true for the time series of the logarithm of the gross growth rates of per capita consumption expenditures on nondurables and services (CNS-Q) calculated in section 6.2.3. Tests were made using the Anderson-Darling statistic[6] A^2 and the Shapiro-Wilk statistic[7] W ($A^2 = .3298$, $W = .9916$). The hypothesis of normally distributed logarithmic growth rates of consumption cannot be rejected. Moreover, the time series exhibits only an insignificantly negative skewness (-.1199) and a low first-order autocorrelation (-.0803). The excess kurtosis (.7362) is positive at a five percent level of significance[8]. Hence, if a Markov state-switching process explains the data

[6] The Anderson-Darling statistic, like the Kolmogorov statistic, is based on the vertical distance between the empirical cumulative density function and the cumulative density function of a normal distribution with known parameters. The larger the test statistic, the stronger the evidence against the normal distribution. In our case the parameters of the normal distribution are estimated from the data. Therefore, the test statistic is adjusted using a factor that depends on the number of observations (see D'Agostino and Stephens (1986, p. 123, table 4.7)). The Anderson-Darling statistic is used because it is more powerful than the Kolmogorov statistic (cf. D'Agostino and Stephens (1986, p. 404)).

[7] The Shapiro-Wilk statistic is the ratio of an estimator of the variance that is valid if the data are normally distributed (based on the square of a linear combination of the order statistics), to the usual corrected sum of squares estimator of the variance ($W \in (0, 1]$). Small values of W provide evidence against the normal distribution.

[8] The significance levels for skewness and excess kurtosis are taken from table 34 panel B and C in Pearson and Hartley (1966).

significantly better than an i.i.d. normally distributed process, then nonstationarities in the time series rather than deviations from a normal distribution or first-order autocorrelation must be the main reason.

The difficulty in estimating the Markov state-switching model with two states and state-dependent normal distributions for consumption growth is that the state variable is not observable. Therefore, the expectation-maximization (EM) algorithm, an iterative technique for computing maximum likelihood estimates with incomplete data[9], is applied. Baum, Petrie, Soules, and Weiss (1970) show that, under certain regularity conditions, the EM-algorithm increases the likelihood function monotonically and that it converges to the maximum likelihood estimates. The application of the EM-algorithm for estimating state-switching models with state-dependent normal distributions is described in Hamilton (1990)[10] (see also Hamilton (1989), (1993), (1996)).

To confirm that a global optimum is found, the estimation is replicated with numerous sets of starting values. Convergence of the algorithm is declared to be obtained when the maximum change in each of the parameters is smaller than 10^{-9}. The algorithm converged to two optima that differ only slightly in the value of the log-likelihood function (474.60 respectively 470.52). From these values one cannot decide which of both optima should be chosen. Therefore, consider the optimum with the log-likelihood of 474.69 first. The maximum likelihood (ML) estimates of this optimum are shown in panel 1 of table A.1 in appendix A.2. Asymptotic standard deviations of the estimates are also reported. Figure A.5 in appendix A.2 graphs the smoothed conditional probability of being in state 1[11] in combination with the time series of consumption growth. This optimum is characterized by a single change in the states over the total period and both states differ only slightly in the mean of consumption growth. Hence, this optimum should be interpreted as identification of a structural change in the time series and not as a characterization of different states of macroeconomic uncertainty[12]. In contrast, the other optimum identifies different states of macroeconomic

[9]Laird (1993) provides an illustrative introduction into the EM-algorithm.

[10]For estimating the Markov state-switching model in this study, the algorithm described in Hamilton (1990) has been implemented in Gauss.

[11]From the parameter estimates of the model, conditional probabilities prob($s_t = i|\lambda_1, \ldots, \lambda_\tau$) can be calculated in a recursive way (cf. Hamilton (1989)). If τ is set equal to t "filter" probabilities about the probable state at time t are obtained. Filter probabilities are based on the information available at time t. For $\tau = T$ "smoothed" probabilities about the probable state at time t are obtained. Smoothed probabilities are more accurate since they are based on the information up to the end of the sample.

[12]This finding suggests that to obtain a higher number of regime changes a state-switching model with four or at least with three states should be estimated. However, with the limited number of observations it is unlikely to obtain good estimates for switching models with more than two states. In the Markov switching model with two states six parameters must be estimated. In a Markov switching model with three (four) states the number of free parameters increases to 12 (20). A possibility to reduce the num-

uncertainty. The estimates for this optimum that is used throughout the study are now described.

The maximum likelihood (ML) estimates of this optimum are shown in the first panel of table 8.1. Asymptotic standard deviations are given in parentheses. The estimates identify an up-state with low volatility of consumption growth (state 2) and a down-state with high volatility (state 1). The generalization to allow the variance to differ across the states thus leads to an estimation result consistent with the empirical finding of increased volatility during recessions, reflecting increased macroeconomic uncertainty. The unconditional probability of the up-state is 76 percent. In 24 percent of the time the economy is in the down-state. The down-state is characterized by an average consumption decrease of about .06 percent. The conditional probability of staying in this state is nearly equal to zero. Hence, the identified down-state can be interpreted as a short and strong crash state. The graphs in figure 8.1 illustrate these results. The first graph exhibits the time series of consumption growth. The second graph shows the time series of the smoothed conditional probability of being in state 1, the down-state of the estimated switching model. It can be seen that the model switches fast between the states and that often the current regime cannot be identified exactly as state 1 or state 2.

8.1.3.2 Evaluation of the Model

Until now the discussion has shown that the ML-estimated model does not separate the regimes exactly. In order to further assess the quality of the switching model the empirically observed first four moments of the distribution and the first-order autocorrelation of the growth rates of consumption are compared with those implied by the switching model (see panel 2 of table 8.1). As can be seen, the implied first and second moments are almost equal to those observed empirically. The implied low negative skewness, excess kurtosis and negative first-order autocorrelation are close to the corresponding historical values. Furthermore, the simulation of the estimated state-switching model in section 8.1.3.3 shows that the hypotheses that the empirically observed first four moments and the first-order autocorrelation are generated by the estimated Markov state-switching model cannot be

---

ber of free parameters is to put restrictions on the parameter vector, for example, to restrict the variances to be equal across states or to restrict the transition matrix to be symmetric. Cecchetti, Lam, and Mark (1990), for example, restrict the variances to be equal across states and explore a three-state model with 10 free parameters. But with the 116 yearly observations they were unable to obtain sensible parameter estimates (cf. Cecchetti, Lam, and Mark (1990, footnote 11)). But one goal of this study is to examine asymmetric market fundamentals and to explicitly allow the variances and therefore the macroeconomic uncertainty to differ across the states.

Table 8.1: Markov state-switching model with two states and state-dependent normal distributions for the logarithm of the quarterly real percentage growth of per capita consumption expenditures on nondurables and services (CNS-Q).

Panel 1:	Maximum Likelihood Estimates[a]			
	$s=1$		$s=2$	
prob_{ss}	.00000	(.15688)[b]	.68848	(.33019)
prob_s[c]	.23753		.76247	
μ_{cs} [%]	-.05917	(.58164)	.74984	(.11541)
σ^2_{cs} [$\%^2$]	.73243	(.33451)	.51346	(.09366)
Panel 2:	**Unconditional Moments**			
	Observed		Implied[d]	
$E\{\ln\tilde{\lambda}\}$ [%]	.55715		.55768	
$\text{var}(\ln\tilde{\lambda})$ [$\%^2$]	.68933		.68400	
Skewness	-.11986		-.25913	
Excess Kurtosis	.73620		.21605	
$\text{corr}(\ln\tilde{\lambda}_t, \ln\tilde{\lambda}_{t-1})$	-.08029		-.05398	
Panel 3:	**LR[e] Statistic and Significance Level**			
LR	2.5407			
Significance level	.50	.10	.05	.01
UTF-LR1[f]	3.8594	9.0612	10.9017	15.3194
UTF-LR2	4.4722	9.9942	11.9823	16.8826

[a] prob_{ss} denotes the conditional probability for staying in state s, prob_s denotes the unconditional probability of state s, μ_{cs} [σ^2_{cs}] denotes the mean [variance] of the state-dependent normal distribution for logarithmic consumption growth $\ln\tilde{\lambda}$.

[b] Standard deviations in parentheses.

[c] The unconditional stationary probabilities of the states are calculated as $\text{prob}_1 = (1-\text{prob}_{22})/(2-\text{prob}_{11}-\text{prob}_{22})$ and $\text{prob}_2 = 1-\text{prob}_1$.

[d] The implied unconditional moments are calculated as follows (see Patel and Read (1982, p. 32)):
$E\{\ln\tilde{\lambda}\} = \mu_c = \sum_{i=1}^{2}\text{prob}_i\mu_{ci}$,
$\text{var}(\ln\tilde{\lambda}) = \sigma_c^2 = \sum_{i=1}^{2}\text{prob}_i(\sigma^2_{ci} + (\mu_{ci}-\mu_c)^2)$,
$\text{Skewness} = \sum_{i=1}^{2}\text{prob}_i(3\sigma^2_{ci}(\mu_{ci}-\mu_c) + (\mu_{ci}-\mu_c)^3)/\sigma_c^3$,
$\text{Excess Kurtosis} = \sum_{i=1}^{2}\text{prob}_i(3\sigma^4_{ci} + 6\sigma^2_{ci}(\mu_{ci}-\mu_c)^2 + (\mu_{ci}-\mu_c)^4)/\sigma_c^4 - 3$,
$\text{corr}(\ln\tilde{\lambda}_t, \ln\tilde{\lambda}_{t-1}) = (\sum_{i=1}^{2}\sum_{j=1}^{2}\sum_{k=1}^{2}\text{prob}_i\text{prob}_{ij}\text{prob}_{jk}\mu_{cj}\mu_{ck} - \mu_c^2)/\sigma_c^2$.
*** (**, *) indicates that the hypothesis that the implied value is equal to the observed value can be rejected at a 1 (5, 10) percent level of significance based on the simulation results presented in section 8.1.3.3.

[e] Likelihood ratio statistic: twice the difference of the maximized loglikelihood value of the Markov state-switching model (H_1) and that using a single normal distribution (H_0).

[f] Upper tail fractiles of the simulated distribution of the LR statistic for the null of an i.i.d. normally distributed process (UTF-LR1) and for the null of a first-order autoregressive process with normal error terms (UTF-LR2), respectively.

Figure 8.1: The upper graph exhibits the time series of quarterly logarithmic real growth rates of per capita consumption expenditures on nondurables and services (CNS-Q). The lower graph shows the time series of conditional smoothed probabilities of state 1, the crash state of the estimated Markov state-switching model for consumption growth.

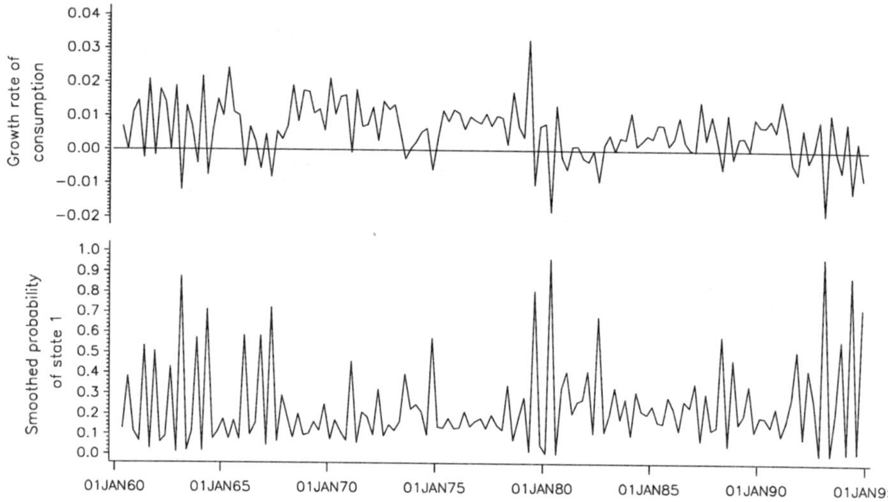

rejected. These results provide evidence for the use of the estimated state-switching model described above[13].

The next step of the evaluation of the estimated model is to test whether the mixture of both normal distributions describes the time series significantly better than a single normal distribution. As can be seen from figure 8.2, the density function of the mixture nearly equals that of a single normal distribution and, thus, it may not substantially improve the fit of the data. However, the excess kurtosis and the skewness of the empirically observed distribution is better represented by the mixture distribution.

The likelihood ratio statistic[14] listed in panel 3 of table 8.1 compares the loglikelihood of the estimated mixture of normal distributions with that of the single normal distribution fitted to the data. Notice, that the transition probabilities are not identified under the null hypothesis of no changes in regime because any value between 0 and 1 leaves the likelihood function unchanged. Therefore, the likelihood ratio statistic does not have a stan-

[13]In contrast, the other optimum implies positive skewness, positive first-order autocorrelation and fails to imply excess kurtosis (see panel 2 of table A.1). However, based on a simulation exercise for this optimum similar to that described in section 8.1.3.3 only for the excess kurtosis the hypothesis that the empirically observed value is generated by the estimated model can be rejected at a ten percent level of significance.

[14]Twice the difference in the maximized loglikelihood values of two alternative models.

Figure 8.2: Empirical density function of the quarterly logarithmic growth rate of per capita consumption expenditures on nondurables and services (CNS-Q) and the density function of the estimated mixture of two normal distributions in comparison with that of a single normal distribution fitted to the data.

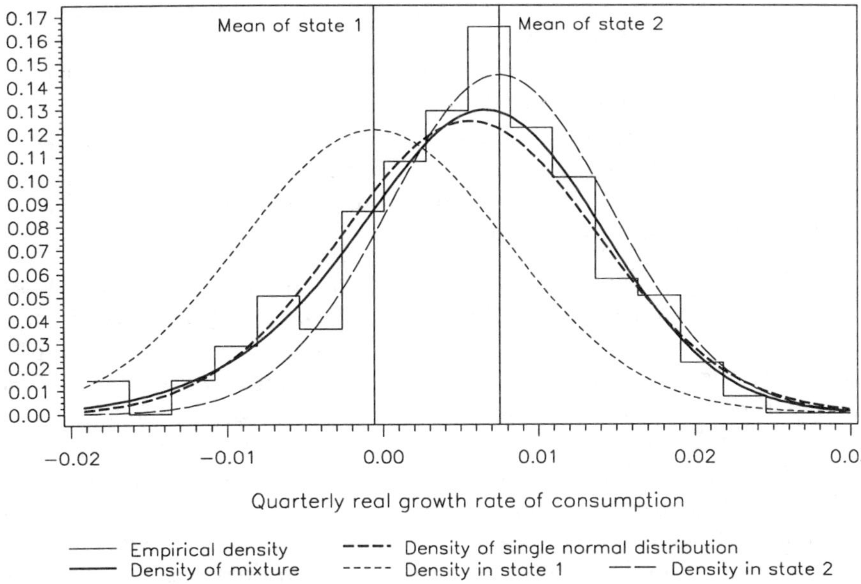

dard chi-squared distribution (cf. Hamilton (1990, p. 61)). Cecchetti, Lam, and Mark (1990, p. 404) simulate the distribution of the likelihood ratio statistic using a Monte Carlo approach. They generate 1,000 time series of i.i.d. normally distributed growth rates with mean and variance set to the empirically observed values and estimate the switching model for each time series. Calculating the likelihood ratio statistic for each realization yields the simulated distribution. Cecchetti, Lam, and Mark (1990) hence test the switching model against independent draws from a single normal distribution.

In a first step a similar approach is applied. 1,500 series of 139 independent draws are taken from a normal distribution with mean and variance set to the empirically observed mean and variance of the logarithmic growth rates of consumption. The sample size of 139 is chosen to correspond to the 139 quarterly historical observations. For each of the 1,500 sequences of 139 draws the Markov state-switching model is estimated and the likelihood ratio statistic is calculated. Upper tail fractiles of the obtained distribution of the likelihood ratio statistic (UTF-LR1) are listed in panel 3 of table 8.1. Within this approach the empirically observed autocorrelation of the growth rate of consumption is not considered, although the possibility to model autocorrelation is an important feature of the state-switching model and autocorrelation in the data affects the estimates of the model. The second step therefore is to investigate if considering the autocorrelation affects the simulated distribution of the likelihood ratio statistic. 1,500 time series are generated using a first-order autoregressive process. The parameter of first-order autocorrelation is set to the empirically observed value of -.08 and the mean and standard deviation of the normally distributed error term are chosen so that the mean and the standard deviation of the unconditional distribution of the generated data equal those of the empirically observed sample. To confirm that the 1,500 realizations are independent, time series with 200 periods are generated and only the last 139 periods are used for the estimation of the Markov state-switching process. Upper tail fractiles of the simulated distribution of the likelihood ratio statistic (UTF-LR2) are listed in the last row of table 8.1.

As can be seen, the upper tail fractiles of the simulated distributions of both LR statistics are quite similar. However, to reject the hypothesis of a first-order autoregressive process with normal error terms against the Markov state-switching process the LR statistic must be slightly larger than to reject the hypothesis of an i.i.d. normally distributed process. For the logarithmic real growth rate of per capita consumption expenditures on nondurables and services the i.i.d. normally distributed process as well as the first-order autoregressive process with normally distributed error terms cannot be rejected against the estimated Markov state-switching process given in table 8.1[15]. Thus, considering first-order autocorrelation does not change the results.

Although the i.i.d. normally distributed process and the first-order autoregressive process with normal error terms cannot be rejected against the estimated Markov state-switching model, the Markov state-switching model captures certain features of the data that alternative models cannot capture. Moreover, it has been demonstrated that the Markov state-switching model is generally a reasonable model for the exogenous endowment process in calibration exercises because it allows to model conditional expectations, nonstationarities, and especially a time-varying one-period risk-free rate of return, and still is theoretically tractable, i.e. it permits closed-form solutions for the implied equilibrium rates of return. Therefore, the estimated Markov state-switching model will be used as the exogenous endowment process in the calibration exercise. Before the calibration results are discussed the estimated endowment process is used to generate a large set of realizations of the time series of the endowment variable. These time series serve as the basis for further simulations throughout the study.

[15]Both processes can be rejected at a ten percent level of significance against the first optimum of the Markov state-switching model (see LR statistic (10.7024) in table A.1 in appendix A.2).

8.1.3.3 Simulation of the Model

In this section the exogenous endowment process is simulated using the Markov state-switching model. The parameters are set equal to the maximum likelihood estimates obtained in the previous section (see table 8.1)[16]. A large set of realizations of the time series of the endowment variable is generated for various reasons: First, comparing the distributions of various moments of the simulated time series with those of the historical time series provides further evidence for or against the estimated data generating process (see this section below). Second, given a utility specification and a set of preference parameters, simulated time series of the implied equilibrium rates of return are calculated from the simulated time series of the endowment variable. By comparing the distributions of moments of these simulated return series with the moments of the historical rates of return, statistical inference is gained concerning the specified equilibrium model (section 8.3.3). Third, given a calibrated equilibrium model, the set of simulated time series of the implied equilibrium rates of return is used for further analyses, for example, for investigating the serial correlation in multiperiod returns on stocks (chapter 9.1).

10,000 time series of the endowment process with 139 observations of the growth rate of consumption are generated. This is done by generating the time series of the state variable and then drawing the random growth rates of consumption from normal distributions conditional on the value of the state variable. To confirm that the time series are independent, time series with length of 200 are generated and the last 139 observations of each time series are taken[17]. Table 8.2 reports the means and fractiles of the simulated distributions of the first four moments and the first-order autocorrelation of the logarithmic real growth rates of per capita consumption expenditures on nondurables and services. Comparing the historical sample moments (see panel 2 of table 8.1) with these simulated moment distributions yields the following results. The historical mean and first-order autocorrelation lie between the 25 and the 50 percent fractile of the corresponding simulated distribution. The variance and skewness lie between the 50 and the 75 percent fractile and the kurtosis lies between the 90 and the 95 percent fractile. Hence, the historical first four moments and the first-order autocorrelation of the logarithmic growth rates of consumption lie at least in the 90 percent confidence interval around the median of simulated distributions of these moments. Thus, analyzing these moments, the hypothesis that the historical time series is generated by the estimated Markov state-switching model cannot be rejected at conventional levels of significance.

[16] A simulation exercise similar to that presented here is conducted for the alternative estimates listed in table A.1 in appendix A.2.

[17] Figure A.6 in appendix A.2 graphs two randomly chosen time series of these 10,000 time series.

Table 8.2: Means and fractiles of the simulated distributions of the first four moments and the first-order autocorrelation of the logarithmic real growth rates of per capita consumption expenditures on nondurables and services (CNS-Q). 10,000 time series with 139 periods are generated using the estimated Markov state-switching model. The star symbols indicate between which fractiles the corresponding sample moments lie.

		Simulation Fractiles						
	Mean	5 %	10 %	25 %	50 %	75 %	90 %	95 %
$E\{\ln\tilde{\lambda}\}$ [%]	.5584	.4472	.4717	.5131 *	.5584	.6032	.6446	.6685
$var(\ln\tilde{\lambda})$ [%^{2}]	.6857	.5500	.5771	.6258	.6818 *	.7413	.7996	.8347
Skewness	-.2457	-.6074	-.5198	-.3865	-.2422 *	-.0994	.0283	.1066
Excess Kurtosis	.1144	-.5074	-.4085	-.2188	.0389	.3616	.7213 *	.9859
$corr(\ln\tilde{\lambda}_t, \ln\tilde{\lambda}_{t-1})$	-.0592	-.1975	-.1658	-.1163 *	-.0603	-.0028	.0486	.0792

8.2 Analysis of the Non-levered Market Portfolio

This section analyzes the implied equilibrium rates of return, dependent on the parameters of the preference function. Graphs are used to illustrate the first moments of the implied risk-free rate and risk premium for different combinations of the preference parameters. Unless these moments match the first moments of the empirically observed risk-free rate and equity premium, higher moments are not considered. The empirically observed rates of return calculated from quarterly averages (Q2) (see table 6.8) are taken as the benchmark returns. Calibrated models, i.e. models that allow to jointly match the first moments of the benchmark returns, are simulated to obtain statistical inference. The calculation of the first and second unconditional moments of the implied risk-free rate of return and the implied risk premium is based on the equations (8.11) to (8.15). For the stochastic exogenous input the estimated mixture of univariate normal distributions described in table 8.1 is used. The implied equilibrium rates of return are analyzed for time-additive expected utility with constant relative risk aversion (section 8.2.1) and for non-expected recursive utility (section 8.2.2).

8.2.1 Time-additive Expected Utility

The implied equilibrium rates of return are calculated for all combinations of discount factor β, $\beta \in \{.9500, .9505, \ldots, 1\}$, and relative risk aversion α, $\alpha \in \{0, .1, \ldots, 10\}$. Figure 8.3 illustrates the implied risk-free rate-risk premium

Figure 8.3: Set of admissible quarterly equilibrium risk-free rates and risk premia using the estimated Markov switching model and **time-additive expected utility**. Discount factor β (b) ≤ 1 and risk aversion α (a) ≤ 10.

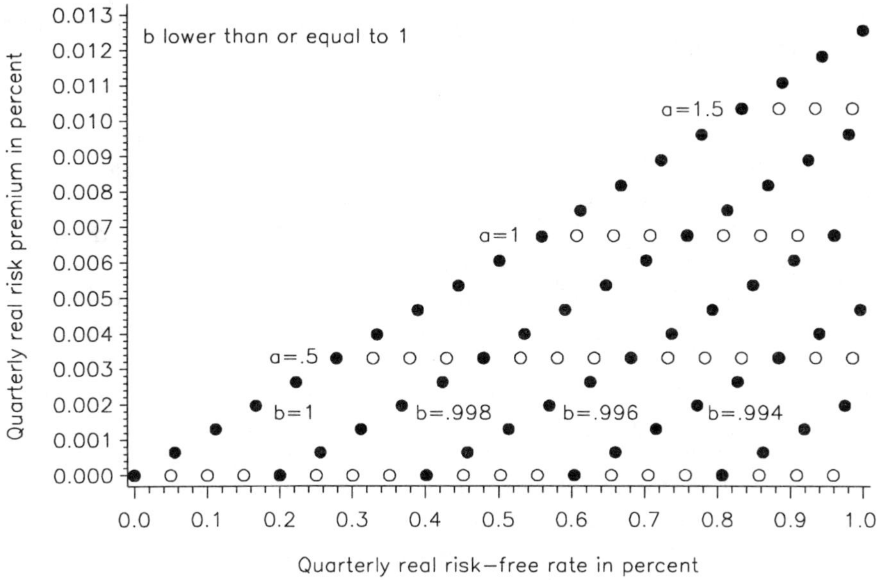

pairs for risk-free rates below one percent per quarter[18]. It is evident that no risk premium larger than .013 percent per quarter can be obtained with the specified equilibrium model for risk-free rates below one percent per quarter. The size of this premium equals only two percent of the empirically observed equity premium. This observation was first reported by Mehra and Prescott (1985, figure 4) for the US, conducting a comparable calibration exercise. It therefore illustrates the "original" Equity Premium Puzzle. Mehra and Prescott observed that with time-additive expected utility and constant relative risk aversion no equity premium beyond .35 percent per annum can be obtained for a risk-free rate below four percent per annum. Their maximum equity premium equals about five percent of their empirically observed value.

Even an increase in the parameter of relative risk aversion, α, does not help to resolve this puzzle. The highest risk premium for risk-free rates below one percent per quarter is obtained using $\alpha = 1.8$ and $\beta = 1$. An increase in α does not only increase the risk premium but also increases the risk-free rate. In other words, for risk-free rates below one percent no plausible risk premium can be obtained, even with implausibly high parameters of risk

[18]Notice that as in the previous chapter Greek characters are replaced by Arabic characters in the figures.

aversion[19]. High risk premia obtained with implausibly high parameters of risk aversion also imply implausibly high risk-free rates. This observation illustrates the previously discussed Risk-free Rate Puzzle. Decreasing β yields higher risk-free rates but does not substantially change the risk premia. Thus, the first moments of the historically observed returns cannot be matched simultaneously.

These results show that for time-additive expected utility an increase in risk aversion does not only increase the risk premium but also increases the risk-free rate. In section 7.2.1 a similar observation has been made applying the variance bound approach. There, it has been argued that this finding is a consequence of the fixed relation between relative risk aversion α and the elasticity of intertemporal substitution σ within this preference function. Since σ is restricted to be the reciprocal of α, an increase in α decreases σ, implying a higher risk-free rate. Hence, an increase in risk aversion α increases the risk-free rate and the risk premium simultaneously. Furthermore, it has been argued that as a consequence of the interrelation of α and σ a discount factor beyond one may be economically plausible and consistent with an equilibrium in this utility framework (see also section 5.2.4). This can now be illustrated. Consider a representative investor with constant time preference and constant elasticity of intertemporal substitution. An increase in his risk aversion α implicitly decreases the elasticity of intertemporal substitution σ and, hence, increases the risk-free rate. But, since the investor's time preference and elasticity of intertemporal substitution are assumed to be constant, the risk-free rate should not be strongly affected by the increase in risk aversion[20]. To keep the risk-free rate constant the discount factor β must be increased beyond one. Thus, for high parameters of risk aversion discount factors larger than one may be plausible because they may contain a compensation for the implicitly chosen low elasticity of intertemporal substitution[21].

---

[19] Parameters of relative risk aversion lower than ten are usually accepted as plausible. For a discussion of the plausible level of risk aversion see, for example, Mehra and Prescott (1985, p. 154).

[20] With constant time preference and constant elasticity of intertemporal substitution, the risk-free rate may be slightly affected by an increase in risk aversion since an investor who is more risk averse will invest less in risky assets. In addition to this well-known effect the precautionary savings effect may affect the risk-free rate (cf. Kimball (1990)). Kimball (1991) distinguishes three kinds of precautionary savings effects. (1) Prudence, which means that investors may accumulate more wealth in response to a risk. (2) Temperance, which means that investors may reduce exposure to other risks in response to an unavoidable risk, even if the risks are statistically independent. (3) Precautionary demand for liquidity, which means that facing a demand for liquidity investors may hold more money in response to a risk.

[21] This result is consistent with earlier findings. Recall that if the elasticity of intertemporal substitution σ is larger than the reciprocal of the parameter of risk aversion α, which is the value implicitly chosen in time-additive expected utility, the investors prefer an early resolution of uncertainty (see section 4.2.3.1). Experimental research has shown

Figure 8.4: Set of admissible quarterly equilibrium risk-free rates and risk premia using the estimated Markov switching model and **time-additive expected utility**. Discount factor β (b) ≤ 1.004 and risk aversion α (a) ≤ 10.

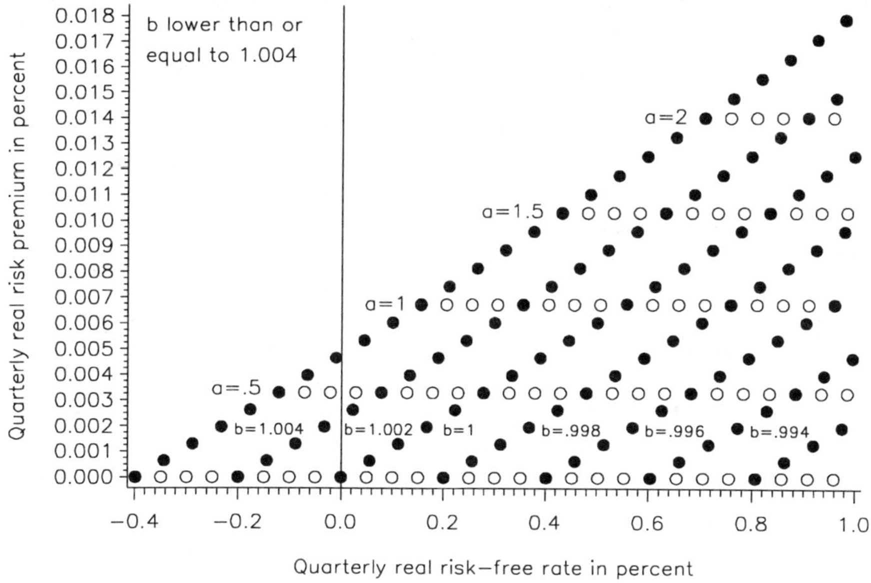

Figure 8.4 depicts the implied equilibrium rates of return for discount factor $\beta \leq 1.004$. The increase in β from one to 1.004 raises the obtainable risk premia by about 40 percent and, hence, risk premia up to .018 percent per quarter can be obtained for risk-free rates less than one percent per quarter. Thus, for a given risk-free rate the risk premium can be raised by simultaneously increasing the coefficient of relative risk aversion α and the discount factor β. For example, with $\alpha = 10$ and $\beta = 1.047$ the model implies a risk-free rate of .750 (1.134)[22] percent per quarter, a risk premium of .092 (1.287) percent per quarter and a correlation between the one-period risk-free rate and the risk premium of -.0169. The risk-free rate matches the empirically observed value for the taxation scenario S1 and also the correlation closely matches the historical value (see table 6.6). However, the risk premium and its standard deviation are still too low to match the empirically observed values. In addition, the standard deviation of the risk-free rate equals twice the historical value.

[22]The standard deviation implied by the model is given in parentheses.

8.2.2 Recursive Non-expected Utility

Now, the implied equilibrium rates of return using recursive non-expected utility are analyzed. Therefore, the implied returns are calculated for different combinations of the parameters of the preference function. At first, the discount factor β is set equal to .999; the relative risk aversion is chosen to be $\alpha \in \{.5, 2, 3, \ldots, 10\}$ and the elasticity of intertemporal substitution is $\sigma \in \{.02, .06, \ldots, 1.22\}$. Figure 8.5 illustrates the implied risk-free rate-risk premium pairs for risk-free rates below one percent per quarter. It is evident that for risk-free rates below one percent per quarter risk premia up to .068 percent per quarter can be obtained with relative risk aversion $\alpha \leq 10$ and discount factor $\beta = .999$.

The graph further shows that for constant elasticity of intertemporal substitution an increase in the level of risk aversion leads to an increase in the risk premium and also affects the risk-free rate. However, this effect is much smaller than for time-additive expected utility and, in contrast, decreases the risk-free rate. A small decrease in the risk-free rate as a result of an increase in risk aversion is plausible, because higher risk aversion may lead to an adjustment in the investor's portfolio to include more risk-free assets and less risky assets, resulting in a simultaneously increasing risk premium and a decreasing risk-free rate[23].

A change in the elasticity of intertemporal substitution clearly influences the risk-free rate and has nearly no effect on the risk premium. A high elasticity of intertemporal substitution implies a low risk-free rate, and vice versa. A high elasticity of intertemporal substitution means that β-adjusted consumption at different times, i.e. consumption levels that are adjusted for the investor's time preference, are nearly perfect substitutes to each other. Hence, considering β-adjusted consumption levels, an investor with a high elasticity of intertemporal substitution is indifferent between a smooth and a volatile consumption stream, whereas an investor with a low elasticity of intertemporal substitution prefers a smoother consumption path. Endowed with a relatively smooth consumption path, an investor with a low elasticity of intertemporal substitution is only willing to unsmooth his consumption by giving up one unit of consumption at date t if he gets more than one additional β-adjusted unit of consumption at date $t + 1$ for sure. Hence, a low elasticity of intertemporal substitution implies higher risk-free rates. Note that for $\beta = .999$ the elasticity of intertemporal substitution cannot be

[23] This effect is related to the precautionary savings effect (see footnote 20 in the previous section). The observation is consistent with the finding of Kimball and Weil (1992) that higher risk aversion tends to increase the strength of the precautionary savings motive using preferences which allow risk aversion and intertemporal substitution to be varied independently. For the precautionary savings motive in non-expected utility specifications see also Mauro (1995).

Figure 8.5: Set of admissible quarterly equilibrium risk-free rates and risk premia using the estimated Markov switching model and **recursive non-expected utility.** Discount factor β (b) = .999, risk aversion α (a) $\leq$ 10 and elasticity of intertemporal substitution σ (s) $\leq$ 1.22.

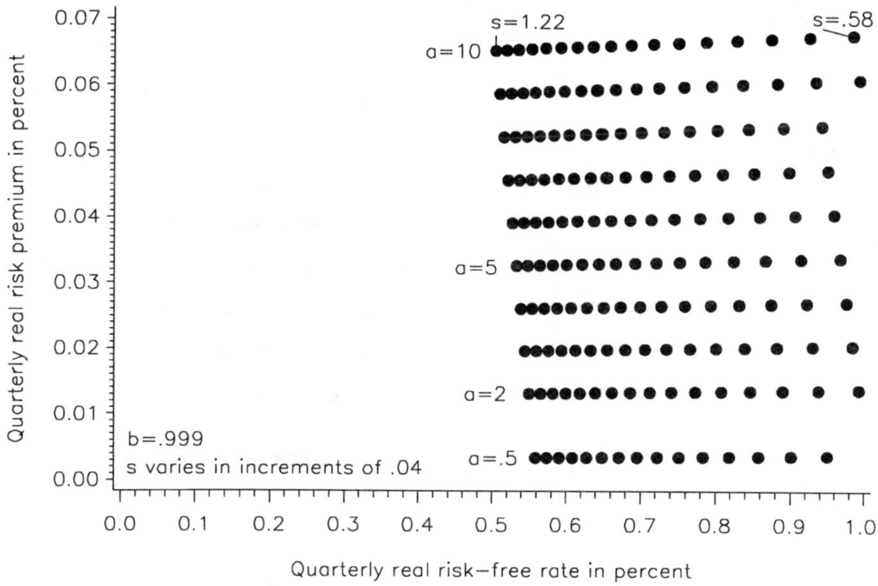

increased beyond 1.22 because then the numerical procedure to solve equation (8.2) does not converge. However, the figure suggests that a further increase in the elasticity of intertemporal substitution would not substantially reduce the risk-free rate. This may indicate that for a higher elasticity of intertemporal substitution no equilibrium exists, given only low positive time preference.

A decrease in β affects the results as expected. Figure 8.6 illustrates the same situation as figure 8.5 for $\beta = .997$ instead of $\beta = .999$[24]. The implied equilibrium risk-free rate increases with the decrease in β, whereas the implied risk premium is almost unaffected. With $\beta = .997$ the numerical procedure converges for values of the elasticity of intertemporal substitution up to 2.25. As can be seen, this increase in σ compensates for the lower discount factor β. Consider, for example, the parameter combination $\alpha = 10$, $\beta = .999$ and $\sigma = 1.15$ in figure 8.5 and the parameter combination $\alpha = 10$, $\beta = .997$ and $\sigma = 2.05$ in figure 8.6. The risk-free rates implied by both parameter combinations are almost equal, .528 percent and .527 percent per quarter, respectively. Hence, the same risk-free rate is implied by different combinations of β and σ. Thus, considering the risk-free rate,

[24]Note that σ now varies in increments of .1.

Figure 8.6: Set of admissible quarterly equilibrium risk-free rates and risk premia using the estimated Markov switching model and **recursive non-expected utility**. Discount factor β (b) = .997, risk aversion α (a) $\leq$ 10 and elasticity of intertemporal substitution σ (s) $\leq$ 2.25.

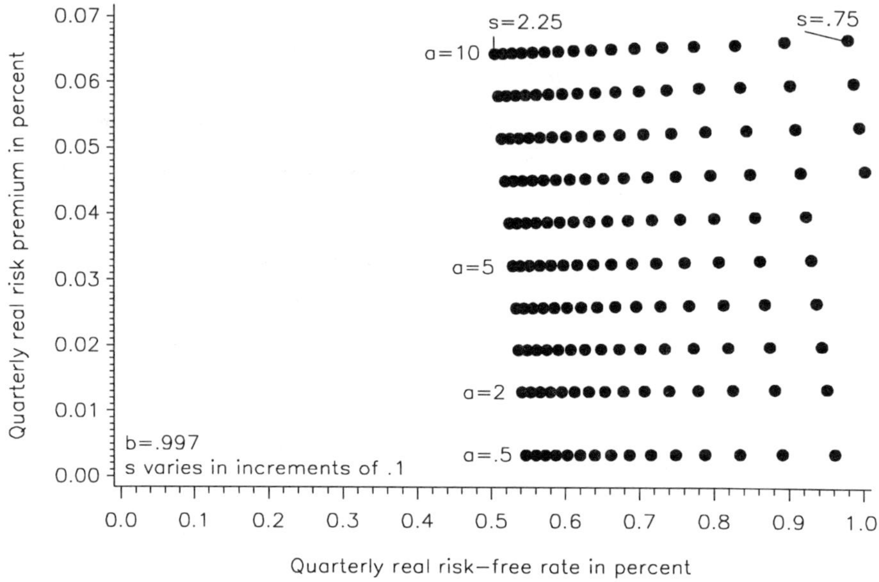

a change in one parameter can always be compensated by a change in the other parameter. However, the implied risk premia may differ across the different combinations of β and σ. The risk-premium implied by the first parameter combination, for example, is slightly larger than that implied by the second parameter combination.

The finding that the same risk-free rate can be obtained by different combinations of β and σ was used in the previous section to justify discount factors larger than one in time-additive expected utility. However, in section 5.2.4 it has been theoretically demonstrated that recursive non-expected utility and time-additive expected utility with an adjusted discount factor which ensures that both imply the same equilibrium risk-free rate, imply the same equilibrium risk premium only if consumption growth is i.i.d.. Therefore, in the analyzed economy with Markov switching consumption growth the risk premia differ. The extent of this effect becomes evident by comparing time-additive expected utility with α = 10 and β = 1.046 and recursive non-expected utility with α = 10, β = .999 and σ = .8. Both combinations imply the same risk-free rate of .73 percent per quarter. The first combination, however, implies a risk premium of .092 percent per quarter, whereas the second implies a risk premium of only .066 percent per quarter.

It has been shown that the Risk-free Rate Puzzle can either be solved by increasing the discount factor beyond one in time-additive expected utility to compensate for the implicitly chosen low elasticity of intertemporal substitution or by using recursive non-expected utility that allows a clear separation of risk aversion and the elasticity of intertemporal substitution. Since consumption growth is not i.i.d. both alternatives yield different equilibrium risk premia. From a theoretical point of view, recursive non-expected utility yields more satisfying results. The parameter of risk aversion determines the risk premium and the elasticity of intertemporal substitution in combination with the discount factor determine the risk-free rate. However, as in the case of time-additive expected utility, with risk aversion $\alpha \leq 10$ the attainable risk premia are far too low to match the empirically observed equity premium.

8.2.3 Summary

Using time-additive expected utility with constant relative risk aversion and a discount factor β below one the implied risk premium does not exceed .013 percent per quarter for risk-free rates below one percent per quarter. This amount equals only two percent of the empirically observed value. Increasing the parameter of risk aversion simultaneously increases the risk premium and the risk-free rate due to the fixed relation of risk aversion and elasticity of intertemporal substitution. This interrelation justifies the use of discount factors beyond one as a compensation for the low elasticity of intertemporal substitution implicitly chosen by high levels of risk aversion. With β larger than one, equity premia up to .09 percent per quarter can be obtained with a parameter of relative risk aversion below ten. But, it has been shown that choosing β beyond one instead of parametrizing time preference and elasticity of intertemporal substitution separately may yield a different risk premium. Therefore, using recursive non-expected utility gives more satisfying results from a theoretical point of view. With recursive non-expected utility a change in risk aversion affects the risk premium but has nearly no effect on the risk-free rate; a change in the discount factor or in the elasticity of intertemporal substitution affects the risk-free rate, leaving the risk premium almost unchanged. However, also with recursive non-expected utility the empirically observed equity premium is still not obtainable with plausible parameters of risk aversion. For a parameter of relative risk aversion of ten the implied risk premium equals only about .07 percent per quarter. Hence, in the specified economy both utility specifications do not allow to simultaneously match the first moments of the empirically observed one-period risk-free rate and the equity premium.

Up to now, the observed equity premium has been compared with the implied equilibrium risk premium on an asset whose dividends equal both consumption and production, i.e. stocks have been modeled as a claim on the exogenous consumption/production stream. The assumption that dividends on equity equal consumption is equivalent to the assumption that aggregate equity, i.e. the aggregate stock market, equals total wealth in the economy. This assumption is unrealistic, because it ignores human capital and debt. Therefore, the model is now generalized by modeling the claim on equity as a levered claim on the market portfolio. This may improve the fit of the model.

8.3 Analysis of the Levered Market Portfolio

8.3.1 Equity Equals Levered Market Portfolio

The analysis in the previous section has shown that the model cannot be calibrated such that the implied equilibrium rate of return on the aggregate wealth portfolio (market portfolio) matches the empirically observed returns on the stock market. However, the return on the stock market represents only the return on equity which is only a fraction of aggregate wealth[25]. The return on human capital and debt, for example, is not represented. Hence, equity does not entitle its owners to aggregate production in reality, but it entitles its owners to the residual claim on aggregate production after all other claims such as wages or interest payments have been deducted. Because the other claims are more or less fixed claims, the aggregate payoff on equity is more risky than the payoff on the aggregate wealth portfolio. This may be a reason for the model's inability to imply risk premia as high as the empirically observed equity premia.

Now, the aggregate payoff on equity is set equal to the aggregate production minus the payments on fixed claims[26]. Wages, representing the payoff on human capital, are assumed to be negotiated before the production is realized and, thus, are fixed claims. Furthermore, the aggregate productive capital, i.e. aggregate wealth minus human capital, is assumed to be financed with debt and equity. Aggregate wealth therefore consists of human capital, equity and debt. Since wages are assumed to be fixed claims such as interest

[25] Notice that subsequently equity does not only represent the stock market but also encloses corporations that are not listed at the stock market as well as unincorporated enterprises. It is assumed that the returns on these firms equal the return on the stock market.

[26] This approach is related to studies including human capital when measuring the return on the aggregate wealth portfolio for tests of CAPM-type models, e.g., Jagannathan, Kubota, and Takehara (1995) and Jagannathan and Wang (1996). These studies find that the inclusion of human capital improves the performance of the models.

payments on debt, human capital is also interpreted as debt. Equity therefore equals the levered market portfolio. It is assumed that at every date t a constant proportion L, $0 \leq L < 1$, of the market portfolio with ex-dividend value P_t^M is financed with one-period debt[27]. The ex-dividend value of equity at date t is therefore given by $P_t^E = P_t^M(1-L)$. Let m_{t+1} again denote the aggregate production (= consumption = payoff on the aggregate wealth portfolio) at date $t+1$. Then, at date $t+1$, the before dividend value of equity equals the before dividend value of the aggregate wealth portfolio, $P_{t+1}^M + m_{t+1}$, minus the value of debt including the aggregate interest payments on debt[28], $L \cdot P_t^M \cdot R_{t+1}^f$.

Before the implied rates of return on equity, i.e. on the levered market portfolio, are analyzed it has to be clarified which level of leverage is reasonable. Hints can be gained from two sources, from macroeconomic data as well as from data on the individual firms' level.

First, have a look at macroeconomic data[29]. The fraction of the German national income from property and entrepreneurial income was 29.9 percent on average over the years 1960 to 1993[30]. However, interest payments are included in this fraction. The aggregate dividend payments of German corporations plus the distributed earnings of unincorporated German enterprises received by German households was equal to 21.3 percent of the national income on average. This fraction partly includes payoffs on human capital because the owners of unincorporated enterprises do not receive wages apart from the firms' earnings. Furthermore, these values are proportions of payoffs rather than of the aggregate wealth. Taking into account that the payoff on equity is more risky than the payoff on aggregate wealth and, therefore, should include a higher risk premium, the fraction of equity on aggregate wealth should be less than 20 percent. Leverage L should therefore be .8 or larger.

Second, examine some data on the micro level. The Deutsches Aktieninstitut (1996, p. FB_04-1) reports an average ratio of equity to total capital of German enterprises of 21.73 percent over the years 1966 to 1994. But what is needed, is the ratio of equity to the firms' total wealth including human capital. Consider that on average the firms' payoffs for wages are 10 to 15 times larger than the firms' interest payments on debt[31]. Hence,

[27]It is assumed that bankruptcy does not occur, so that the interest rate on debt equals the risk-free rate. Hence, in contrast to the model without leverage, where the risk-free asset has been in zero net supply, the risk-free asset is now in positive net supply.

[28]Note that debt includes human capital and that the interest payments therefore include wages.

[29]The macroeconomic data are taken from the Statistisches Bundesamt. East German data are not included.

[30]This value is comparable to the ones for the USA and Japan reported by Jagannathan and Wang (1996, p. 13), respectively Jagannathan, Kubota, and Takehara (1995, p. 4).

[31]These values can be confirmed, for example, by comparing the average fraction of

assuming that the return on human capital equals the return on debt, the firms' human capital should amount 10 to 15 times the firms' debt. From these values the proportion of equity on the firms' total wealth can be approximated. Assume, for example, a firm with 25 equity and 75 debt. This firm's human capital equals at least $10 \cdot 75 = 750$. Therefore, conservatively calculated, equity accounts for only three percent ($25/850 \approx 0.03$) of the firm's total wealth[32] [33]. Thus, the leverage parameter L should be .95 or higher. Considering this, two different levels of leverage, L, will be used, 80 percent and 95 percent, respectively.

8.3.2 Implied Equilibrium Return on Equity

For the analyzed state-switching economy, the expected gross return on equity from state i to state j, $\mathrm{E}\{\tilde{R}^E_{ij}\}$, is given by

$$\begin{aligned} \mathrm{E}\{\tilde{R}^E_{ij}\} &= \mathrm{E}\{\tilde{R}^M_{ij}\} + (\mathrm{E}\{\tilde{R}^M_{ij}\} - R^f_i)\frac{L}{1-L} \\ & \quad \text{for} \quad i,j \in \{1,2\} \quad \text{and} \quad 0 \le L < 1. \end{aligned} \tag{8.16}$$

L is the proportion of the market portfolio (aggregate wealth portfolio) financed with debt, $\mathrm{E}\{\tilde{R}^M_{ij}\}$ is the implied equilibrium gross return on the market portfolio in a period from state i to state j as calculated in equation (8.5) and R^f_i is the implied equilibrium risk-free rate of return as calculated in equation (8.9). The unconditional expected return on equity then is

$$\mathrm{E}\{\tilde{R}^E\} = \sum_{i=1}^{2}\sum_{j=1}^{2} \mathrm{prob}_i \mathrm{prob}_{ij} \mathrm{E}\{\tilde{R}^E_{ij}\}. \tag{8.17}$$

The conditional variance of the return on equity from state i to state j is calculated as

$$\mathrm{var}(\tilde{R}^E_{ij}) = (1-L)^{-2}\mathrm{var}(\tilde{R}^M_{ij}). \tag{8.18}$$

The unconditional variance of the return on equity is

$$\mathrm{var}(\tilde{R}^E) = \sum_{i=1}^{2}\sum_{j=1}^{2} \mathrm{prob}_i \mathrm{prob}_{ij} \left(\mathrm{var}(\tilde{R}^E_{ij}) + \left(\mathrm{E}\{\tilde{R}^E_{ij}\} - \mathrm{E}\{\tilde{R}^E\}\right)^2\right). \tag{8.19}$$

the firms' sales paid off for interests and the average fraction paid off for wages reported by Herrmann (1996, table 27) for the years 1990 to 1992.

[32] This value corresponds to the finding of Cecchetti, Lam, and Mark (1993) that "during the twentieth century total dividends have averaged between 3% and 5% of aggregate consumption" (Cecchetti, Lam, and Mark 1993, p. 22). Also Jagannathan, Kubota, and Takehara (1995, p. 4) and Jagannathan and Wang (1996, p.13) report that dividends amount to only three percent of income in the USA and Japan.

[33] Less conservatively calculated, for example with 20 equity, 80 debt and $15 \cdot 80 = 1200$ human capital, equity accounts for only 1.5 percent of the firm's total wealth.

The conditional expected equity premium from state i to state j is given by

$$\mathrm{E}\{\tilde{p}_{ij}^E\} = \mathrm{E}\{\tilde{R}_{ij}^E\} - R_i^f \quad \text{for} \quad i,j \in \{1,2\}.$$

The unconditional expectation of the equity premium follows as

$$\mathrm{E}\{\tilde{p}^E\} = \sum_{i=1}^{2}\sum_{j=1}^{2} \mathrm{prob}_i \mathrm{prob}_{ij} \mathrm{E}\{\tilde{p}_{ij}^E\}. \tag{8.20}$$

The conditional variance of the equity premium, $\mathrm{var}(\tilde{p}_{ij}^E)$, equals the conditional variance of the return on equity, $\mathrm{var}(\tilde{R}_{ij}^E)$. The unconditional variance of the equity premium then is

$$\mathrm{var}(\tilde{p}^E) = \sum_{i=1}^{2}\sum_{j=1}^{2} \mathrm{prob}_i \mathrm{prob}_{ij} \left(\mathrm{var}(\tilde{R}_{ij}^E) + \left(\mathrm{E}\{\tilde{p}_{ij}^E\} - \mathrm{E}\{\tilde{p}^E\}\right)^2\right). \tag{8.21}$$

The unconditional covariance of the equity premium and the one-period risk-free rate is given by

$$\mathrm{cov}(\tilde{p}^E, \tilde{R}^f) = \sum_{i=1}^{2} \mathrm{prob}_i \left(\mathrm{E}\{\tilde{p}_i^E\} - \mathrm{E}\{\tilde{p}^E\}\right) \left(R_i^f - \mathrm{E}\{\tilde{R}^f\}\right). \tag{8.22}$$

8.3.3 Calibration Results

8.3.3.1 Time-additive Expected Utility

The analysis is started using time-additive expected utility. Figure 8.7 exhibits the implied equilibrium rates of return for discount factor $\beta = 1$ and leverage L equal to 0, .8 and .95, respectively. Since the highest risk premia are obtained for discount factor $\beta = 1$, only those implied returns are presented. Holding risk aversion α and discount factor β constant, it is evident that an increase in leverage leads to higher risk premia. However, with leverage $L \leq .95$ and $\beta \leq 1$ no risk premium larger than .26 percent per quarter can be obtained for risk-free rates below one percent per quarter. Thus, exclusively introducing leverage in time-additive expected utility is not sufficient to explain the empirically observed equity premium.

In addition to leverage, the discount factor β is now allowed to increase beyond one. Figure 8.8 depicts the implied equilibrium rates of return for discount factor $\beta = 1.03$ and leverage L equal to 0, .8 and .95, respectively. The empirically observed risk-free rate-equity premium pairs for the four taxation scenarios are represented by heart symbols. It is evident that the scenarios S1, S2 and S4 exhibit lower risk premia than those obtainable with

Figure 8.7: Set of admissible equilibrium risk-free rates and risk premia using the estimated Markov switching model, **time-additive expected utility** and **leverage**. Three levels of leverage are compared for α (a) ≤ 10 and β (b) $= 1$.

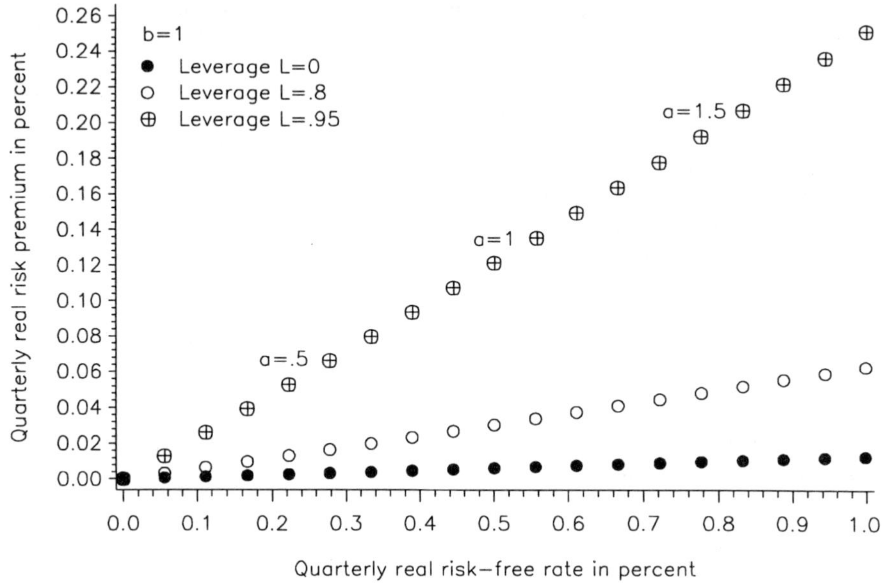

$\beta = 1.03$ and $L = .95$. Scenario S3 can be matched after a further increase in β. A further increase in β would also allow to match scenario S4 with the lower level of leverage, $L = .8$. The risk premia of the other scenarios cannot be obtained with $L = .8$ and $\alpha \leq 10$. The finding that with leverage and discount factors larger than one the Equity Premium Puzzle can be resolved is similar to the results of Benninga and Protopapadakis (1990).

Now, as the first moments of the empirically observed risk-free rate and equity premium can be matched by the implied values, it is interesting to compare the second moments. Table 8.3 lists the first and second moments and the correlation of the one-period risk-free rate and the equity premium for both the empirical observations calculated from quarterly averages (Q2) and the values implied by the model with time-additive expected utility and leverage. The preference parameters are chosen such that the first moments of the implied equilibrium rates of return exactly match those of the empirically observed rates of return. Comparing row 1 with row 2, for example, shows that the average risk-free rate and equity premium of taxation scenario S1 can exactly be obtained with relative risk aversion $\alpha = 4.26$, discount factor $\beta = 1.0159$ and leverage $L = .95$ (model UE1). In general, for $L = .95$, reasonable values of relative risk aversion between 2.00 and 6.18 are sufficient to generate the historical equity premia. For taxation

Figure 8.8: Set of admissible equilibrium risk-free rates and risk premia using the estimated Markov switching model, **time-additive expected utility** and **leverage.** Three levels of leverage are compared for α (a) $\leq$ 10 and β (b) = 1.03. The empirically observed quarterly average returns (Q2) for the four taxation scenarios are represented by heart symbols.

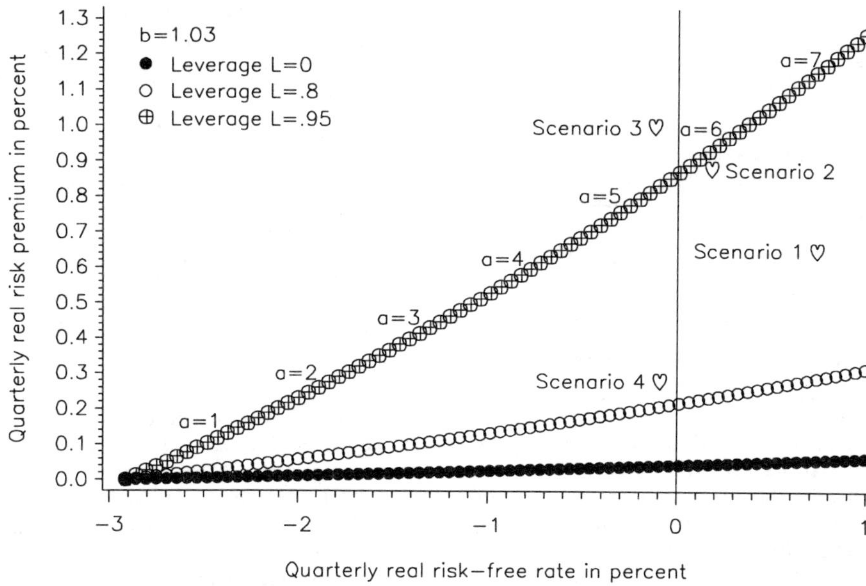

scenario S1 the implied standard deviation of the risk-free rate is close to that empirically observed, for the scenarios S2 and S3 it is about 50 percent larger and for scenario S4 it is smaller than the observed value. For all scenarios, the implied standard deviation of the equity premium is far in excess of the empirically observed value. The models imply a slightly negative correlation between the risk-free rate and the equity premium, whereas, except for scenario S1, the empirically observed correlation is slightly positive. Overall, considering the second moments it seems as if scenario S1 can be better matched than the other scenarios. Comparing the moments of scenario S4 implied by model UE4a (L = .95) with those implied by model UE4b (L = .8) shows that the implied standard deviation of the equity premium is much smaller for L = .8 than for L = .95. However, it is still larger than the empirically observed value.

So far it has been found that models calibrated to match the first moments of the empirically observed risk-free rate and the empirically observed equity premium do not exactly match the empirically observed second moments and the empirically observed correlation between the risk-free rate and the equity premium. However, this does not necessarily mean that the observed rates of return are inconsistent with the calibrated models and the un-

Table 8.3: First and second moments of real rates of return as implied by equilibrium models with **time-additive expected utility** and **leverage** in comparison with the corresponding moments of empirically observed quarterly average rates of return (Q2). The rates of return are given in percent per quarter.

Scenario[a]		Model parameter[b]			Risk-free rate		Equity premium		corr
		α	β	L	$E\{\tilde{r}^f\}$	$std(\tilde{r}^f)$	$E\{\tilde{p}^E\}$	$std(\tilde{p}^E)$	$(\tilde{r}^f, \tilde{p}^E)$
S1	observed				.74	.54	.65	7.58	-.032
UE1	implied[c]	4.26	1.0159	.95	.74	.47***	.65	19.12***	-.006
S2	observed				.18	.42	.88	7.52	.076
UE2	implied[c]	5.54	1.0285	.95	.18	.61***	.88	20.32***	-.008
S3	observed				-.12	.40	1.00	7.50	.123
UE3	implied[c]	6.18	1.0350	.95	-.12	.68***	1.00	20.96***	-.010
S4	observed				-.09	.40	.28	3.57	.098
UE4a	implied[c]	2.00	1.0120	.95	-.09	.22***	.28	17.05***	-.002
UE4b	implied[c]	6.80	1.0380	.80	-.09	.75***	.28	5.42***	-.011

[a]Scenario S1 is the zero tax scenario, S2 is the scenario of a private investor with a marginal tax rate of 36 percent, S3 is the scenario of a private investor with the maximum marginal tax rate and S4 is the scenario of a corporate investor. The observed rates of return for these scenarios are taken from table 6.8. UE1 to UE4b denote the calibrated models; U: univariate exogenous process, E: time-additive expected utility, 1 to 4: the matched taxation scenario.

[b]α is the parameter of relative risk aversion, β is the discount factor and L is the level of leverage.

[c]***, (**, *) indicates that the hypothesis that the observed value is equal to the implied value can be rejected at a 1 (5, 10) percent level of significance on the basis of 10,000 simulations of the calibrated model.

derlying state-switching endowment process. To obtain statistical inference Monte Carlo simulations are conducted employing the 10,000 simulated time series of the state variable and the endowment variable generated in section 8.1.3.3[34]. For each of the five calibrated models given in table 8.3 the corresponding 10,000 time series of the risk-free rate and the equity premium are calculated. This is done as follows. (1) Given the time series of the state variable and the state-dependent risk-free rates for the calibrated model (see equation (8.9)), the simulated time series of the risk-free rate is obtained. (2) Given Θ_1 and Θ_2 for the calibrated model (see equations (8.2) and (8.4)) and given the time series of the state variable and of the endowment variable λ, the time series of the return on the market is calculated using equation (8.1). (3) Given the level of leverage, in each period the return on equity

[34]Simulation methods for testing calibrated equilibrium asset pricing models have also been implemented by Cecchetti, Lam, and Mark (1990), Gregory and Smith (1991), and Bonomo and Garcia (1994), (1996).

Table 8.4: Means and fractiles of the simulated distributions of the first and second moments and the correlation between the risk-free rate and the equity premium implied by the calibrated model UE1. The simulation consists of 10,000 time series with 139 periods.

		Simulation Fractiles						
	Mean	.5 %	2.5 %	5 %	50 %	95 %	97.5 %	99.5 %
$E\{\tilde{r}^f\}$ [%]	.7405	.6626	.6786	.6946	.7426	.7906	.7986	.8146
$std(\tilde{r}^f)$ [%]	.4700	.4129	.4267	.4395	.4726	.4989	.5027	.5098
$E\{\tilde{p}^E\}$ [%]	.6706	-3.4746	-2.5282	-1.9983	.6845	3.3193	3.8150	4.7409
$std(\tilde{p}^E)$ [%]	19.0977	16.2170	16.8804	17.2155	19.0871	21.0057	21.3678	22.0378
$corr(\tilde{r}^f, \tilde{p}^E)$	-.0051	-.1941	-.1470	-.1235	-.0041	.1145	.1377	.1774

is calculated from the simulated return on the market and the simulated risk-free rate using a formula similar to equation (8.16). (4) Subtracting the simulated risk-free rate from the simulated return on equity in each period yields the simulated time series of the equity premium. The distributions of the first and second moments and the correlation of the risk-free rate and the equity premium are calculated from these 10,000 simulated time series. For illustration, the means and fractiles of these simulated distributions are listed in table 8.4 for the calibrated model UE1. Comparing the simulated distributions with the historical values yields uniform results for all five models. The hypotheses that the first moments and the correlation of the risk-free rate and the equity premium are generated by the calibrated model and the underlying state-switching endowment process cannot be rejected, whereas for both second moments the hypothesis can be rejected at a one percent level of significance.

To summarize, using time-additive expected utility with leverage and discount factors beyond one the model can be calibrated to exactly imply the first moments of both the empirically observed risk-free rate and equity premium. With leverage equal to .95 the required levels of relative risk aversion fall into a reasonable range between two and seven. With a lower level of leverage the parameter of relative risk aversion required to match the first moment of the equity premium rises beyond 15, except for taxation scenario S4. The implied correlation between the risk-free rate and the equity premium does not significantly differ from the empirically observed value. On the other hand, the implied second moment of the risk-free rate as well as the implied second moment of the equity premium are significantly different from the empirically observed values. Especially the implied second moment of the equity premium is unreasonably large. In all cases the discount factor is larger than one. It has already been discussed that discount factors beyond one may be plausible in the time-additive expected utility framework

Figure 8.9: Set of admissible equilibrium risk-free rates and risk premia using the estimated Markov switching model, **recursive non-expected utility** and **leverage.** α (a) $\leq$ 10, β (b) = .999, σ (s) $\leq$ 1.22 and L = .8. The empirically observed quarterly average returns (Q2) for the four taxation scenarios are represented by heart symbols.

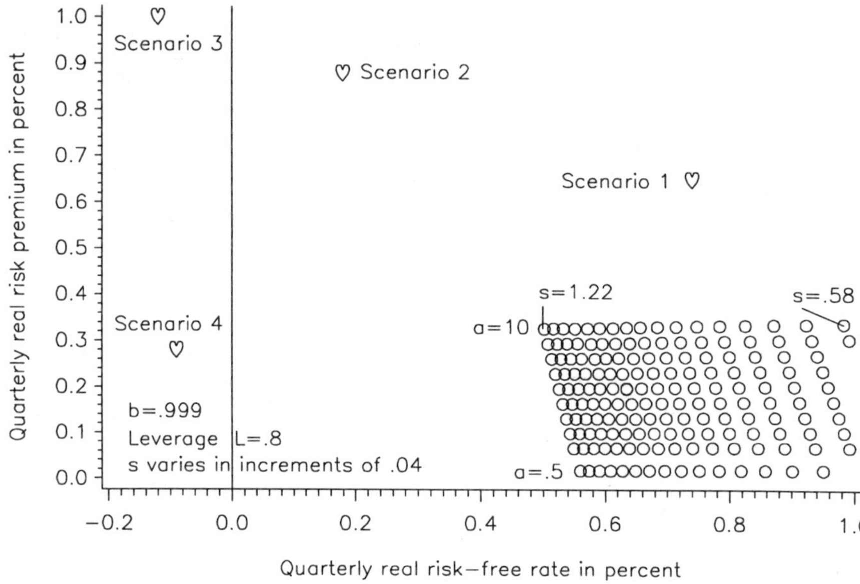

to compensate for the low elasticity of intertemporal substitution implicitly chosen by the level of risk aversion. But, the elasticity of intertemporal substitution now implicitly chosen by the combination of the parameter of risk aversion and the discount factor beyond one may also be unreasonable. To investigate this, the model with recursive non-expected utility is analyzed next.

8.3.3.2 Recursive Non-expected Utility

Now, the recursive non-expected utility specification is applied. Figure 8.9 exhibits the implied rates of return for leverage $L = .8$ and discount factor $\beta = .999$. The heart symbols represent the empirically observed risk-free rate-equity premium pairs for the four taxation scenarios. Although leverage increases the risk premia (see figure 8.5 for the case with $L = 0$), with $L = .8$ and relative risk aversion below ten the maximum attainable risk premium is only about .3 percent per quarter. This value suffices only to match the equity premium of taxation scenario S4. With an elasticity of intertemporal substitution below 1.22 no risk-free rate below .5 percent per quarter can

Figure 8.10: Set of admissible equilibrium risk-free rates and risk premia using the estimated Markov switching model, **recursive non-expected utility** and **leverage.** α (a) $\leq$ 10, β (b) = .999, σ (s) $\leq$ 1.22 and L = .95. The empirically observed quarterly average returns (Q2) for the four taxation scenarios are represented by heart symbols.

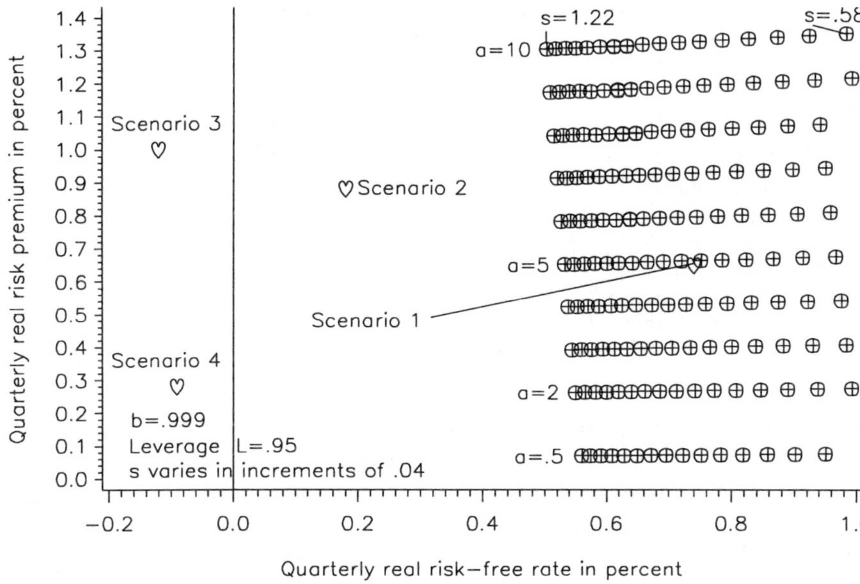

be obtained, so that also the risk-free rate-equity premium pair of scenario S4 cannot be generated.

Figure 8.10 exhibits the situation with leverage $L = .95$. As in the case of time-additive expected utility, considerably larger risk premia are obtained using $L = .95$ as compared to $L = .8$. The equity premia of all scenarios can be matched with parameters of relative risk aversion below ten. However, as in the cases with $L = 0$ and $L = .8$, an elasticity of intertemporal substitution σ below 1.22 implies risk-free rates above .5 percent per quarter. Hence, only the risk-free rate-equity premium pair of scenario S1 lies in the admissible region. As noted in section 8.2.2, a further increase in the elasticity of intertemporal substitution would not substantially reduce the risk-free rate. Thus, with recursive non-expected utility and leverage the empirically observed rates of return of the scenarios S2, S3 and S4 cannot be matched, given a positive time preference. As in the case of time-additive expected utility, a discount factor larger than one may help to reduce the equilibrium risk-free rate[35]. However, unlike the situation with time-additive

[35]This finding is consistent with the estimation results of Epstein and Zin (1991, tables 2-5). They directly estimate the parameters of the recursive non-expected utility

Table 8.5: First and second moments of real rates of return as implied by equilibrium models with **recursive non-expected utility** and **leverage** in comparison with the corresponding moments of empirically observed quarterly average rates of return (Q2). The rates of return are given in percent per quarter.

Scenario[a]		Model parameter[b]				Risk-free rate		Equity premium		corr
		α	β	σ	L	$E\{\tilde{r}^f\}$	std($\tilde{r}^f$)	$E\{\tilde{p}^E\}$	std($\tilde{p}^E$)	($\tilde{r}^f,\tilde{p}^E$)
S1	observed					.74	.54	.65	7.58	-.032
UN1a	implied[c]	5.00	.9970	1.20	.95	.74	.09***	.65	16.41***	-.005
UN1b	implied[c]	4.90	.9990	.83	.95	.74	.13***	.65	16.64***	-.005
UN1c	implied[c]	4.85	1.0000	.72	.95	.74	.15***	.65	16.76***	-.005

[a]See footnote *a* in table 8.3. UN1a to UN1c denote the calibrated models; U: univariate exogenous process, N: recursive non-expected utility, 1: the matched taxation scenario.
[b]See footnote *b* in table 8.3. σ is the elasticity of intertemporal substitution.
[c]See footnote *c* in table 8.3.

expected utility, with recursive non-expected utility that clearly separates risk aversion from the elasticity of intertemporal substitution no economic reason can be found to justify this step. Instead, these results indicate that the discount factors beyond one chosen for time-additive expected utility may be too large to be solely justified as compensation for the implicitly chosen low elasticity of intertemporal substitution.

The analysis of the second moments is confined to the zero tax scenario (S1) because, as discussed above, the first moments of the other scenarios cannot be matched with recursive non-expected utility and leverage. Table 8.5 compares the second moments of the implied rates of return with those of the empirically observed rates of return. The implied values are given for three different parameter combinations, which allow to exactly match the first moments of the risk-free rate and the equity premium. In one of these calibrated models the discount factor is $\beta = .997$ (UN1a), in one it is $\beta = .999$ (UN1b) and in the third it is $\beta = 1$ (UN1c). The required parameter of relative risk aversion is 5, 4.9 and 4.85 and the elasticity of intertemporal substitution is 1.2, .83 and .72, respectively. The levels of relative risk aversion seem reasonable. Similar to the findings in section 8.2.2, different combinations of β and σ imply the same risk-free rate. But, given a constant level of risk aversion, these combinations do not imply the same equity premium. Instead, the level of risk aversion must be adjusted to imply the same equity premium. It must be clarified which combination of β and σ is the most reasonable. Taking into account the discussion in sec-

specification from US data applying the GMM estimation technique and find that the estimated time preference is often significantly less than zero.

tion 7.2.3, only values of the elasticity of intertemporal substitution below one can be accepted as reasonable. Hence, in combination with the reasonability of a slightly positive time preference, the model UN1b seems most reasonable. In contrast to the time-additive expected utility, the implied standard deviation of the risk-free rate is too small to match the empirically observed value. The implied standard deviation of the equity premium is slightly smaller than for time-additive expected utility but it is still more than twice the observed value. Similar to the time-additive expected utility a slightly negative correlation between the risk-free rate and the equity premium is implied, which corresponds with the empirically observed correlation. Statistical inference is gained by conducting Monte Carlo simulations similar to those in the previous section. As for the case with time-additive expected utility, the hypotheses that the second moments of the empirically observed rates of return are generated by the calibrated model and the underlying state-switching process can be rejected at a one percent level of significance.

8.3.4 Summary

The implied rates of return on equity have been analyzed. Equity was defined as a levered market portfolio, that means equity was set equal to a constant fraction of the aggregate wealth portfolio. The remaining parts of the aggregate wealth portfolio, i.e. human capital and debt, were assumed to be fixed claims with rates of return equal to the one-period risk-free rate.

For both, time-additive expected utility and recursive non-expected utility, leverage substantially increases the implied risk premia. With leverage equal to .95 the parameters of relative risk aversion required to generate the empirically observed equity premium lie in a plausible range between two and six. Matching the observed risk-free rates with time-additive expected utility necessitates discount factors in a range between 1.01 and 1.035. With recursive non-expected utility and a discount factor of .999 only the risk-free rate of scenario S1 can be matched with a reasonable value of the elasticity of intertemporal substitution. The risk-free rates of the other scenarios cannot be matched. This indicates that although discount factors beyond one in time-additive expected utility may be plausible, those necessary to imply the risk-free rates of the scenarios S2 to S4 may be too high to be justified.

Further analyses of calibrated models that allow to exactly match the first moments of the empirically observed rates of return show that the models with leverage are not capable to match the first and second moments and the correlation of the risk-free rate and the risk premium simultaneously, neither with time-additive expected utility and discount factors beyond one, nor with recursive non-expected utility. For both utility specifications the implied standard deviation of the risk premium is far in excess of the empirically observed value. For recursive non-expected utility the implied standard deviation of the risk-free rate is too small, whereas for time-additive expected utility it exhibits quite reasonable values. Nevertheless, Monte Carlo simulations show that the hypotheses that the empirically observed second moments are generated by the calibrated model and the underlying state-switching endowment process can be rejected at a one percent level of significance in all cases.

So far, the sum of human capital and debt has been assumed to equal a constant fraction L of the market portfolio, which yields the one-period risk-free rate. Although this model allows to relax the assumption of the equality of consumption, i.e. the payoff on the aggregate wealth portfolio, and the aggregate payoff on equity, it imposes a strong structure. For example, the rates of return on human capital and on debt are restricted to be equal and, hence, to be perfectly correlated. But more important, the low variability of the risk-free rate together with the low correlation between the risk-free rate and the return on equity implied by the model restricts the return on the aggregate wealth portfolio and the return on equity to be strongly correlated. To illustrate this, consider again the parameter combinations in table 8.3 and table 8.5 which exactly match the first moments of the implied rates of return to those of empirically observed returns. The correlation between the return on the market and the return on equity implied by these parameter combinations varies in a range between .87 and .99[36]. Moreover, four out of the eight values lie above .97. This strong correlation between the return on equity and the return on the aggregate wealth portfolio implies a strong correlation between aggregate production and the aggregate payoff on equity. In reality, however, such a strong correlation does not seem reasonable. Wages, for example, may be negatively correlated with the payoff on equity because for a given aggregate production, increasing wage costs reduce corporate earnings and therefore reduce the payoff on equity, and vice versa. Therefore, it is now examined whether a model that does not presume a strong correlation between the payoff on the aggregate wealth portfolio and the aggregate payoff on equity allows to better match the moments of the observed rates of return.

[36]The correlation between the return on the aggregate wealth portfolio and the return on equity can be calculated as follows: $\text{corr}(\tilde{R}^M, \tilde{R}^E) = (L \cdot \text{corr}(\tilde{R}^E, \tilde{R}^f)\text{std}(\tilde{R}^f) + (1 - L)\text{std}(\tilde{R}^E))/\text{std}(\tilde{R}^M)$, where $\text{corr}(\tilde{R}^E, \tilde{R}^f) = (\text{corr}(\tilde{p}^E, \tilde{R}^f)\text{std}(\tilde{p}^E) + \text{std}(\tilde{R}^f))/\text{std}(\tilde{R}^E)$. $\text{std}(\tilde{R}^M)$ is the standard deviation of the risky asset for $L = 0$. If the risk-free rate is constant, the return on the aggregate wealth portfolio is perfectly correlated with the return on equity.

8.4 Separating Consumption and Dividends

8.4.1 The Bivariate Model

Both models that have been analyzed until now, the model without leverage and the model with leverage, imply that the aggregate production (= consumption) and the aggregate payoff on equity are strongly correlated. But, as argued at the end of the previous section, such a strong correlation is not observable in reality because other components of income, i.e. wages and interest payments on debt, are not perfectly correlated with the aggregate payoff on equity. The assumption of a strong correlation between aggregate production and the aggregate payoff on equity should therefore be relaxed. Besides, observed consumption data may be smoothed in comparison with aggregate production. Consumption smoothing can result from stock-keeping of the consumption good or from exports and imports[37]. Such effects further weaken the relation between aggregate consumption and the aggregate payoff on equity.

To lower the correlation between aggregate production (= consumption) and the aggregate payoff on equity, the aggregate payoff on equity and the aggregate payoff on human capital and debt are now assumed to be two separate random variables. Aggregate production is the sum of both payoffs[38]. The claim on equity is assumed to be traded at the stock market. The exogenous aggregate payoff on equity $\tilde{e}_t$ exhibits a random gross growth rate $\tilde{\gamma}_{t+1} = \tilde{e}_{t+1}/e_t$. The date $t+1$ gross return on equity is given by $\tilde{R}^E_{t+1} = (\tilde{e}_{t+1} + \tilde{P}^E_{t+1})/P^E_t$, where P^E_t denotes its ex-dividend market value at date t. Since equity, human capital and debt together pay off aggregate production $\tilde{m}_t$, the aggregate payoff on human capital and debt equals $\tilde{m}_t - \tilde{e}_t$[39]. Total exogenous production, $\tilde{m}_t$, again exhibits the random gross growth rate $\tilde{\lambda}_{t+1} = \tilde{m}_{t+1}/m_t$. Although aggregate production and the aggregate payoff on equity need not be equal or strongly correlated within this framework, the relation between both can be derived[40].

---

[37]The assumption of a closed economy is a critical assumption especially for countries like Germany, which are small in size and have a relatively large volume of exports and imports of goods and services.

[38]For a comparable approach distinguishing stocks from other assets see Barsky (1989, p. 1141). Abel (1994) provides an equilibrium model where production (= consumption) is the sum of labor income and capital income.

[39]Note that the aggregate payoff on human capital and debt may also enclose the effects of stock-keeping of the consumption good and of exports and imports.

[40]At date t the market weight of equity equals $v_t = P^E_t/P^M_t$, the weight of human capital and debt equals $v'_t = 1 - v_t$. Let $\tilde{\gamma}'_{t+1}$ denote the random gross growth rate of the payoff on human capital and debt. Then, it follows that $\mathrm{E}\{\tilde{\lambda}_{t+1}\} = v_t \mathrm{E}\{\tilde{\gamma}_{t+1}\} + v'_t \mathrm{E}\{\tilde{\gamma}'_{t+1}\}$ and $\mathrm{var}(\tilde{\lambda}_{t+1}) = v_t^2 \mathrm{var}(\tilde{\gamma}_{t+1}) + v_t'^2 \mathrm{var}(\tilde{\gamma}'_{t+1}) + 2 v_t v'_t \mathrm{corr}(\tilde{\gamma}_{t+1}, \tilde{\gamma}'_{t+1}) \mathrm{std}(\tilde{\gamma}_{t+1}) \mathrm{std}(\tilde{\gamma}'_{t+1})$. Hence, the growth rate of aggregate production, $\tilde{\lambda}_{t+1}$, and the growth rate of the aggregate

As before (see section 8.1.1), a Markov state-switching model with two states is explored. At each date t the economy is in one of both states. The state variable s_t again follows an ergodic Markov switching process with a 2×2 transition matrix and the transition probabilities

$$\text{prob}_{ij} = \text{prob}(s_{t+1} = j | s_t = i) \qquad \text{for} \qquad i, j \in \{1, 2\}.$$

In both states $\ln \tilde{\lambda}_t$, the logarithmic growth rate of aggregate production (= consumption), and $\ln \tilde{\gamma}_t$, the logarithmic growth rate of the aggregate payoff on equity, are assumed to be bivariate normally distributed and serially uncorrelated. The bivariate normal distribution in state s is given by $N_s = N(\mu_{cs}, \mu_{es}, \sigma^2_{cs}, \sigma^2_{es}, \sigma_{ces})$, $s \in \{1, 2\}$, where $\mu_{cs} = \text{E}\{\ln \tilde{\lambda}_t | s_t = s\}$, $\mu_{es} = \text{E}\{\ln \tilde{\gamma}_t | s_t = s\}$, $\sigma^2_{cs} = \text{var}(\ln \tilde{\lambda}_t | s_t = s)$, $\sigma^2_{es} = \text{var}(\ln \tilde{\gamma}_t | s_t = s)$ and $\sigma_{ces} = \text{cov}(\ln \tilde{\lambda}_t, \ln \tilde{\gamma}_t | s_t = s)$. This approach will be referred to as the "bivariate model". In contrast, the models that have been analyzed until now will be referred to as "univariate models".

Similar to the argument in section 5.2.1 and section 8.1.1, the ex-dividend market value of aggregate equity, P_t^E, is proportional to the current level of the payoff on equity, e_t. The proportionality factor is again state-dependent, i.e. $P_t^E = \Psi_i \cdot e_t$, $i \in \{1, 2\}$[41]. It depends on the expected future growth of consumption, because the preferences still exhibit constant relative risk aversion (CRRA). As described above, the growth rates of consumption are exogenous and i.i.d. in each state. The gross return on equity in a period from state i at date t to state j at date $t + 1$ follows as

$$\tilde{R}_{ij}^E = \frac{\tilde{\gamma}_j e_t + \tilde{P}_{j,t+1}^E}{P_{i,t}^E} = \frac{\tilde{\gamma}_j e_t + \Psi_j \tilde{\gamma}_j e_t}{\Psi_i e_t} = \tilde{\gamma}_j \frac{1 + \Psi_j}{\Psi_i} \quad \text{for } i, j \in \{1, 2\}. \tag{8.23}$$

For the gross return on the aggregate wealth portfolio which consists of equity, debt and human capital equation (8.1) still holds.

---

payoff on equity, $\tilde{\gamma}_{t+1}$, are related through the market weight of aggregate equity, v_t, and the growth rate of the aggregate payoff on human capital and debt, $\tilde{\gamma}'_{t+1}$. A low variance of $\tilde{\gamma}'_{t+1}$ and/or a negative correlation between $\tilde{\gamma}_{t+1}$ and $\tilde{\gamma}'_{t+1}$ are consistent with the empirically observed low volatility of consumption growth in combination with the high volatility of the aggregate payoff on equity.

41 Ψ_i can be interpreted as a state-dependent price-dividend- or price-earnings-ratio of the stock market. By multiplying the current payoff on equity with this ratio the current price of the equity is obtained. Subsequently, the equilibrium price-dividend- (resp. price-earnings-) ratios are derived. As intuitively expected these equilibrium ratios depend on the preference function, the risk-free rate and the expected future growth of consumption and the payoff on equity.

8.4.2 Implied Equilibrium Return on Equity

The derivation of the implied equilibrium return on the market portfolio and the implied one-period risk-free rate is similar to that in section 8.1.2 and is therefore not repeated in this section. Now, the implied equilibrium return on equity is derived. The Euler equation (5.5) for the return on equity, $\tilde{R}^E_{t+1}$, is

$$1 = \mathrm{E}\left\{\left[\beta\left(\frac{\tilde{c}_{t+1}}{c_t}\right)^{-\rho}\right]^{\frac{1-\alpha}{1-\rho}} (\tilde{R}^M_{t+1})^{\frac{1-\alpha}{1-\rho}-1}\tilde{R}^E_{t+1}\right\}. \tag{8.24}$$

Using the equations (8.1) and (8.23), the transition probabilities, prob_{ii} and prob_{ij}, and $\tilde{c}_{t+1}/c_t = \tilde{\lambda}_{t+1}$ the Euler equation (8.24) for an economy in state i yields a system of two linear equations with the unknowns Ψ_1 and Ψ_2

$$\begin{aligned} 1 &= \sum_{j=1}^{2}\mathrm{prob}_{ij}\mathrm{E}\left\{[\beta\tilde{\lambda}_j^{-\rho}]^{\frac{1-\alpha}{1-\rho}}\left(\tilde{\lambda}_j\frac{1+\Theta_j}{\Theta_i}\right)^{\frac{1-\alpha}{1-\rho}-1}\tilde{\gamma}_j\frac{1+\Psi_j}{\Psi_i}\right\} \\ \Leftrightarrow \Psi_i &= \sum_{j=1}^{2}\mathrm{prob}_{ij}\beta^{\frac{1-\alpha}{1-\rho}}\mathrm{E}\{\tilde{\lambda}_j^{-\alpha}\tilde{\gamma}_j\}\left(\frac{1+\Theta_j}{\Theta_i}\right)^{\frac{1-\alpha}{1-\rho}-1}(1+\Psi_j) \qquad (8.25) \\ & \quad \text{for} \quad i \in \{1,2\}, \end{aligned}$$

which can be solved analytically. Solving this system yields

$$\Psi_i = \frac{q_{ii}(1-q_{jj}) + q_{ij}(1+q_{ji})}{(1-q_{ii})(1-q_{jj}) - q_{ij}q_{ji}} \quad \text{for} \quad i,j \in \{1,2\} \quad \text{and} \quad i \neq j, \tag{8.26}$$

where $q_{ij} = \mathrm{prob}_{ij}\beta^{\frac{1-\alpha}{1-\rho}}\mathrm{E}\{\tilde{\lambda}_j^{-\alpha}\tilde{\gamma}_j\}((1+\Theta_j)/\Theta_i)^{\frac{1-\alpha}{1-\rho}-1}$.

The expected gross return on equity from state i to state j follows from equation (8.23) as

$$\mathrm{E}\{\tilde{R}^E_{ij}\} = \mathrm{E}\{\tilde{\gamma}_j\}\frac{1+\Psi_j}{\Psi_i} \quad \text{for} \quad i,j \in \{1,2\}. \tag{8.27}$$

Substituting Ψ_1 and Ψ_2 from equation (8.26) in (8.27) and using $\mathrm{E}\{\tilde{\gamma}_j\} = \exp[\mu_{ej} + 0.5\sigma^2_{ej}]$ and $\mathrm{E}\{\tilde{\lambda}_j^{-\alpha}\tilde{\gamma}_j\} = \exp[-\alpha\mu_{cj} + \mu_{ej} + 0.5\alpha^2\sigma^2_{cj} + 0.5\sigma^2_{ej} - \alpha\sigma_{cej}]$,[42] $\mathrm{E}\{\tilde{R}^E_{ij}\}$ can be calculated. The unconditional expected gross return on equity is given by

$$\mathrm{E}\{\tilde{R}^E\} = \sum_{i=1}^{2}\sum_{j=1}^{2}\mathrm{prob}_i\mathrm{prob}_{ij}\mathrm{E}\{\tilde{R}^E_{ij}\}. \tag{8.28}$$

[42] See appendix A.1 for the derivation of the expected value of the product of bivariate lognormally distributed random variables.

The variance of the return on equity from state i to state j follows from equation (8.23) as

$$\mathrm{var}(\tilde{R}^E_{ij}) = \left(\frac{1+\Psi_j}{\Psi_i}\right)^2 \mathrm{var}(\tilde{\gamma}_j) \quad \text{for} \quad i,j \in \{1,2\}, \tag{8.29}$$

where $\mathrm{var}(\tilde{\gamma}_j) = \exp[2\mu_{ej} + \sigma^2_{ej}](\exp[\sigma^2_{ej}] - 1)$.

The unconditional variance of the return on equity, $\mathrm{var}(\tilde{R}^E)$, can be calculated using an equation similar to that for the market portfolio (equation (8.8)). The conditional and unconditional expected risk premium on equity, $\mathrm{E}\{\tilde{p}^E_{ij}\}$ and $\mathrm{E}\{\tilde{p}^E\}$, the unconditional variance of the risk premium on equity, $\mathrm{var}(\tilde{p}^E)$, and its covariance with the one-period risk-free rate, $\mathrm{cov}(\tilde{p}^E, \tilde{R}^f)$, are calculated similarly to the values for the risk premium on the market portfolio in section 8.1.2.3.

8.4.3 Growth Rates of Payoff on Equity

In the bivariate model the growth rate of aggregate consumption, respectively of aggregate production, and the growth rate of the aggregate payoff on equity are separately modeled. For the empirical investigation thus a quarterly time series of the real growth rate of the payoff on equity is needed in addition. Therefore, it must be clarified which time series should be used to represent the payoff on equity.

First, exclusively using data either of corporations traded at the stock market or of all corporations may be misleading due to changes in the weights of corporations and unincorporated firms during the investigated time period (1960 to 1994). For example, many unincorporated firms have become corporations. Therefore, figures for the whole corporate sector including both corporations and unincorporated firms should be used. Second, the aggregate payoff on equity could be represented by the historically observed aggregate dividend payments on the whole corporate sector, i.e. the aggregate dividends paid on the stock market plus the distributed earnings of the unincorporated firms. But dividend payments are usually smoothed in comparison with earnings because firms routinely pay out only a fraction of earnings. Hence, aggregate earnings may better reflect the market value of aggregate equity than aggregate dividends. On the other hand, it must be taken into account that using earnings in a present value context is incorrect because earnings are partly reinvested raising future earnings in a double counting[43]. Third, since the model is based on the consumption

[43] Shiller (1981) discusses the alternative use of dividends and earnings in his famous "excess volatility"-article. He further argues that earnings are figures conceived by accountants which are supposed to provide an indicator of how well a company is doing, and that there is a great deal of latitude for the definition of earnings (Shiller 1981, p. 429).

and investment decision of a representative agent and since consumption is represented by the consumption expenditures of all private households, only those payoffs on equity received by private households should be used. Using aggregate dividend payments or aggregate earnings may bias the results because some of the dividend payments or earnings are received by investors others than private households. Since no clear answer is obtained, two alternative time series are used. First, a time series of aggregate earnings of all corporations and unincorporated firms. Second, a time series of the dividend payments of all corporations plus the distributed earnings of all unincorporated firms, both received by private households. These time series will be referred to as "corporate earnings" and "corporate dividends", respectively.

Unfortunately, these time series are only available on an annual basis. Therefore, quarterly time series must be approximated from the yearly time series. The yearly time series of corporate earnings (E-Y) and of corporate dividends (D-Y) are provided by the Statistisches Bundesamt[44]. One possibility is to interpolate the quarterly time series (E-Q and D-Q) by assuming that earnings and dividends develop continuously over each year. Then, however, their quarterly growth rates exhibit no dynamic structure within the years. Therefore, the dynamic structure of a quarterly time series that comes closest to both time series is imposed on them. For this purpose the quarterly time series measuring the aggregate entrepreneurial and property income[45] (EPI-Q) provided by the Deutsches Institut für Wirtschaftsforschung is used[46]. The construction of the quarterly corporate earnings and corporate dividends proceeds in three steps. In the first step the ratio of corporate earnings (corporate dividends) over the aggregate entrepreneurial and property income is calculated on a yearly basis, E'-Y=E-Y/EPI-Y (D'-Y=D-Y/EPI-Y). The second step is to interpolate the quarterly ratios (E'-Q and D'-Q) from the yearly ratios. A continuous curve is fitted to the yearly ratios by connecting successive straight line segments. Interval midpoints, i.e. July 1 of each year, are used as the break points and ordinates are chosen so that the average of the four estimated quarterly ratios of each year equals the yearly input ratio. Finally, in the third step, the interpolated quarterly ratios are multiplied by the quarterly aggregate entrepreneurial and property income to obtain approximations of the quarterly corporate earnings (E-Q) and the quarterly corporate dividends (D-Q), E-Q=E'-Q·EPI-Q and D-Q=D'-Q·EPI-Q.

---

[44] Again, East German data are not included. Schüler and Spies (1991) provide a description of the calculation of these time series.

[45] "Einkommen aus Unternehmertätigkeit und Vermögen der Inländer".

[46] To adjust for seasonalities, a seasonally adjusted time series is calculated using the SAS-adaptation of the US Bureau of the Census X-11 Seasonal Adjustment Program.

Table 8.6: Descriptive statistics of the real quarterly growth rates and the real quarterly logarithmic growth rates of per capita entrepreneurial and property income (EPI-Q), of per capita corporate earnings (E-Q) and of per capita corporate dividends (D-Q) for the period 1960 to 1994 (139 observations).

	Growth rate[a]		Logarithmic growth rate				
	Mean	Std	Mean	Std	Skw	Kur	Acorr
Entr. and prop. inc. (EPI-Q)	$.436^{*}_{\circ}$	(2.655)					
Corporate earnings (E-Q)	$.424^{*}_{\circ}$	(2.759)	$.386^{*}_{\circ}$	(2.741)	.1582	$.9799^{*}$	$.2632^{***}$
Corporate dividends (D-Q)	$.695^{***}_{\circ\circ\circ}$	(2.535)	$.662^{***}_{\circ\circ\circ}$	(2.510)	.2304	.2510	.0862

[a]Std denotes standard deviation, Skw denotes skewness, Kur denotes kurtosis and Acorr denotes the first-order autocorrelation. For Mean *** (**, *) [∘ ∘ ∘ (∘∘, ∘)] indicates that the hypothesis of zero mean can be rejected at a 1 (5, 10) percent level of significance on the basis of the parametric t-statistic [the nonparametric Wilcoxon-Sign-Rank statistic]. For the other values *** (**, *) indicates that the hypothesis of zero skewness, zero kurtosis, respectively zero first-order autocorrelation can be rejected at a 1 (5, 10) percent level of significance. For the skewness and kurtosis the percentage points are taken from Pearson and Hartley (1966, table 34 B and C). For the autocorrelation the percentage points are based on the parametric t-statistic.

In order to express the corporate earnings (corporate dividends) in per capita terms, the time series are divided by the quarterly average West German population taken from the monthly population data published by the Statistisches Bundesamt. From both time series discrete growth rates are calculated and deflated using the growth rate of the quarterly average consumer price index (CPI-Q2). Table 8.6 contains descriptive statistics of these time series. For comparison, summary statistics for the quarterly real growth rate of per capita entrepreneurial and property income are reported, too.

In the investigated time period corporate dividends have grown faster than corporate earnings. This finding results from the fact that a larger fraction of the earnings has been retained at the beginning of the time period than at the end of the period. For example, the proportion of distributed earnings received by the households over total earnings has continuously grown from 57.5 percent in 1960 to 85 percent in 1993. As a consequence of this special development in the investigated time period, the observed growth rate of corporate dividends may be biased upwards. Observed corporate earnings may better reflect the realized level of growth.

The standard deviation of the growth rate of corporate dividends is slightly smaller than that of the growth rate of corporate earnings. But the difference between the standard deviations may be underestimated because both time series are assumed to exhibit the dynamic structure of the aggregate entrepreneurial and property income. Evidence for this argument is provided by the finding that on the yearly basis the standard deviation of the growth rate of corporate earnings (corporate dividends) equals 1.15 (.85) times the standard deviation of the growth rate of the aggregate entrepreneurial and property income, whereas it equals only 1.04 (.95) times the standard deviation of the growth rate of the aggregate entrepreneurial and property income on the quarterly basis (see table 8.6). Thus, the approximation procedure seems to underestimate the volatility of the growth rate of corporate earnings and to overestimate the volatility of the growth rate of corporate dividends.

The bivariate approach was motivated by the argument that aggregate consumption and the aggregate payoff on equity should not be assumed to be equal or perfectly correlated. Comparing the time series in table 8.6 with those for consumption in table 6.9 confirms the argument that consumption growth is smoothed; the standard deviations of the growth rate of corporate earnings (E-Q) and of the growth rate of corporate dividends (D-Q) are about three times larger than that of the growth rate of consumption expenditures on nondurables and services (CNS-Q). Furthermore, the correlation between the logarithmic growth rate of CNS-Q and the logarithmic growth rate of E-Q (D-Q) is only .32922 (.34625). Since both correlations are nearly equal, the correlation with consumption growth provides no clear argument for the use of one over the other time series.

So far, only the special development of the payout ratio would justify the use of the corporate earnings instead of the corporate dividends. Therefore, two bivariate Markov state-switching models are estimated in the next section, one using corporate earnings and the other using corporate dividends. As the mixture of univariate normal distributions, the mixture of bivariate normal distributions allows to capture skewness and excess kurtosis; the state-switching process allows to model first-order autocorrelation. However, as for the logarithmic growth rate of consumption (see section 8.1.3.1) the hypothesis of normally distributed logarithmic growth rates cannot be rejected for corporate earnings and corporate dividends. The Anderson-Darling statistic A^2 and the Shapiro-Wilk statistic W for the logarithmic growth rate of corporate earnings are $A^2 = .2398$ and $W = .9915$. For the logarithmic growth rate of corporate dividends these statistics are $A^2 = .2761$ and $W = .9868$. Furthermore, only the excess kurtosis and the first-order autocorrelation of the logarithmic growth rate of corporate earnings is significantly positive (see table 8.6).

8.4.4 The Bivariate Markov State-switching Model

8.4.4.1 Estimation of the Model

In section 8.1.3.1 a univariate Markov state-switching model with state-dependent normal distributions for the logarithmic growth rate of consumption was estimated using the expectation-maximization (EM) algorithm. Now, the EM-algorithm is applied to estimate a bivariate Markov state-switching model with state-dependent normal distributions for the joint process of the logarithmic growth rate of consumption and the logarithmic growth rate of corporate earnings, respectively corporate dividends. As for the univariate state-switching model the estimation is replicated with numerous sets of starting values to confirm that a global optimum is found[47].

The estimates of the bivariate state-switching model using corporate earnings are shown in the first panel of table 8.7[48]. Asymptotic standard deviations of the estimates are given in parentheses. The estimates identify an up-state that is characterized by high values and low volatilities of the growth rates of consumption and corporate earnings (state 2) and a down-state with low values and high volatilities of the growth rates (state 1). The variance of consumption growth is almost two times larger in the down-state than in the up-state. For the growth rate of corporate earnings this difference is smaller. The covariance of both growth rates differs strongly across the states. The correlation between both growth rates is .6955 in state 1 and .0530 in state 2, i.e. in the recession state changes in corporate earnings are strongly correlated with changes in consumption, whereas in the up-state consumption growth and growth of corporate earnings are almost independent. Thus, similar to the findings for the univariate model, the generalization allowing the variance and covariance to differ across the states leads to an estimation result consistent with the empirical finding of increased volatilities during recessions, reflecting macroeconomic uncertainty. The unconditional probability of the up-state is 73 percent. In 27 percent of the time the economy is in the down-state. The density functions of the estimated state-dependent bivariate normal distributions are illustrated in the upper two graphs in figure 8.11. Especially the higher volatility and the

[47]Convergence of the algorithm is declared to be obtained when the maximum change in each of the parameters is smaller than 10^{-9}

[48]It turns out that the structure of the estimated bivariate model using corporate earnings is similar to that of the estimated bivariate model using corporate dividends. But the first model fits the joint process of consumption growth and growth of corporate earnings better than the second model fits the joint process of consumption growth and growth of corporate dividends. Therefore, taking into account the special character of corporate dividends discussed in the previous section, only the results for the bivariate model using corporate earnings are presented in this section and used further on. For comparison, the estimation results for the bivariate model using corporate dividends are presented in table A.2, figure A.7 and figure A.8 in appendix A.2.

Table 8.7: Markov state-switching model with two states and state-dependent bivariate normal distributions for the logarithm of the quarterly real percentage growth of per capita consumption expenditures on nondurables and services (CNS-Q) and the logarithm of the quarterly real percentage growth of per capita corporate earnings (E-Q).

Panel 1:	Maximum Likelihood Estimates[a]			
	$s = 1$		$s = 2$	
prob_{ss}	.82344	(.09636)	.93531	(.04116)
prob_s	.26813		.73187	
μ_{cs} [%]	.26536	(.25139)	.65680	(.09007)
μ_{es} [%]	-1.99099	(.73917)	1.19753	(.32098)
σ^2_{cs} [%²]	.98450	(.28395)	.54287	(.08107)
σ^2_{es} [%²]	6.27614	(1.86316)	5.27374	(.79801)
σ_{ces} [%²]	1.72878	(.66868)	.08969	(.20258)
Panel 2:	**Unconditional Moments**			
	$\ln\tilde{\lambda}$		$\ln\tilde{\gamma}$	
	Observed	Implied[b]	Observed	Implied
E{·} [%]	.55715	.55185	.38585	.34261
var(·) [%²]	.68933	.69135	7.51221	7.53757
Skewness	-.11986	-.18653	.15820	-.23348*
Excess Kurtosis	.73620	.31582	.97991	.04502**
$\text{corr}(\cdot_t, \cdot_{t-1})$	-.08029	.03300	.26323	.20083
$\text{corr}(\ln\tilde{\lambda}_t, \ln\tilde{\gamma}_t)$	.32922	.33910		
Panel 3:	**LR[c] Statistic and Significance Level**			
LR	21.3235			
Significance level	.50	.10	.05	.01
UTF-LR[d]	7.9736	15.1816	18.0101	24.1912

[a] prob_{ss} denotes the conditional probability for staying in state s, prob_s denotes the unconditional probability of state s. The unconditional probabilities are calculated as $\text{prob}_1 = (1 - \text{prob}_{22})/(2 - \text{prob}_{11} - \text{prob}_{22})$ and $\text{prob}_2 = 1 - \text{prob}_1$. μ_{cs} [σ^2_{cs}] denotes the mean [variance] of the state-dependent normal distribution for logarithmic consumption growth $\ln\tilde{\lambda}$, μ_{es} [σ^2_{es}] denotes the mean [variance] of the state-dependent normal distribution for the logarithmic growth rate of corporate earnings $\ln\tilde{\gamma}$, σ_{ces} denotes the state-dependent covariance of both growth rates. Standard deviations are given in parentheses.

[b] The calculation of the implied unconditional moments is described in the footnote of table 8.1. The implied unconditional correlation of consumption growth and dividend growth is calculated as follows:
$\text{corr}(\ln\tilde{\lambda}_t, \ln\tilde{\gamma}_t) = \sum_{i=1}^{2} \text{prob}_i(\sigma_{cei} + (\mu_{ci} - \mu_c)(\mu_{ei} - \mu_e))/(\sigma_c\sigma_e)$. *** (**, *) indicates that the hypothesis that the implied value is equal to the observed value can be rejected at a 1 (5, 10) percent level of significance based on the simulation results presented in section 8.4.4.3.

[c] Likelihood ratio statistic: twice the difference of the maximized loglikelihood value of the bivariate Markov state-switching model (H_1) and that using a single bivariate normal distribution (H_0).

[d] Upper tail fractiles of the simulated distribution of the LR statistic for the null of an i.i.d. bivariate normally distributed process.

Figure 8.11: The upper part exhibits the density functions of the two estimated state-dependent bivariate normal distributions of the Markov state-switching model for consumption growth and growth of corporate earnings. The lower part shows the density function of the mixture of the estimated state-dependent bivariate normal distributions in comparison with that of a single bivariate normal distribution fitted to the data.

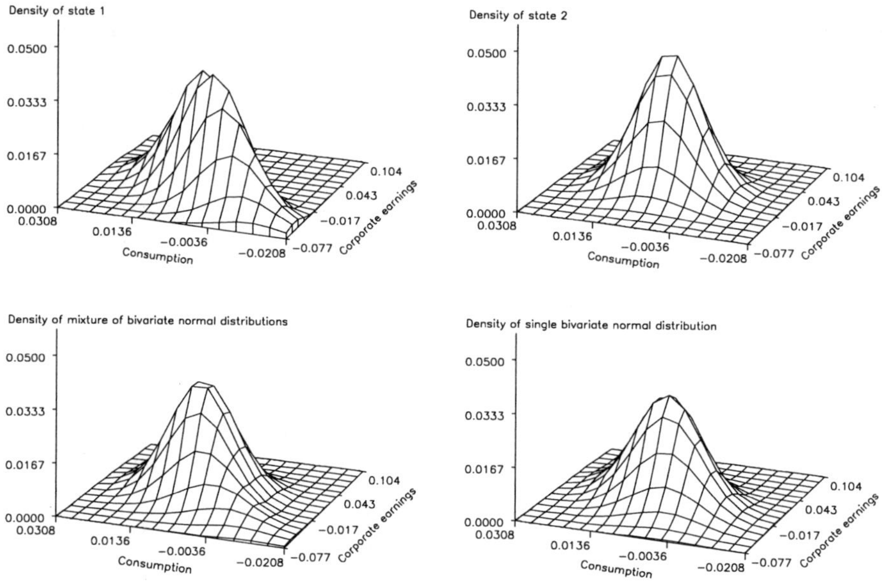

higher correlation of both growth rates in state 1 as compared to state 2 can be seen.

Comparing the state-dependent growth of consumption with the state-dependent growth of corporate earnings illustrates that consumption growth is smoothed. The average quarterly growth of corporate earnings switches between -1.99 percent and 1.20 percent, whereas the average quarterly consumption growth switches only between .27 percent and .66 percent. Moreover, the variance of the growth rate of corporate earnings is in both states more than six times larger than the variance of the growth rate of consumption.

The estimates of the bivariate model are now compared with those obtained for the univariate model (section 8.1.3.1). The down-state in the univariate model is characterized by an average consumption decrease of .06 percent, whereas the down-state identified by the bivariate model exhibits a low increase in consumption of .27 percent. Thus, the univariate model identifies

Figure 8.12: The upper two graphs exhibit the time series of quarterly logarithmic real growth rates of per capita consumption expenditures on nondurables and services (CNS-Q) and of per capita corporate earnings (E-Q). The lower graph shows the time series of the conditional smoothed probabilities of state 1, the recession state of the estimated bivariate Markov state-switching model for consumption growth and growth of corporate earnings.

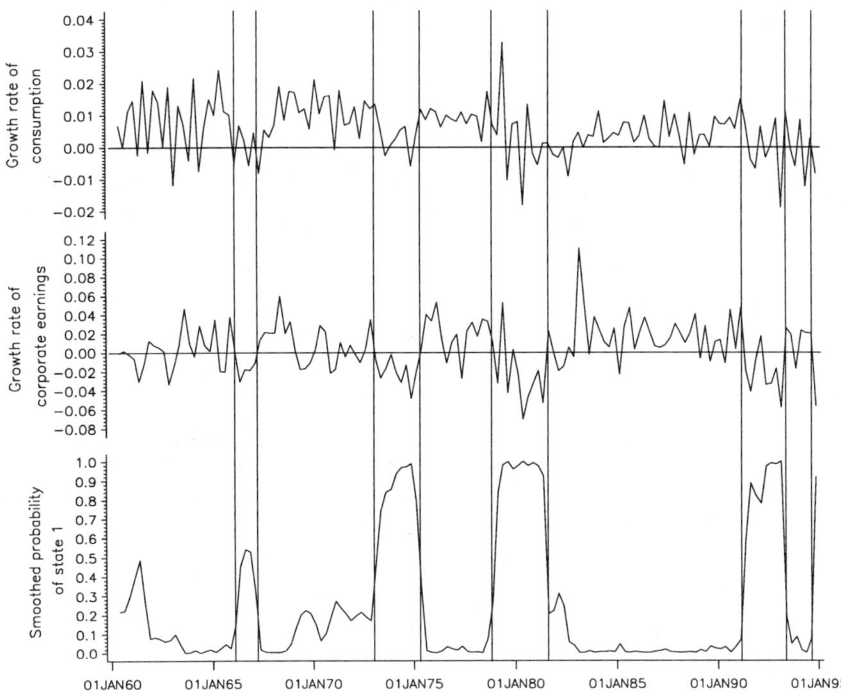

a stronger down-state which occurs with a slightly lower stationary probability (24 percent) as compared to the down-state of the bivariate model. The conditional probability of staying in the same state is much smaller in the univariate model than in the bivariate model. Hence, the states identified by the bivariate model can be characterized as longer-lasting boom or depression regimes in contrast to the short and strong crash- and up-states identified by the univariate model. Comparing the graphs for the univariate model in figure 8.1 with those for the bivariate model in figure 8.12 illustrates these results. The upper two graphs in figure 8.12 exhibit the time series of consumption growth and growth of corporate earnings. The lower graph shows the time series of the smoothed conditional probability of being in state 1, the recession state of the bivariate model. In contrast to the univariate model (lower graph in figure 8.1) which switches fast between the states and which often cannot exactly identify the current regime as state 1 or state 2, the bivariate model produces long up- and down-regimes and the current regime can be identified nearly always exactly as state 1 or 2. The vertical lines in figure 8.12 separate the recession and boom phases identified by the bivariate model. Comparing the growth rates in the upper two graphs across these regimes illustrates the result obtained above. During the recession phases the growth rates are mostly negative and more volatile than during the boom phases.

8.4.4.2 Evaluation of the Model

To further assess the quality of the bivariate switching model the empirically observed first four moments and the first-order autocorrelation of the logarithmic growth rates of consumption and of corporate earnings, and the correlation of both growth rates are compared with the values implied by the switching model (see panel 2 of table 8.7). As can be seen, the implied first and second moments are almost equal to those empirically observed. In addition, the model captures the negative skewness and excess kurtosis of the growth rate of consumption and the positive first-order autocorrelation of corporate earnings. The model fails to imply the positive skewness and excess kurtosis of the growth rate of corporate earnings. The results of the Monte Carlo simulation of the estimated state-switching model in section 8.4.4.3 show that the hypothesis that the observed skewness and excess kurtosis of the growth rate of corporate earnings is generated by the bivariate model can be rejected at a ten, respectively five, percent level of significance. The implied correlation between both time series is almost equal to that empirically observed.

The next step of the evaluation of the estimated model is to test whether the mixture of the state-dependent bivariate normal distributions fits the joint distribution of consumption growth and growth of corporate earnings significantly better than a single bivariate normal distribution. The density functions of both distributions are illustrated in the lower two graphs of figure 8.11. Since these distributions are not equal, the mixture distribution may fit the data better than the single normal distribution. This finding is confirmed by the large value of the likelihood ratio statistic (LR=21.324) given in panel 3 of table 8.7. As in the case of the univariate switching model the likelihood ratio statistic is not chi-squared distributed under the null hypothesis of no change in regime. To provide statistical inference the distribution of the likelihood ratio statistic is simulated. For this purpose the Monte Carlo approach used for the univariate model (see section 8.1.3.2) is generalized to the bivariate case[49]. 1,500 series of 139 independent draws

[49] In the univariate case the likelihood ratio statistic was simulated for the null hy-

of the logarithmic growth rate of consumption and the logarithmic growth rates of corporate earnings are generated from a single bivariate normal distribution with means, variances and correlation set to the empirically observed values. To generate the random numbers of the bivariate normal distribution with the prespecified correlation the algorithm described in Boswell, Gore, Patil, and Tallie (1993, pp. 708-9) is used. To confirm that the realizations are independent, time series with 200 periods are generated and the last 139 observations of each time series are taken. For each series of draws the Markov switching model is estimated and the likelihood ratio statistic is computed. Upper tail fractiles of the obtained distribution of the likelihood ratio statistic (UTF-LR) are listed in panel 3 of table 8.7. It can be seen that the hypothesis that the joint process of consumption growth and growth of corporate earnings is an i.i.d. bivariate normally distributed process can be rejected at a five percent level of significance against the hypothesis that it is the estimated Markov state-switching process with state-dependent bivariate normal distributions. Thus, in contrast to the univariate state-switching model, the bivariate state-switching model explains the data significantly better than a single bivariate distribution[50].

Overall, in addition to the theoretical advantage that the bivariate model allows to separate consumption and the payoff on equity, it is found that the bivariate model identifies boom and recession states significantly better than the univariate model. Before the calibration results are discussed the estimated bivariate process is used to generate a large set of realizations of the joint time series of the logarithmic growth rate of consumption and corporate earnings. As in the case of the univariate switching model, these time series serve as a basis for further simulations throughout the study.

8.4.4.3 Simulation of the Model

In this section the exogenous joint process of the growth rates of consumption and the growth rates of the payoff on equity is simulated using the estimated bivariate Markov state-switching model (see table 8.7)[51]. 10,000

pothesis of i.i.d. normally distributed consumption growth and, alternatively, for the null of a first-order autoregressive process with normal error terms. It was found that the simulated distributions of the statistics do not differ substantially. Therefore, the distribution of the likelihood ratio statistic is now simulated only for the null of i.i.d. bivariate normally distributed consumption growth and growth of corporate earnings.

[50] A simulation of the likelihood ratio statistic similar to that presented here is also conducted for the estimated bivariate Markov state-switching model using corporate dividends instead of corporate earnings. For this model, however, the hypothesis that the joint process of consumption growth and growth of corporate dividends is an i.i.d. bivariate normally distributed process cannot be rejected (see panel 3 of table A.2 in appendix A.2).

[51] A simulation exercise similar to that presented here is also conducted for the estimated bivariate Markov state-switching model using corporate dividends listed in table

Table 8.8: Means and fractiles of the simulated distributions of the first four moments, the first-order autocorrelation and the correlation between the logarithmic real growth rates of per capita consumption expenditures on nondurables and services (CNS-Q) and per capita corporate earnings (E-Q). 10,000 time series with 139 periods are generated using the estimated bivariate Markov state-switching model. The star symbols indicate between which fractiles the corresponding sample moments lie.

	Mean	Simulation Fractiles 5 %	10 %	25 %	50 %	75 %	90 %	95 %
Panel 1:	Growth rate of consumption (CNS-Q)							
$E\{\ln\tilde{\lambda}\}$ [%]	.5506	.4170	.4473	.4990	.5516 *	.6046	.6504	.6781
$var(\ln\tilde{\lambda})$ [$\%^2$]	.6894	.5371	.5656	.6171	.6841 *	.7536	.8231	.8643
Skewness $\ln\tilde{\lambda}$	-.1609	-.5640	-.4694	-.3128	-.1550 *	-.0013	.1354	.2216
Ex. Kurtosis $\ln\tilde{\lambda}$	.1649	-.5015	-.4037	-.2024	.0697	.4262 *	.8389	1.1425
$corr(\ln\tilde{\lambda}_t, \ln\tilde{\lambda}_{t-1})$	.0222	-.1245	-.0948 *	-.0376	.0220	.0825	.1377	.1677
Panel 2:	Growth rate of corporate earnings (E-Q)							
$E\{\ln\tilde{\gamma}\}$ [%]	.3387	-.2977	-.1536	.0904	.3577 *	.6030	.8121	.9282
$var(\ln\tilde{\gamma})$ [$\%^2$]	7.4623	5.7966	6.1530	6.7470	7.4238 *	8.1409	8.8139	9.2392
Skewness $\ln\tilde{\gamma}$	-.2060	-.5402	-.4622	-.3410	-.2027	-.0676	.0475	.1159 *
Ex. Kurtosis $\ln\tilde{\gamma}$	-.0458	-.5948	-.5083	-.3386	-.1105	.1634	.4883	.7185 *
$corr(\ln\tilde{\gamma}_t, \ln\tilde{\gamma}_{t-1})$	.1811	.0159	.0525	.1162	.1842	.2483 *	.3036	.3388
$corr(\ln\tilde{\lambda}_t, \ln\tilde{\gamma}_t)$	.3302	.1438	.1848	.2581 *	.3349	.4075	.4679	.4978

time series with 139 observations of the growth rate of consumption and the growth rate of corporate earnings are generated. As in the univariate case (see section 8.1.3.3), the time series of the state variable are generated and then the random growth rates are drawn from bivariate normal distributions conditional on the value of the state variable[52]. Table 8.8 reports the means and fractiles of the simulated distributions of the first four moments, the first-order autocorrelation and the correlation of the logarithmic real growth rates of per capita consumption expenditures on nondurables and services (CNS-Q) and of per capita corporate earnings (E-Q).

Comparing the empirically observed values (see panel 2 of table 8.7) with these simulated distributions yields the following results. The observed means and variances of both time series and the observed skewness of the growth rate of consumption lie between the 50 and the 75 percent fractile

A.2 in the appendix A.2.

[52]To generate the bivariate normally distributed random numbers in both states again the algorithm proposed by Boswell, Gore, Patil, and Tallie (1993, pp. 708-9) is used. To confirm that the time series are independent, time series with 200 periods are generated and the last 139 observations of each time series are taken.

of the corresponding simulated distributions. The observed excess kurtosis of the growth rate of consumption lies between the 75 and the 90 percent fractile. This is also the case for the first-order autocorrelation of the growth rate of corporate earnings. The first-order autocorrelation of the growth rate of consumption lies between the ten and the 25 percent fractile and the correlation between both time series lies between the 25 and the 50 percent fractile. Thus, for these values the hypothesis that the observed moments are generated by the estimated bivariate switching model cannot be rejected at a ten percent level of significance. The results for the skewness and the excess kurtosis of the growth rate of corporate earnings are not comparably good. The skewness lies between the 95 and the 97.5 percent fractile and the excess kurtosis lies between the 97.5 and the 99.5 percent fractile. Hence, the hypothesis that these two values are generated by the estimated bivariate switching model can be rejected at a ten, respectively five percent level of significance. Overall, with the exception of the excess kurtosis of the growth rate of corporate earnings, the hypothesis that the observed first four moments, the first-order autocorrelation and the correlation of the logarithmic growth rate of consumption and corporate earnings are generated by the estimated bivariate Markov state-switching model cannot be rejected at a five percent level of significance. The generated set of 10,000 time series of the joint growth rates of consumption and corporate earnings will be used for further analyses (see section 8.4.5 and chapter 9).

8.4.5 Calibration Results

As before, the implied equilibrium risk-free rate and risk premium are calculated for time-additive expected utility with constant relative risk aversion and, alternatively, for recursive non-expected utility. For relative risk aversion below ten the attainable equilibrium rates of return for risk-free rates below one percent per quarter are graphed in the figures 8.13 and 8.14, respectively. It is evident from these figures that the implied risk premia are smaller than for the univariate model with leverage (see figures 8.7 and 8.10). On the other hand, the implied risk premia are larger than for the univariate model without leverage. Comparing figure 8.14 with the equivalent figure for the model without leverage (figure 8.5) reveals that for recursive non-expected utility the risk premia obtained with the same level of risk aversion are about three to four times larger in the bivariate model. But with relative risk aversion below ten, the empirically observed equity premia still cannot be matched. For time-additive expected utility the implied risk premia are only slightly higher than in the univariate model without leverage (compare figures 8.13 and 8.3). The reason for the different results for both utility specifications is that in contrast to the univariate model a decrease in the elasticity of intertemporal substitution decreases the risk premium in the bivariate model. The low levels of elasticity of intertemporal

Figure 8.13: Set of admissible quarterly equilibrium risk-free rates and risk premia using the estimated **bivariate** Markov switching model and **time-additive expected utility**. Risk aversion α (a) ≤ 10 and discount factor β (b) ≤ 1.

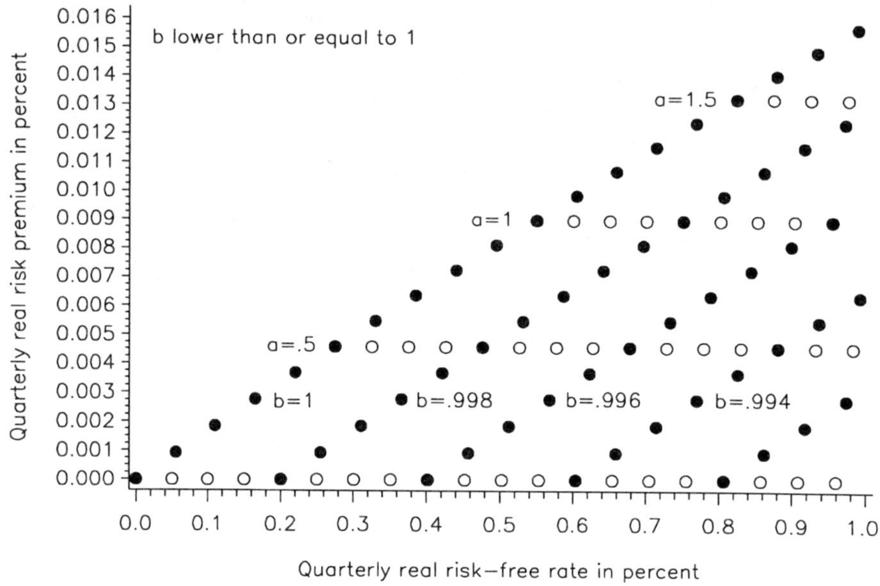

substitution implicit in time-additive expected utility hence lead to lower risk premia. For both utility specifications the risk-free rates implied by the bivariate model are almost the same as in the univariate models. Thus, as in the univariate case, only the risk-free rate of the zero tax scenario (S1) can be generated with a plausible level of the elasticity of intertemporal substitution and a discount factor below one.

What are the reasons for the implied low equity premia as compared to the univariate model with leverage? Besides the fact that the constructed time series of the growth rate of corporate earnings may underestimate the true variability of corporate earnings (see discussion is section 8.4.3) two other aspects may account for this failure. First, the observed low correlation between the growth rate of aggregate consumption and the growth rate of corporate earnings. Second, the inclusion of payoffs on human capital in the time series of corporate earnings. The effect of the first aspect is analyzed first.

Recall that in contrast to the univariate model with leverage the bivariate model does not assume a strong correlation between the growth rate of consumption and the growth rate of the aggregate payoff on equity. The implied strong correlation between aggregate consumption and the aggregate payoff on equity in the univariate models raises the risk premium for given vari-

Figure 8.14: Set of admissible quarterly equilibrium risk-free rates and risk premia using the estimated **bivariate** Markov switching model and **recursive non-expected utility**. Risk aversion α (a) $\leq$ 10, discount factor β (b) = .999 and elasticity of intertemporal substitution σ (s) $\leq$ 1.22. The empirically observed quarterly average returns (Q2) for the four taxation scenarios are represented by heart symbols.

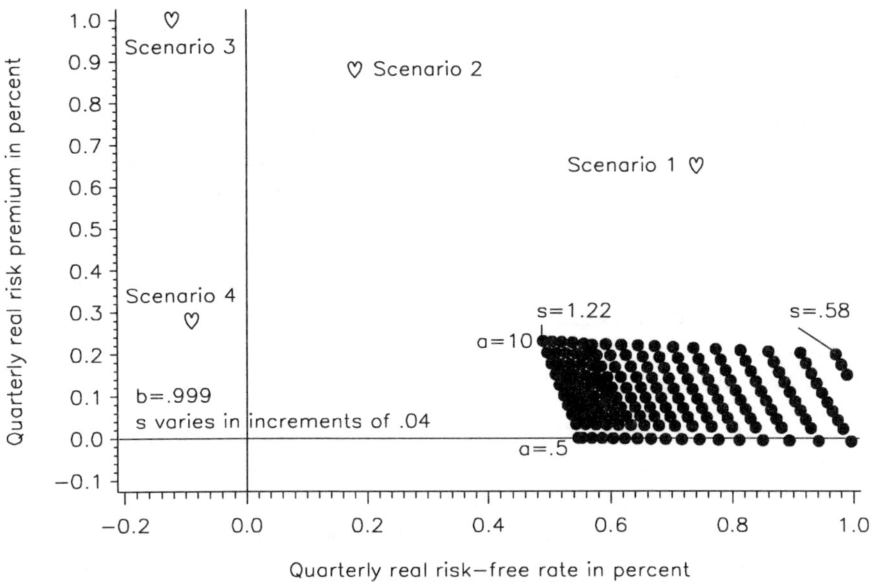

ances of consumption and payoffs on equity. An asset with high payoffs in states of high consumption and low payoffs in states of low consumption is more risky than one with payoffs which are lower correlated with consumption. Thus, the bivariate model may produce lower risk premia than the univariate model with leverage because the correlation between the growth rate of consumption and the growth rate of the aggregate payoff on equity is set equal to the empirically observed low value of .329 (see table 8.7). This low correlation may result, first, from the fact that only a fraction of aggregate equity is held by private households and, second, from the fact that only a small fraction of all households own equity and that households holding claims on equity may consume differently than households that do not hold claims on equity. Therefore, aggregate consumption may not be the right measure of the consumption of equityholders[53]. Empirical observations

[53]Shiller (1982, pp. 218) notes "It may be that only a small percent of the population invests in stocks, perhaps due to the fixed cost of acquiring information. Aggregate consumption may, accordingly, be a poor proxy for the consumption of the generally wealthy minority who hold stocks. It may be influenced substantially by the behavior of a liquidity-constrained segment of the population."

support this argument. The fraction of all German stocks held by German private households decreased from 30.3 percent in 1960 to 15.2 percent in 1994 (Deutsches Aktieninstitut 1996, pp. FB_08.1-2 and FB_08.1-3). Furthermore, this fraction of all stocks is not equally held across all households. Instead, only about 5.5 percent of the private households hold stocks at all (Deutsches Aktieninstitut 1996, p. FB_08.6-1-b). Hence, the time series of aggregate consumption should be substituted by a time series measuring the consumption behavior of households holding claims on equity. Unfortunately, such data does not exist for Germany[54]. Mankiw and Zeldes (1991) and Jacobs (1996) provide some evidence for the US. Mankiw and Zeldes (1991) construct a time series of aggregate consumption of stockholders and a time series of aggregate consumption of non-stockholders using household panel data. They find that the consumption of stockholders is more volatile and more highly correlated with the excess return on the stock market than the consumption of non-stockholders, but that the differentiation between stockholders and non-stockholders does not fully resolve the Equity Premium Puzzle. Jacobs (1996) investigates Euler equations directly on the household level. His results also indicate that the Equity Premium Puzzle does not result from aggregation problems.

To investigate the effect of the low correlation between consumption and corporate earnings in the bivariate model, a modified bivariate model is constructed. The covariances in both states of the bivariate Markov state-switching model are modified such that the growth rate of consumption and the growth rate of corporate earnings are perfectly correlated in both states. Figure 8.15 exhibits the implied equilibrium rates of return for recursive non-expected utility using this modified model. It is evident that with perfect correlation the risk premia are higher than in the original bivariate model. But only the equity premium observed for the taxation scenario S4 can be matched; the equity premia of the other scenarios are still too high to be generated. This result indicates that using a time series of consumption growth that is more highly correlated with the growth rate of the aggregate payoff on equity does not help to generate sufficiently high equity premia.

The second aspect to be analyzed is the inclusion of the payoff on the human capital of the owners of unincorporated firms into the time series measuring the aggregate payoff on equity. Recall that the owners of unincorporated firms do not receive wages apart from their firms' earnings but that their human capital could otherwise yield a risk-free return. Interpreting a fraction of the earnings as a fixed claim on this human capital, again justifies the introduction of leverage. At the beginning of section 8.3 it was demonstrated that equity accounts for only three percent of aggregate wealth, whereas the distributed earnings of unincorporated firms and the dividend payments of

[54]The German Socio-Economic Panel (GSOEP) which yearly surveys the same private households since 1984 does no contain information about their consumption behavior.

Figure 8.15: Set of admissible quarterly equilibrium risk-free rates and risk premia using a **modified bivariate** Markov switching model and **recursive non-expected utility**. The Markov switching model is modified such that consumption growth and growth of corporate earnings are perfectly positively correlated in both states. Discount factor β (b) = .999, risk aversion α (a) $\leq$ 10 and elasticity of intertemporal substitution σ (s) $\leq$ 1.22. The empirically observed quarterly average returns (Q2) for the four taxation scenarios are represented by heart symbols.

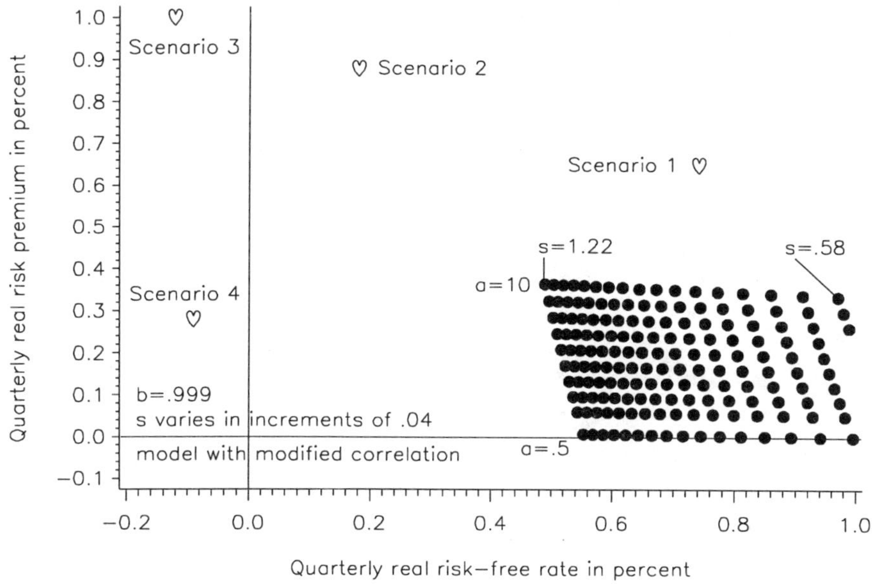

corporations, both received by private households, amount to over 20 percent of the national income. Even if the return on equity is assumed to be twice the return on human capital these figures indicate that about 75 percent of the aggregate earnings may be attributed to the human capital of the owners of unincorporated firms. Therefore, leverage L equal to .75 is now introduced into the bivariate model. The implied equilibrium return on equity is calculated from the return on equity implied by the bivariate model without leverage similarly to the introduction of leverage described in section 8.3.2.

The implied equilibrium rates of return for the bivariate model with leverage (L = .75) are illustrated in the figures 8.16 and 8.17 for time-additive expected utility and for recursive non-expected utility, respectively. The model with time-additive expected utility does not generate sufficiently high risk premia to match the empirically observed equity premia. In contrast, the model with recursive non-expected utility generates sufficiently high risk premia to match the empirically observed equity premia for all taxation

Figure 8.16: Set of admissible quarterly equilibrium risk-free rates and risk premia using the estimated **bivariate** Markov switching model, **time-additive expected utility** and **leverage.** α (a) ≤ 10, β (b) $= 1.03$ and leverage $L = .75$. The empirically observed quarterly average returns (Q2) for the four taxation scenarios are represented by heart symbols.

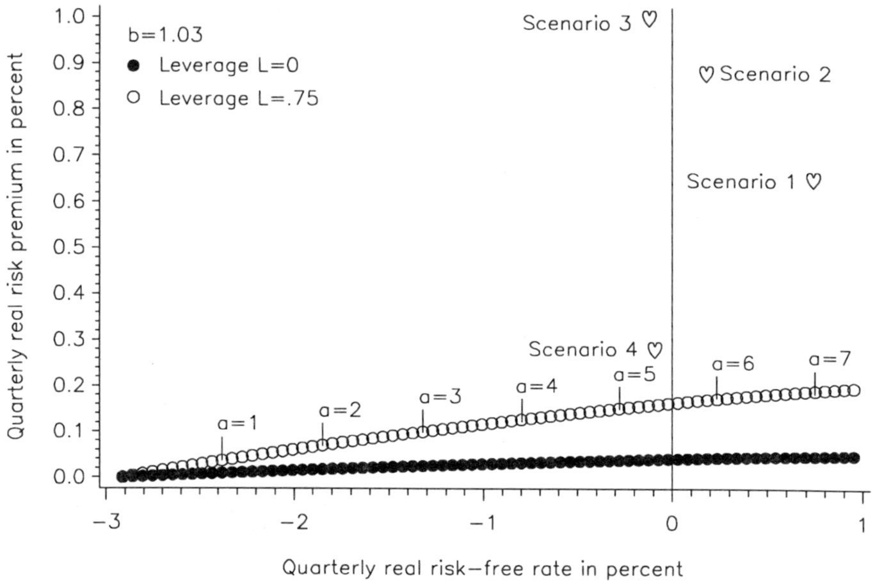

scenarios. As before, with recursive non-expected utility only the risk-free rate of the zero tax scenario can be matched. The difference in the results for both utility specifications can be explained as follows. As noted above, the elasticity of intertemporal substitution affects the risk premium; a decrease in σ reduces the risk premium. Therefore, an increase in the discount factor β cannot compensate for the implicitly chosen low elasticity of intertemporal substitution in time-additive expected utility, because it only reduces the risk-free rate but does not increase the risk premium. Hence, the potential of the recursive non-expected utility to distinguish between risk aversion and the elasticity of intertemporal substitution becomes clear in the bivariate model.

Finally, the second moments of the implied rates of return are analyzed for the bivariate model with leverage. Since the empirically observed risk-free rate and equity premium can jointly be generated only with recursive non-expected utility and only for the zero tax scenario (S1), only these results are listed in table 8.9. As for the univariate model with leverage (see table 8.3), three alternative parameter combinations are given that allow to exactly match the first moments of the risk-free rate and of the equity premium. In the first of these calibrated models the discount factor

Figure 8.17: Set of admissible quarterly equilibrium risk-free rates and risk premia using the estimated **bivariate** Markov switching model, **recursive non-expected utility** and **leverage.** α (a) $\leq$ 10, β (b) = .999, σ (s) $\leq$ 1.22 and L = .75. The empirically observed quarterly average returns (Q2) for the four taxation scenarios are represented by heart symbols.

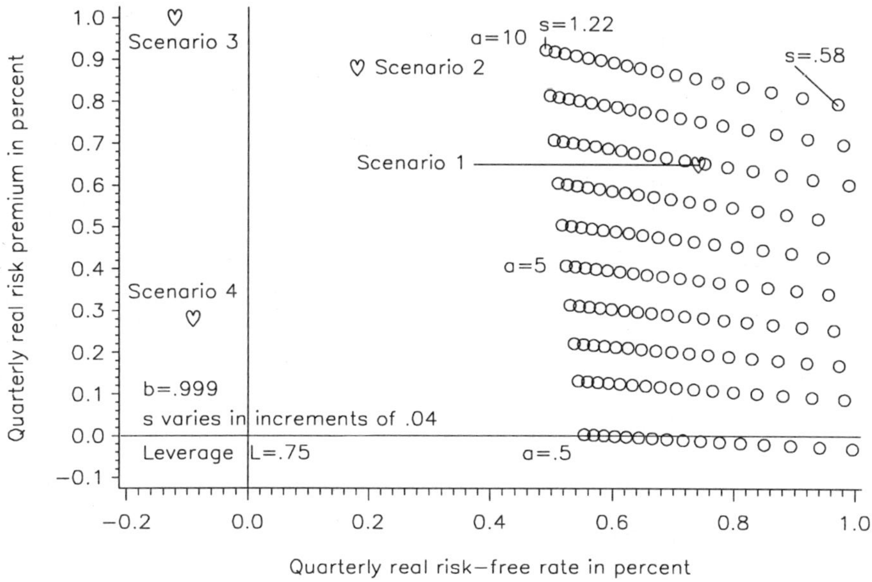

is $\beta = .997$ (BN1a), in the second it is $\beta = .999$ (BN1b) and in the third it is $\beta = 1$ (BN1c). The required elasticity of intertemporal substitution ranges between .69 and 1.13. As discussed above, different combinations of β and σ imply the same risk-free rates but imply different risk premia so that the parameter of risk aversion must be adjusted to obtain the empirically observed equity premium. The effect of different combinations of β and σ on the risk premium is much stronger for the bivariate model than for the univariate model. This can be seen by comparing the amount of the required adjustment in the parameter of risk aversion. For the bivariate model the required parameters of relative risk aversion lie between 7.56 and 8.16 yielding a range of .6, whereas this range was only .15 (4.85 to 5.00, see table 8.3) for the univariate model. The implied standard deviations of the one-period risk-free rate are slightly higher for the bivariate models than for the univariate models and therefore come closer to the empirically observed value. The implied standard deviations of the equity premium are again larger than the empirically observed value and do not differ substantially from the values implied by the univariate models. The implied one-period risk-free rate and equity premium are stronger correlated for the bivariate models than for the univariate models. This correlation closely matches the

Table 8.9: First and second moments of real rates of return as implied by **bivariate** equilibrium models with **recursive non-expected utility** and **leverage** in comparison with the corresponding moments of empirically observed quarterly average rates of return (Q2). The rates of return are given in percent per quarter.

Scenario[a]		Model parameter[b]				Risk-free rate		Equity premium		corr
		α	β	σ	L	$E\{\tilde{r}^f\}$	$std(\tilde{r}^f)$	$E\{\tilde{p}^E\}$	$std(\tilde{p}^E)$	$(\tilde{r}^f, \tilde{p}^E)$
S1	observed					.74	.54	.65	7.58	-.032
BN1a	implied[c]	7.56	.997	1.13	.75	.74	.13***	.65	17.22***	-.023
BN1b	implied[c]	7.96	.999	.79	.75	.74	.18***	.65	16.78***	-.023
BN1c	implied[c]	8.16	1.000	.69	.75	.74	.21***	.65	16.58***	-.024

[a]Scenario S1 is the zero tax scenario. The observed rates of return for this scenario are taken from table 6.8. BN1a to BN1c denote the calibrated models; B: bivariate exogenous process, N: recursive non-expected utility, 1: the matched taxation scenario.

[b]α is the parameter of relative risk aversion, β is the discount factor, σ is the elasticity of intertemporal substitution and L is the level of leverage.

[c]***, (**, *) indicates that the hypothesis that the observed value is equal to the implied value can be rejected at a 1 (5, 10) percent level of significance on the basis of 10,000 simulations of the calibrated model.

empirically observed value.

To obtain statistical inference again Monte Carlo simulations are conducted for the calibrated models. For each of the three calibrated models given in table 8.5 10,000 time series with 139 observations of the risk-free rate and the equity premium are calculated from the 10,000 time series of the endowment variables and the state variable generated in section 8.4.4.3. This is done as follows. (1) Given the time series of the state variable and the state-dependent risk-free rates for the calibrated model (see equation (8.9)), the simulated time series of the risk-free rate is obtained. (2) Given Ψ_1 and Ψ_2 for the calibrated model (see equation (8.26)) and given the time series of the state variable and of the endowment variable γ, the time series of the return on equity before leverage is calculated using equation (8.23). (3) Given the level of leverage, in each period the return on equity is calculated from the simulated return on equity before leverage and the simulated risk-free rate using a formula similar to equation (8.16). (4) Subtracting the simulated risk-free rate from the simulated return on equity in each period yields the simulated time series of the equity premium. The distributions of the first and second moments and the correlation of the risk-free rate and the equity premium are calculated from the 10,000 simulated time series. The means and percentage points of these simulated distributions for the model BN1b are listed in table 8.10. Comparing the simulated distributions with the empirically observed values yields the same results for all three

Table 8.10: Means and fractiles of the simulated distributions of the first and second moments and the correlation between the risk-free rate and the equity premium implied by the calibrated model BN1b. The simulation consists of 10,000 time series with 139 periods.

		Simulation Fractiles						
	Mean	.5 %	2.5 %	5 %	50 %	95 %	97.5 %	99.5 %
$E\{\tilde{r}^f\}$ [%]	.7424	.6243	.6569	.6688	.7460	.8053	.8142	.8320
$std(\tilde{r}^f)$ [%]	.1761	.0902	.1201	.1316	.1805	.2041	.2050	.2055
$E\{\tilde{p}^E\}$ [%]	.6390	-3.0621	-2.1095	-1.6602	.6809	2.7795	3.1413	3.8768
$std(\tilde{p}^E)$ [%]	16.7336	12.0005	13.1484	13.7801	16.7674	19.5751	20.0733	20.9692
$corr(\tilde{r}^f, \tilde{p}^E)$	-.0408	-.3343	-.2480	-.2092	-.0409	.1301	.1635	.2460

models. The hypotheses that the observed first moments and the correlation of the risk-free rate and the equity premium are generated by the calibrated model and the underlying state-switching endowment process cannot be rejected, whereas the hypotheses that the second moments are generated by the model can be rejected at a one percent level of significance.

8.4.6 Summary

The univariate model with leverage restricts aggregate consumption, i.e. the payoff on the aggregate wealth portfolio, and the aggregate payoff on equity to be strongly correlated. The bivariate model relaxes this restriction. The growth rate of aggregate consumption and the growth rate of the aggregate payoff on equity are modeled by separate random variables. Both growth rates are assumed to be bivariate normally distributed in each state of a Markov state-switching model. The payoff on equity is measured by the earnings of corporations and unincorporated firms. The standard deviation of the growth rate of the payoff on equity is three times larger than the standard deviation of consumption growth. The correlation between both growth rates is slightly above .30. The estimated bivariate Markov state-switching model identifies up- and down-states more clearly than the estimated univariate Markov state-switching model. As for the univariate model, the generalization allowing the variances and the covariances to differ across the states yields estimation results consistent with the empirical finding of increased volatilities during recessions. The evaluation of the estimated bivariate Markov state-switching model shows that in contrast to the univariate switching model the bivariate switching model explains the data significantly better than a single distribution.

The equilibrium risk premia implied by the bivariate model are larger than those implied by the univariate model without leverage and smaller than those implied by the univariate model with leverage. The low correlation between the payoff on equity and aggregate consumption, that may result from the fact that aggregate consumption is not a good proxy for the consumption of equityholders, is not the sole reason for the low implied equity premia. The introduction of leverage, which is justifiable because the earnings figures partly include payoffs on human capital, allows to increase the implied risk premia. But only with recursive non-expected utility the empirically observed equity premia can be generated. The low risk-free rates of the taxation scenarios S2 to S4 cannot be generated with plausible values of the elasticity of intertemporal substitution. Thus, only the risk-free rate and the equity premium of the zero tax scenario can jointly be matched. The correlation between the risk-free rate and the equity premium is close to the empirically observed value and the hypothesis that the observed value is generated by the model cannot be rejected. The implied standard deviation, however, differs from the empirically observed value. The implied standard deviation of the one-period risk-free rate is too low, whereas the implied standard deviation of the equity premium is too high. For both values the hypothesis that the empirically observed value is generated by the model can be rejected.

8.5 Summary of Chapter 8

State-switching model economies have been calibrated in order to match the first and second moments of the empirically observed German real risk-free rate and real equity premium. Two methods for relaxing the assumption of aggregate consumption being equal to the aggregate payoff on equity have been compared: the leverage approach and the usage of a bivariate stochastic process. The univariate and the bivariate Markov state-switching model for the stochastic processes of consumption, respectively consumption and the payoff on equity, have been estimated directly from the data and evaluated by Monte Carlo simulations. The parametric models of the market pricing kernel derived from time-additive expected utility with constant relative risk aversion and, alternatively, from recursive non-expected utility have been applied.

The bivariate model is advantageous because it models aggregate consumption and the aggregate payoff on equity separately and thus allows to relaxe the assumption of the equality or strong correlation between both. In contrast to the estimated univariate Markov switching model the estimated bivariate Markov switching model clearly identifies boom and recession states. Moreover, it fits the empirically observed data significantly better than a single bivariate normal distribution. A similar result could not be obtained for the univariate switching model. The generalization to allow the variances and covariance to differ across the states of the exogenous growth processes yields for both models estimation results consistent with the empirical finding of increased volatilities during recessions.

The calibration results favor recursive non-expected utility which clearly distinguishes between risk aversion and the elasticity of intertemporal substitution. In contrast, time-additive expected utility restricts the elasticity of intertemporal substitution to equal the reciprocal of the parameter of risk aversion. In section 5.2.4 it has been demonstrated that an implicitly chosen elasticity of intertemporal substitution that differs from the true value can be adjusted by solely changing the value of the discount factor, if the endowment process is i.i.d.. The calibration results have shown that this is not true for the state-switching endowment process. The discount factor exclusively changes the risk-free rate, whereas the elasticity of intertemporal substitution may also affect the risk premium, especially in the bivariate model.

Only the historical risk-free rate observed for the zero tax scenario can be explained with plausible levels of the elasticity of intertemporal substitution around .8 and a discount factor of .999. Thus, although recursive non-expected utility helps to resolve the Risk-free Rate Puzzle, the finding that the tax adjusted risk-free rates are too low to be explained with reasonable values of the elasticity of intertemporal substitution partly revives this puzzle. Without leverage the univariate as well as the bivariate model do not allow to match the empirically observed equity premia with plausible values of relative risk aversion below ten. But introducing leverage is justifiable in both models. With reasonable levels of leverage both models generate sufficiently high risk premia with values of relative risk aversion below eight. Hence, leverage helps to resolve the Equity Premium Puzzle. Thus, using recursive non-expected utility and leverage the first moments of the risk-free rate and the equity premium of the zero tax scenario can jointly be matched with a discount factor below one and with reasonable values of risk aversion and elasticity of intertemporal substitution. Seven different models have been specified that allow to exactly match the first moments of the returns in the zero tax scenario (see tables 8.3, 8.5 and 8.9). The hypotheses that the empirically observed first moments and the correlation of the risk-free rate and the equity premium are generated by the model and the underlying state-switching endowment process can be rejected for none of these calibrated models. But the hypotheses that the empirically observed second moments are generated by these models must be rejected at the one percent level of significance. Especially the implied standard deviation of the equity premium is much higher than the empirically observed standard deviation.

Does this observation raise a new puzzle, an "Equity Premium Volatility Puzzle" stating that the empirically observed volatility of the equity premium is too low to be explained? One could argue this way. But, this observation can also be attributed to the Equity Premium Puzzle. Although the first moment of the observed equity premium can be explained with reasonable levels of risk aversion, it can only be explained in combination with an unrealistically high volatility of the equity premium. Hence, the observed equity premium is too high to be explained with reasonable levels of risk aversion, given its "low" empirically observed volatility. Thus, the specified models that allow to match the first moments of the risk-free rate and the equity premium seem to solve the Equity Premium Puzzle, but by implying a high volatility of the equity premium the puzzle arises again. Therefore, this study cannot claim to have really solved the Equity Premium Puzzle[55].

Nevertheless, the calibrated models do match the empirically observed first moments and the correlation of the risk-free rate and the equity premium, and except for the second moments they may also be consistent with other properties of the observed time series such as skewness, kurtosis, first-order autocorrelation or mean reversion of multiperiod stock returns. Furthermore, the state-switching structure of the model may also allow to explain the empirically observed time-variation in the risk-free rate and the equity premium. These aspects are analyzed in the following chapter.

[55]Cecchetti, Lam, and Mark (1993) and Hung (1994) do not find the implied high volatility of the equity premium because they use an incorrect measure of the unconditional standard deviation implied by the model (see footnote 5 in this chapter). Cecchetti, Lam, and Mark (1993) find that the implied standard deviation is too low, whereas Hung (1994) claims that his model is consistent with all five first and second moments of the risk-free rate and the risk premium in the US data.

Chapter 9

Evaluating the Calibrated Equilibrium Models

This chapter is concerned with a further assessment of the calibrated equilibrium models. Recall that within this study a calibrated model is a model that exactly implies the empirically observed average one-period risk-free rate and average equity premium. In the previous chapter it has already been shown that calibrated equilibrium models are able to imply the empirically observed correlation between the risk-free rate and the equity premium but that they are not able to generate the empirically observed standard deviations. While the primary focus has been on the unconditional first and second moments of the one-period risk-free rate and the equity premium in the previous chapter, the performance of the calibrated equilibrium models is now measured along other dimensions.

The analysis follows different directions. Section 9.1 investigates additional unconditional moments and properties of the implied rates of return. First, it is analyzed whether or not the implied rates of return match the empirically observed rates of return with respect to the first-order autocorrelation, the skewness and the kurtosis (section 9.1.1). Second, it is investigated whether or not the calibrated models are consistent with the widely observed mean reversion in long-horizon (multiperiod) returns on stock price indices (section 9.1.2). Section 9.2 uses unconditional moments to provide further evidence for or against the calibrated equilibrium models and the underlying Markov state-switching endowment processes. All analyses within this chapter are conducted for the calibrated models UE1 (see table 8.3), UN1b (see table 8.5) and BN1b (see table 8.9). On the one hand, this selection includes models with time-additive expected utility and models with recursive non-expected utility. On the other hand, it includes models with the univariate exogenous process and models with the bivariate exogenous process.

9.1 Additional Unconditional Properties of Implied Rates of Return

9.1.1 Properties of One-period Returns

The purpose of this section is to analyze whether or not the calibrated equilibrium models are consistent with stylized facts of empirically observed one-period rates of return others than the first and second moments. In particular, it is investigated whether the models UE1, UN1b and BN1b are consistent with the skewness, kurtosis and first-order autocorrelation of asset returns, i.e. whether the models imply these properties. In addition to the risk-free rate and the equity premium the return on equity is now explicitly analyzed.

Panel 1 of table 9.1 lists the first four moments, the first-order autocorrelation and the correlation with the one-period risk-free rate of the empirically observed quarterly one-period risk-free rates, of the observed quarterly returns on equity and of the observed quarterly equity premia for the zero-tax scenario (S1) in the period 1960 to 1994. Except for the first-order autocorrelation the values are calculated from quarterly average returns (Q2) in order to correspond with the values used in the calibration exercises in the previous chapter. The values for the first-order autocorrelation are calculated from end-of-period returns (Q1) because the averaging within the periods used to calculate Q2-returns implies an artificial first-order autocorrelation of .25 (cf. Working (1960) and the discussion in section 6.2.1.2). Higher-order autocorrelations are not affected by the averaging.

The analysis whether or not the three calibrated equilibrium models imply comparable figures is conducted similarly to the analyses in the sections 8.3.3 and 8.4.5. There, 10,000 time series of the equilibrium risk-free rate, the equilibrium return on equity and the equilibrium equity premium have been generated for each calibrated model in order to evaluate the fit of the second moments implied by the calibrated models. Now, for each of the 10,000 time series the first four moments, the first-order autocorrelation and the correlation with the risk-free rate are calculated for each of the three models. The tabulation of these statistics yields the Monte Carlo distributions of these statistics from which inference can be drawn. For illustration, table A.3 in appendix A.2 lists the means and fractiles of these distributions for model BN1b. Panel 2 of table 9.1 exhibits how these distributions correspond with the empirically observed values listed in panel 1. For each of the three models so-called p-values giving the percentage of the simulated distribution with values less than the empirically observed values are given. The closer the p-value is to .5, the stronger the support for the hypothesis that the observed value is implied by the model. A low p-value indicates that the value implied

Table 9.1: First four moments, first-order autocorrelation and correlation with the real risk-free rate of real rates of return as implied by the calibrated equilibrium models UE1, UN1b and BN1b in comparison with the corresponding empirically observed values.

		E{·} [%]	std(·) [%]	Skewness	Kurtosis	corr(·$_{t-1}$, ·$_t$)	corr($\tilde{r}^f$, ·)
Panel 1:		Empirically observed quarterly rates of return[a] (1960-1994)					
	$\tilde{r}^f$	.740	.538	.037	-.678	.371	
S1	$\tilde{r}^E$	1.390	7.579	.214	1.289	.035	.039
	$\tilde{p}^E$	.651	7.577	.193	1.312	.040	-.032
Panel 2:		p-values[b] of implied equilibrium rates of return[c]					
	$\tilde{r}^f$	.476	1.000^{---}	.000^{+++}	.317	1.000^{---}	
UE1	$\tilde{r}^E$	.492	.000^{+++}	.998^{---}	.989^{--}	.739	.605
	$\tilde{p}^E$	.492	.000^{+++}	.996^{---}	.991^{--}	.706	.355
	$\tilde{r}^f$	.370	1.000^{---}	.000^{+++}	.317	1.000^{---}	
UN1b	$\tilde{r}^E$	.492	.000^{+++}	.971^{-}	.984^{--}	.696	.673
	$\tilde{p}^E$	.493	.000^{+++}	.965^{-}	.986^{--}	.705	.364
	$\tilde{r}^f$	.436	1.000^{---}	.986^{--}	.557	.003^{+++}	
BN1b	$\tilde{r}^E$	.490	.000^{+++}	.813	.054	.684	.756
	$\tilde{p}^E$	.492	.000^{+++}	.785	.048^{+}	.722	.537

[a]The quarterly average returns before taxation (Q2, S1) are used. Only the value for the first-order autocorrelation is taken from end-of-period returns (Q1). $\tilde{r}^f$ denotes the one-period risk-free rate, $\tilde{r}^E$ denotes the return on equity and $\tilde{p}^E$ denotes the equity premium.

[b]The p-value is the percentage of the simulated distribution with values less than the empirically observed value. $- - -$ ($--$, $-$) [$+ + +$ ($++$, $+$)] indicates that the value implied by the model is too low [high] and that the hypothesis that the observed value is drawn from the simulated distribution can be rejected at a 1, (5, 10) percent level of significance.

[c]UE1 is the calibrated model with the univariate endowment process and time-additive expected utility ($\alpha = 4.26$, $\beta = 1.0159$, $L = .95$). UN1b is the calibrated model with the univariate endowment process and recursive non-expected utility ($\alpha = 4.9$, $\beta = .999$, $\sigma = .83$, $L = .95$). BN1b is the calibrated model with the bivariate endowment process and recursive non-expected utility ($\alpha = 7.96$, $\beta = .999$, $\sigma = .79$, $L = .75$).

by the model is too high, and vice versa. p-values without "+" or "−" signs indicate that the empirically observed value lies in the 90 percent confidence interval and that therefore the hypothesis that the observed value is drawn from the distribution implied by the model cannot be rejected. The signs "+" and "−" indicate that the hypothesis that the observed value is drawn from the distribution can be rejected. The more of these signs are given, the stronger the evidence against the model.

First, consider the expected value and the standard deviation. None of the models can be rejected with respect to the first moments. This is not surprising since the three models have been calibrated to match the first moments of the empirically observed rates of return. As discussed in the previous chapter the calibrated models are not able to yield the standard deviations of observed returns. The implied standard deviation of the risk-free rate is too low, whereas the implied standard deviations of the return on equity and of the equity premium are too high. Second, consider the skewness and the kurtosis. The models with the univariate exogenous process (UE1 and UN1b) do not capture the excess kurtosis and the slightly positive skewness of the return on equity and the equity premium, whereas the model with the bivariate exogenous process (BN1b) is consistent with the empirically observed values. All models are consistent with the kurtosis of the risk-free rate. However, the skewness of the risk-free rate implied by the models with the univariate exogenous process is too high and that implied by the model with the bivariate exogenous process is too low. Third, consider the first-order autocorrelation. All models are consistent with the slightly positive first-order autocorrelation of the return on equity and the equity premium, whereas the positive autocorrelation of the risk-free rate is not captured. For the risk-free rate the models with the univariate exogenous process imply a lower autocorrelation, whereas the model with the bivariate exogenous process implies a higher autocorrelation. Finally, all models are consistent with the empirically observed correlation between the risk-free rate and the return on equity, respectively the equity premium.

Overall, the results for the skewness, the kurtosis, the first-order autocorrelation and the correlation with the risk-free rate of return are more favorable for the calibrated models than the results for the second moments. The calibrated equilibrium models with the Markov state-switching exogenous process are consistent with at least some of these empirically observed properties. Especially the model with the underlying bivariate Markov state-switching process captures with exception of the standard deviation all unconditional properties of the equity return and the equity premium. This indicates that a Markov state-switching endowment process may explain at least some of the stylized facts of empirically observed asset returns that cannot be explained with models implying constant conditional moments of the asset returns.

9.1.2 Mean Reversion in Multiperiod Returns

In this section it is analyzed whether the calibrated equilibrium models explain the empirically observed mean reversion in stock market prices at long horizons. This observation is well established in the literature since the studies of Fama and French (1988a) and Poterba and Summers (1988). Mean reversion in prices or returns implies autocorrelation of returns at some investment horizons. Using different approaches, these studies provide evidence that three- to ten-year returns on stock price indices exhibit negative serial correlation, whereas the returns are positively autocorrelated at shorter horizons. The evidence is broad. Poterba and Summers (1988) analyze 18 different countries[1] and Fama and French (1988a) analyze different industry- and size-portfolios of US stocks. Furthermore, Poterba and Summers (1988) show that this structure of the serial correlation in multiperiod returns is also observable for stock returns in excess of a risk-free rate and for returns on individual stocks in excess of a market index[2]. A general problem of such analyses is that the rejection of the random walk hypothesis, i.e. the hypothesis of no serial correlation, is complicated by the small number of non-overlapping observations available for multiperiod returns. The standard deviations of the estimated autocorrelation measures are typically so large that the random walk can hardly be rejected, and, as pointed out by Poterba and Summers (1988), sample autocorrelations of returns are biased towards negative autocorrelation in small samples. This effect is known as the "small sample bias".

Mean reversion in stock prices may be explained either by variations in expected returns and risks over time or by departures from the fundamental values that are corrected in the long-run (cf. Poterba and Summers (1988, p. 28)). But if individuals are rational and the market is efficient with respect to information stock prices should not depart from their fundamental values. Hence, models assuming rational individuals must allow for variation in expected returns and risks to have any ability to explain the autocorrelation of stock returns. The calibrated equilibrium models are based on a Markov state-switching economy and therefore allow for variation in expected returns and risks[3]. The question is whether these models imply a

[1]For 24- and 36-month returns on a German stock price index in the period 1957 to 1986 they report a positive serial correlation. The serial correlation of returns over 48 and more (up to 96) months is negative and declines with an increasing number of months (Poterba and Summers 1988, table 4).

[2]This structure of autocorrelation in the excess returns on individual assets is also found by studies analyzing the profitability of contrarian investment strategies. For example, DeBondt and Thaler (1985) find that a portfolio of loser stocks whose value has declined in the last three to five years, outperforms the market in the next three to five years, whereas a comparable portfolio of winner stocks whose value has risen in the past underperforms the market in the future. Evidence for a similar behavior at the German stock market is given in Meyer (1994), (1995).

[3]An equilibrium model with a state-switching endowment process may imply mean

serial correlation in multiperiod equilibrium asset returns comparable with the empirically observed serial correlation.

Previous studies have shown that equilibrium models based on state-switching economies may imply negative serial correlation in multiperiod equilibrium asset returns. Cecchetti, Lam, and Mark (CLM) (1990)[4] analyze a univariate two-state Markov switching model in combination with time-additive expected utility and constant relative risk aversion. Their model produces negative serial correlation in multiperiod returns comparable with empirically observed values. A model with a geometric Brownian motion as endowment process should not imply serial correlation so that the serial correlation implied by such a model is a measure for the small sample bias. CLM (1990) use this fact to show that the small sample bias is considerably smaller than the serial correlation implied by the model with the state-switching endowment. Kandel and Stambaugh (KS) (1990), (1991) report similar results for a univariate symmetric four-state switching model. Alternatively using time-additive expected utility and recursive non-expected utility they show that the negative serial correlation is not sensitive to the parameter of risk aversion but to the elasticity of intertemporal substitution. With declining elasticity of intertemporal substitution the negative serial correlation becomes stronger, which suggests that consumption smoothing implies the mean reversion in stock prices. Bonomo and Garcia (1994) argue that the favorable results of CLM (1990) and KS (1990) may result from a misspecification of their state-switching models. CLM (1990) restrict the variances to be equal in their two-state model and the four-state model of KS (1990) is restricted to two means and two variances. Bonomo and Garcia (1994) provide evidence that a two-state model with one mean but different variances implies substantially smaller serial correlation although it fits the underlying endowment variable significantly better than the models of CLM (1990) and KS (1990). In their subsequent study, Bonomo and Garcia (1996) apply a bivariate[5] three-state model with no restrictions. Using time-additive expected utility their model produces some negative autocorrelation in real returns, but it fails to produce any autocorrelation in excess returns. Moore and Schaller (1996) investigate whether the assumption that the current state of the Markov state-switching model is known to the representative agent affects the implied mean reversion. They argue that when the agent must learn about the state this alters the timing of the equilibrium

---

reversion in asset prices for two reasons. First, the individuals may substitute intertemporally when faced with alternating high and low growth rates in their endowment, i.e. when the means differ across the states of the switching model. Second, when the variances differ across the states the expected asset returns may alternate between high and low values as a consequence of the alternating levels of risk, which in turn may imply mean reversion in asset prices (cf. Bonomo and Garcia (1994, p. 27)).

[4] See also Cecchetti and Mark (1990).

[5] As in the model used in section 8.4, they separate dividend growth from consumption growth.

price response and, as a result, it may change the serial correlation properties of returns. Although they find that learning can significantly alter the autocorrelation in equilibrium returns, they conclude that it does not rule out Markov state-switching as an explanation for the empirically observed mean reversion. All these studies refer to US data and none has analyzed recursive non-expected utility in combination with a state-switching model that is restricted neither in the means nor in the variances. Therefore, this section provides evidence for Germany and for more general specifications of the equilibrium model.

Besides directly estimating autoregressive models, two different but related methods are usually applied to analyze the serial correlation in multiperiod returns, namely variance-ratio tests and regression tests. Variance ratios compare the relative variability of returns over different horizons. Variance-ratio tests proposed by Lo and MacKinlay (1988) are based on the fact that if logarithmic returns follow a random walk the variance of the returns should be proportional to the investment horizon, i.e. $\text{var}(\tilde{r}_{t,t+k}) = k\text{var}(\tilde{r}_{t,t+1})$, where $\tilde{r}_{t,t+k}$ denotes the compound return from date t to date $t+k$. If this proportionality is not satisfied in large samples the returns are autocorrelated at some lags. The regression tests proposed by Fama and French (1988a) simply regress multiperiod returns on lagged multiperiod returns and use the estimated slope coefficient as a measure of the serial correlation of the multiperiod returns. Poterba and Summers (1988, pp. 29-31) show how both measures of serial correlation depend on the autocorrelation coefficients of different lags. As a consequence of the different weights both measures place on these autocorrelation coefficients they differ in the power to reject the random walk hypothesis. However, the purpose of this study is to analyze whether the serial correlation in empirically observed multiperiod returns is also implied by the calibrated equilibrium models and not to analyze the random walk hypothesis. Therefore, the more intuitive regression test is applied. The serial correlation is measured by the slope coefficient, b_k, of a regression of multiperiod returns on lagged multiperiod returns,

$$\tilde{r}_{t,t+k} = a_k + b_k \tilde{r}_{t-k,t}, \tag{9.1}$$

where $\tilde{r}_{t,t+k}$ is the compound rate of return from date t to date $t+k$:

$$\tilde{r}_{t,t+k} = \left(\prod_{\tau=1}^{k} (\tilde{r}_{t+\tau} + 1) \right) - 1. \tag{9.2}$$

b_k is calculated for horizons between one and 40 quarters, i.e. $k \in \{1, 2, \ldots, 40\}$. As in Fama and French (1988a) overlapping observations are used in the regression[6].

[6]For statistical tests the standard deviation of the estimated regression coefficient

Table 9.2: Serial correlation in multiperiod real returns on equity and in multiperiod real equity premia as implied by the calibrated equilibrium models UE1, UN1b and BN1b in comparison with the corresponding empirically observed values.

Observed values[c]		Simulated values[a]						Simulated values "large sample"[b]					
		UE1		UN1b		BN1b		UE1		UN1b		BN1b	
k	b_k	b_k	p	b_k	p	b_k	p	b_k	p	b_k	p	b_k	p
Panel 1:		Multiperiod return on equity $\tilde{r}^E_{t,t+k}$											
1	.035	-.018	.739	-.009	.683	-.008	.695	-.012	.965*	-.007	.925	-.003	.916
2	.089	-.018	.842	-.022	.841	-.014	.840	-.005	.993**	-.011	.993**	-.001	.992**
3	.074	-.030	.793	-.034	.793	-.028	.792	-.004	.957*	-.012	.963*	-.002	.949
4	-.036	-.044	.520	-.045	.524	-.042	.514	-.005	.280	-.013	.337	-.004	.264
8	-.194	-.094	.287	-.085	.286	-.091	.288	-.008	.005**	-.015	.007**	-.009	.005**
12	-.041	-.141	.686	-.132	.655	-.140	.678	-.014	.381	-.013	.386	-.012	.376
16	-.154	-.185	.556	-.190	.551	-.192	.559	-.020	.072	-.018	.089	-.017	.074
20	-.154	-.228	.634	-.234	.621	-.236	.632	-.025	.090	-.020	.119	-.023	.104
24	-.135	-.271	.720	-.283	.710	-.282	.723	-.031	.161	-.025	.180	-.029	.163
28	-.208	-.324	.668	-.333	.669	-.339	.673	-.037	.047*	-.030	.074	-.035	.061
32	-.315	-.357	.554	-.370	.568	-.377	.572	-.041	.005**	-.035	.010**	-.039	.008**
36	-.554	-.378	.311	-.394	.332	-.399	.333	-.048	.001***	-.044	.000***	-.046	.000***
40	-.310	-.388	.565	-.399	.575	-.406	.577	-.051	.012**	-.047	.016**	-.050	.017**
Panel 2:		Multiperiod equity premium $\tilde{p}^E_{t,t+k}$											
1	.040	-.003	.705	-.014	.722	-.005	.705	.003	.927	-.012	.968*	.001	.929
2	.089	-.014	.833	-.030	.856	-.014	.837	-.001	.991**	-.020	.997***	-.001	.991**
3	.061	-.027	.759	-.044	.788	-.027	.761	-.001	.915	-.022	.959	-.001	.914
4	-.064	-.042	.433	-.055	.476	-.041	.436	-.003	.124	-.023	.233	-.003	.123
8	-.258	-.092	.173	-.093	.191	-.090	.176	-.007	.000***	-.024	.001***	-.009	.000***
12	-.114	-.140	.548	-.138	.542	-.140	.546	-.013	.114	-.020	.154	-.012	.120
16	-.222	-.185	.429	-.193	.451	-.191	.448	-.019	.012**	-.023	.022**	-.017	.014**
20	-.224	-.228	.509	-.235	.522	-.236	.521	-.025	.019**	-.024	.034*	-.023	.022**
24	-.195	-.272	.635	-.283	.634	-.283	.640	-.031	.049*	-.029	.073	-.029	.061
28	-.274	-.325	.577	-.332	.586	-.339	.593	-.037	.010**	-.034	.021**	-.035	.015**
32	-.385	-.359	.466	-.369	.477	-.377	.491	-.041	.000***	-.038	.002***	-.039	.001***
36	-.631	-.380	.240	-.393	.253	-.399	.258	-.047	.000***	-.046	.000***	-.046	.000***
40	-.359	-.391	.526	-.396	.531	-.405	.537	-.051	.004***	-.049	.006**	-.049	.007**

[a] b_k is the median of the Monte Carlo distribution of the multiperiod serial correlations obtained from the 10,000 simulated time series of equity returns, respectively equity premia. p is the p-value, i.e. the percentage of the simulated distribution with values less than the observed value. *** (**, *) indicates that the hypothesis that the observed value is drawn from the simulated distribution can be rejected at a 1, (5, 10) percent level of significance. For a description of the models see the footnotes in table 9.1.

[b] "large sample" means that 10,000 time series with 1,390 instead of 139 observations are generated and that non-overlapping regressions are used instead of overlapping regressions.

[c] Serial correlation in empirically observed k-quarter returns on equity and equity premia. k-quarter returns are compound end-of-quarter returns on stocks before taxation (Q1, S1). k-quarter equity premia are k-quarter stock returns minus the compound risk-free rate over the k quarters. Serial correlation is measured by the regression coefficient b_k of a regression of k-quarter returns on lagged k-quarter returns.

Column two of table 9.2 lists the regression coefficients b_k estimated from the empirically observed returns on equity (panel 1) and from the empirically observed equity premia (panel 2) in the period 1960 to 1994. End-of-quarter returns before taxation (Q1, S1) are used for the estimation[7]. To save space the regression coefficients are not listed for all horizons. For both the returns on equity and the equity premia the estimated serial correlation is positive for horizons shorter than one year and negative for longer horizons. The negative serial correlation becomes stronger with increasing length of the horizon. To analyze the serial correlation implied by the calibrated equilibrium models the previously used 10,000 time series of the equilibrium equity returns and equity premia generated for each calibrated model are again employed. Estimating the regression coefficients b_k for each time series yields Monte Carlo distributions of the regression coefficients for each of the three models. To investigate the effects of the small sample bias additional 10,000 time series with 1,390 observations instead of 139 observations are simulated and the regression coefficients are estimated for these time series using non-overlapping regressions[8]. These regression coefficients are referred to as the "large sample" values.

Table 9.2 presents the median and the p-values of the Monte Carlo distributions of the regression coefficients for each of the three calibrated models. As in the previous section, p-values are the percentage of the simulated distribution with values less than the empirically observed value. First, consider the results for the model UE1. The medians of the simulated distributions of the regression coefficients indicate that the model implies a negative serial correlation of multiperiod returns on equity and of multiperiod equity premia, but that it fails to imply the positive serial correlation for short horizons. For horizons longer than one year the size of the implied negative serial correlation is comparable with that observed empirically. The p-values suggest that for none of the horizons the hypothesis of the empirically observed serial correlation being drawn from the implied distribution can be rejected. The results for model UE1 are illustrated in figure 9.1. The upper graph illustrates that the empirically observed serial correlation of multiperiod equity returns lies in the 50 percent confidence interval of the simulated distribution for almost all horizons. The same result is obtained for the serial correlation of multiperiod equity premia illustrated in

---

must be adjusted when overlapping periods are used. Here, the purpose is simply to compare the empirically observed values with the simulated distribution of implied values. Therefore, the standard deviation of the regression coefficient is not calculated.

[7]Zero tax returns are chosen because the models to be analyzed are calibrated to the zero tax scenario (S1). Alternatively using tax adjusted returns does not substantially alter the results. End-of-quarter returns (Q1) are chosen because at short horizons the serial correlation of period average returns (Q2) is artificially increased by the averaging (cf. Working (1960)).

[8]A comparable approach is applied by Cecchetti, Lam, and Mark (1990) and Bonomo and Garcia (1994), (1996).

Figure 9.1: Empirically observed serial correlation in multiperiod real returns on equity and in multiperiod real equity premia for various investment horizons in comparison with the serial correlation implied by the calibrated equilibrium model UE1.

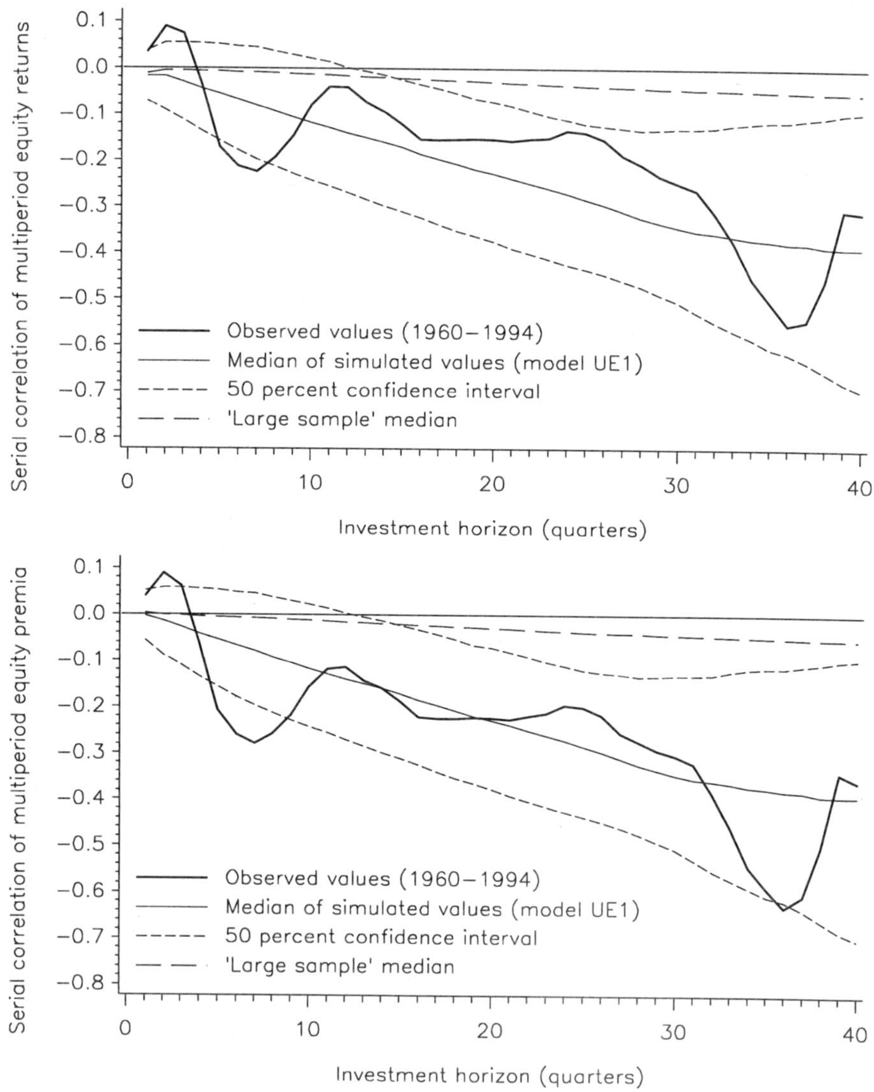

the lower graph. Both graphs also exhibit the large sample medians. It is evident from the graphs that in the large sample substantially less negative serial correlation is found as compared to the normal sample. This indicates that most of the negative serial correlation may result from the small sample bias. However, for the question whether the empirically observed values are consistent with the equilibrium model, this aspect is irrelevant because the small sample bias affects both the observed values and the simulated values. Nevertheless, the large sample medians indicate that the model also implies negative serial correlation in large samples, which is the equilibrium explanation of mean reversion in asset prices. For the models UN1b and BN1b similar results are obtained[9]. For none of the horizons the hypothesis of the empirically observed serial correlation being drawn from the distribution implied by model UN1b respectively BN1b can be rejected. Considering the large sample results reveals that again most of the serial correlation found in the observed returns on equity, the observed equity premia and the corresponding simulated values results from the small sample bias. The amount of the implied serial correlation is almost equal across the three calibrated equilibrium models.

Overall, the results show that the calibrated equilibrium models are able to explain the empirically observed mean reversion in stock market prices at long horizons. This result depends neither on one of the two endowment processes – univariate or bivariate – nor on one of the two utility specifications – time-additive expected utility or recursive non-expected utility. All three models produce as much negative serial correlation in multiperiod returns on equity and in multiperiod equity premia as is empirically observed. Thus, in contrast to Bonomo and Garcia (1996) the models analyzed here do not fail to produce negative serial correlation in excess returns. Large sample simulations indicate that most of the mean reversion in both the observed and the simulated equilibrium returns may be attributed to the small sample bias. This, however, does not change the result that the equilibrium models are consistent with the observed mean reversion.

[9]The figures A.9 and A.10 in appendix A.2 illustrate the results for the models UN1b and BN1b.

9.2 Using Conditional Moments

While the focus was on unconditional moments and properties of empirically observed and implied rates of return in the calibration exercises in chapter 8 and in the evaluation of the calibrated equilibrium models in section 9.1, this section is concerned with the models' implications about expected returns and volatilities conditional upon the state of the underlying Markov state-switching process. The purpose of analyzing conditional moments is twofold. First, it is examined whether and to what extent expected returns and volatilities vary over time according to the calibrated equilibrium models UE1, UN1b and BN1b. The structure of these three models is related to the three two-period equilibrium models used by Barsky (1989) to theoretically explore the impact of increased risk and reduced economic growth on the prices of stocks and bonds. Therefore, the predictions of the models UE1, UN1b and BN1b, given that the economy is currently in a down- respectively an up-state, are discussed in the light of Barsky's (1989) theoretical results. Second, the forecasts of the models are empirically investigated by comparing them with subsequently realized returns.

Implied Conditional Moments Given the estimated univariate and bivariate Markov state-switching models, the conditional expected rates of return and the conditional variance can be calculated for the three calibrated equilibrium models. The one-period risk-free rate conditional upon being in state s, r_s^f, $s \in \{1,2\}$, follows from equation (8.9). The expected return on equity conditional upon being in state s, $\mathrm{E}_s\{\tilde{r}^E\}$, is calculated by adding the probability-weighted expected returns $(\mathrm{prob}_{s1} \cdot \mathrm{E}\{\tilde{r}^E_{s1}\})$ and $(\mathrm{prob}_{s2} \cdot \mathrm{E}\{\tilde{r}^E_{s2}\})$ (see equation (8.16)). $\mathrm{prob}_{ss'}$ is the transition probability for moving from state s to state s', and $\tilde{r}^E_{ss'}$ is the expected return on equity from state s to state s'. Equivalently, the conditional variance of the return on equity, $\mathrm{var}_s(\tilde{r}^E)$, is obtained as the sum of $(\mathrm{prob}_{s1} \cdot \mathrm{var}(\tilde{r}^E_{s1}))$ and $(\mathrm{prob}_{s2} \cdot \mathrm{var}(\tilde{r}^E_{s2}))$ (see equation (8.18)). These moments and the conditional equity premium, $\mathrm{E}_s\{\tilde{p}^E\}$, are listed in table 9.3. The models UE1 and UN1b, based on the univariate Markov state-switching process (see table 8.1), are given in panel 1; model BN1b, based on the bivariate process (see table 8.7), is given in panel 2.

The discussion starts with the models UE1 and UN1b. Given that the economy is in the down-state ($s = 1$), both models imply a higher risk-free rate and a lower equity premium than in the up-state ($s = 2$). The low equity premium in the down-state corresponds with the low standard deviation of the return on equity. For both models the variation in the conditional risk-free rate is larger than in the conditional equity premium. Therefore, the expected return on equity is higher in the down-state than in the up-state despite the lower equity premium. Hence, the risk-free rate and the return on equity move together, whereas the risk-free rate and the equity premium move in opposite directions. The later effect previously became evident by the implied negative correlation between the risk-free rate and the equity premium (see tables 8.3 and 8.5).

The observation of higher expected returns in the down-state than in the up-state may be surprising at first sight. However, it is a consequence of the low persistence of the down-state of the estimated univariate Markov state-switching model. The probability of staying in the down-state is almost zero ($\text{prob}_{11} = .000$, see table 8.1), i.e., given that the economy is in the down-state, it is almost certain that it will be in the up-state in the next period. The up-state is characterized by high growth rates of consumption and a low volatility of consumption growth. On the other hand, in the up-state there exists a considerable probability of changing to the down-state ($\text{prob}_{21} = .312$), which is characterized by low consumption growth and high volatility. Thus, moving from the down-state to the up-state equals an increase in risk and a reduction in expected consumption growth, and vice versa. Barsky (1989) theoretically explores the impact of increased risk and reduced economic growth on the behavior of bond and stock prices in a simple two-period equilibrium model[10]. The present findings are consistent with Barsky's results. Using expected utility he shows that in response to an increase in risk or a reduction in expected economic growth the price of the risk-free asset always rises but stock prices fall (rise) when risk aversion $\alpha < 1$ ($\alpha > 1$) (Barsky 1989, p. 1139). Hence, within this model the values of stocks and risk-free assets move together for reasonable values of risk aversion ($\alpha > 1$). Only for $\alpha < 1$ the values of stocks and risk-free assets move in opposite directions. Here, the model with expected utility (UE1) has $\alpha > 1$. Thus, the prices of stocks and of the risk-free asset rise by moving from the down-state to the up-state which in turn simultaneously reduces the risk-free rate and the expected return on equity.

In model UN1b risk aversion α and the elasticity of intertemporal substitution σ are separately specified by using recursive non-expected utility. Barsky (1989) applies Selden's (1978) OCE-preferences to separate α and σ in his two-period world. He shows that the risk-free rate always falls as risk increases and that this effect is increasing in α and decreasing in σ (Barsky 1989, p. 1141). This is consistent with the observed large variation in the risk-free rate for model UE1 as compared to model UN1b. The parameters of risk aversion are almost equal in both models but UE1 implicitly assumes $\sigma = 1/4.26$ which is substantially smaller than the value $\sigma = .83$ which is explicitly chosen in model UN1b. In the present model the intuition for the high variability of the risk-free rate for low values of σ is as follows:

[10] Abel (1988) derives closed-form solutions for stock and bond prices in an infinite horizon model with time-varying risk and productivity growth and shows that the results of Barsky (1989) are robust with respect to the extension to the infinite horizon case.

Table 9.3: Conditional moments of real rates of return as implied by the calibrated equilibrium models UE1, UN1b and BN1b in comparison with the corresponding empirically observed conditional moments.

		State s	prob$_s$	r_s^f [%]	E$_s\{\tilde{r}^E\}$ [%]	E$_s\{\tilde{p}^E\}$ [%]	std$_s(\tilde{r}^E)$ [%]
Panel 1:		Univariate Markov state-switching process					
Moments implied by calibrated model[a]	UE1	1	.2375	1.583	2.027	.444	7.097
		2	.7625	.479	1.193	.715	13.373
	UN1b	1	.2375	.984	1.492	.508	7.055
		2	.7625	.669	1.364	.695	13.404
Observed moments (discrete approach)[b]	F-prob	1	.2319	.888	1.747	.859	6.433
		2	.7681	.694	1.204	.510	7.909
	S-prob	1	.2391	.775	1.075	.300	7.526
		2	.7609	.728	1.410	.682	7.623
Observed moments (regression approach)[c]	F-prob	1	.2341	1.114	3.399	2.294	
		2	.7659	.625	.698	.073	
		R^2		.032**	.005	.003	
	S-prob	1	.2346	1.016	3.004	1.988	
		2	.7654	.655	.817	.162	
		R^2		.019	.004	.002	
Panel 2:		Bivariate Markov state-switching process					
Moments i. b. c. m.[d]	BN1b	1	.2681	.443	1.742	1.299	5.145
		2	.7319	.853	1.266	.413	7.977
Observed moments (discrete approach)[b]	F-prob	1	.2609	.959	-1.090	-2.048	5.892
		2	.7391	.662	2.184	1.522	7.934
	S-prob	1	.2754	.940	-1.174	-2.114	6.193
		2	.7246	.663	2.282	1.618	7.856
Observed moments (regression approach)[c]	F-prob	1	.2437	1.026	-.710	-1.737	
		2	.7563	.647	1.987	1.341	
		R^2		.047**	.012	.015	
	S-prob	1	.2497	1.032	-1.549	-2.582	
		2	.7503	.642	2.288	1.647	
		R^2		.063***	.031**	.037**	

[a]UE1 is the calibrated model with time-additive expected utility ($\alpha = 4.26$, $\beta = 1.0159$, $L = .95$). UN1b is the calibrated model with recursive non-expected utility ($\alpha = 4.9$, $\beta = .999$, $\sigma = .83$, $L = .95$).

[b]F-prob (S-prob) denotes that the current state is inferred from filter (smoothed) probabilities. The state-variable is set equal to 1 if the probability for state one exceeds 30 percent. Otherwise, the state-variable is set equal to 2. The moments are calculated from empirically observed quarterly average returns before taxation (Q2, S1) over the periods following state 1 respectively state 2.

[c]The empirically observed quarterly average returns before taxation (Q2, S1) are regressed on the conditional filter (F-prob) respectively smoothed (S-prob) probabilities of state 1 at the beginning of the period. The moments for state 2 then equal the estimated intercept. The moments for state 1 equal the sum of the intercept and the slope coefficient. R^2 denotes the R^2-statistic from the regression. *** (**, *) indicates that the F-statistic is positive at a 1 (5, 10) percent level of significance.

[d]Conditional moments implied by the calibrated model BN1b, the model with recursive non-expected utility ($\alpha = 7.96$, $\beta = .999$, $\sigma = .79$, $L = .75$).

Faced with the alternating levels and risks of the exogenous endowment, individuals with strong preferences for a smooth consumption path (low σ) must face a large fluctuation in the risk-free rate to be satisfied with the exogenous endowment and not to have the desire to substitute intertemporally. Barsky (1989, p. 1141) further shows that the return on equity falls (rises) with increasing risk when $\sigma < 1$ ($\sigma > 1$). This is consistent with the finding that for model UN1b the expected return on equity is smaller in the up-state than in the down-state.

Next, consider the model BN1b. In contrast to the previous results the risk-free rate is now lower in the down-state than in the up-state. This is a consequence of the strong persistence of both states of the bivariate model. Given that the economy is in the down-state (up-state), it is likely that it will be in the down-state (up-state) in the next period, too. Therefore, expected economic growth is now lower and expected risk is higher in the down-state than in the up-state, leading to a lower risk-free rate in the down-state. The standard deviation of the return on equity does not vary as much as for the univariate endowment process. Nonetheless, the equity premia differ strongly across states. As for the univariate model the conditional equity premium and the risk-free rate move in opposite directions, but in contrast to the univariate model this is now also true for the return on equity and the risk-free rate. This is a consequence of the strong variation in the conditional equity premium. Interpreting the findings for model BN1b one must take into account that within this model the return on equity and economic growth are not strongly correlated. This situation corresponds to Barsky's (1989) model with two risky assets, one of which is what he calls "corporate captial"[11]. For corporate capital he does not obtain comparably clear results as for the stocks in the previous models because now the covariance between consumption and the return on corporate capital plays a crucial role. But, he illustrates "that very reasonable values of ... [α] are consistent with movements of debt and equity values in opposite directions" (Barsky 1989, p. 1143), which is the case in the present model BN1b, too.

Overall, these results indicate that the three models imply a considerable variation in conditional expected returns and risks. The structure of this variation is strongly affected by the structure of the underlying endowment process. For example, the short persistence of the down-state of the univariate model implies that the up-state is actually the state with higher uncertainty about the future and lower expected economic growth. Model BN1b, based on the bivariate process, implies that the values of stocks and risk-free assets move in opposite directions, whereas the models UE1 and UN1b, based on the univariate process, imply that the values of stocks and risk-free assets move together.

[11]Unfortunately, Barsky (1989) analyzes this model only for expected utility and not for the OCE-preferences.

Empirically Observed Conditional Moments Now, it is analyzed whether the conditional moments implied by the models correspond with the empirically observed conditional moments, i.e. the conditional forecasts of the calibrated models are compared with subsequently realized returns. Regarding how hard it has been to account for the unconditional moments of asset returns using consumption-based equilibrium asset pricing models, explaining conditional moments, which comes close to explaining time series behavior of asset returns, is not an easy task[12].

Determining the empirically observed conditional moments is a two-step procedure. The first step is to infer the state of the economy at each date because the state variable is not directly observable. For this task filter probabilities (F-prob) and, alternatively, smoothed probabilities (S-prob) both obtained from the EM-algorithm, can be used[13]. Given these state probabilities, the second step is to calculate the sample moments conditional on the state. Two alternative approaches are applied.

The first approach is to assume that the current state is the down-state when F-prob respectively S-prob of the down-state exceeds a fixed value. Intuitively, one would fix this value at .5. But since the probabilities are often between zero and .5 this would result in only few realizations of the down-state. Therefore, the value is set to .3 to match the unconditional probability of the down-state to the model's 23.75 percent. Since the economy by assumption switches between discrete states this approach is referred to as the "discrete approach". The observed conditional moments of the rates of return are calculated from the quarterly average returns before taxation (Q2, S1)[14] over all quarters following a down-state respectively an up-state between 1960 and 1994. Table 9.3 lists these moments for the univariate and the bivariate state-switching model.

The second approach explicitly captures the fact that the state variable is not directly observable and that therefore the probability of being in the down-state may not only be zero or one but may also be somewhere in between. In this case the conditional distribution of the rates of return is assumed to be a mixture of the conditional distribution in the down-state

---

[12]This may be the reason why, as to my knowledge, Kandel and Stambaugh (1990) is the only study that also investigates conditional moments implied by calibrated equilibrium models. Kandel and Stambaugh use a slightly different approach than applied here. This will be discussed later on in this section.

[13]Filter probabilities as well as smoothed probabilities are the probabilities of being in the down- respectively the up-state at time t, given the estimated Markov state-switching model. The filter probabilities are inferred from information up to date t only, whereas the smoothed probabilities are inferred from the information of the full sample. For predictions the filter probabilities should be used because they do not include future information (see also footnote 11 on page 167). For the univariate (bivariate) model the smoothed probabilities are graphed in figure 8.1 on page 170 (figure 8.12 on page 205).

[14]Alternatively using end-of-quarter returns before taxation (Q1, S1) does not qualitatively change the results discussed below.

and the conditional distribution in the up-state. The empirically observed means of these two distributions are obtained by regressing the empirically observed rates of return respectively excess returns on F-prob respectively S-prob of the down-state at the beginning of the period, i.e. $\tilde{r}_{t+1} = a + b \cdot \text{prob}_{1,t} + \varepsilon_t$. Within this "regression approach" the observed conditional mean return following an up-state equals a and the observed conditional mean return following a down-state equals $(a + b)$[15]. Table 9.3 contains these conditional moments and the R^2-statistics.

Let us first discuss the results for the univariate Markov state-switching model. As predicted by the models UE1 and UN1b, the conditional risk-free rate is smaller in the up-state than in the down-state for both approaches, but the observed conditional risk-free rates are closer to the predicted values when the regression approach is used. Using the regression approach or the discrete approach and the filter probabilities the mean return on equity following an up-state is smaller than that following a down-state. Thus, in three cases the empirical findings are consistent with the predicted comovement of the conditional expected return on stocks and the conditional risk-free rate. Using the regression approach, the variation in the observed conditional mean returns on equity is much stronger than predicted by the equilibrium models. The conditional equity premium is smaller in the down-state as predicted by the models only for the discrete approach and smoothed probabilities. In the other cases the conditional risk-free rate and the conditional equity premium move together which is in contrast to the models. The conditional standard deviation of the return on equity is only estimated for the discrete approach. It turns out that the conditional standard deviations are almost equal across states. This is inconsistent with the implications of the equilibrium models. Nevertheless, the findings are more promising than those obtained by Kandel and Stambaugh (1990). Here, the ex-post realized risk-free rate and the return on equity are consistent with the predictions of the equilibrium models, whereas Kandel and Stambaugh (1990) find that "the pattern of the real rate is at odds not only with the numerical example of the model ..., but also with any other parametrization of the Lucas framework that uses conditional moments of consumption as state variables." (Kandel and Stambaugh 1990, p. 224). However, as is discussed in the next paragraph, the results for the bivariate Markov state-switching model are as discouraging as those of Kandel and Stambaugh (1990).

[15]The approach of Kandel and Stambaugh (1990) is related to the regression approach applied here. Instead of regressing the empirically observed rates of return on the state variable at the beginning of the period, they regress the observed rates of return on the lagged conditional means and standard deviations of quarterly consumption growth which are estimated as linear functions of three predetermined financial variables.

For the bivariate Markov state-switching model the empirically observed conditional risk-free rate and mean return on equity respectively the equity premium move in opposite directions as predicted by the model. But the empirically observed risk-free rate is low in the up-state and not in the down-state and the empirically observed mean return on equity respectively the equity premium is low in the down-state and not in the up-state. An explanation of this finding may be as follows. In contrast to the univariate model the bivariate model may not identify states of economic growth relevant for the risk-free rate. The state of the bivariate state-switching model is mainly determined by the growth rate of the payoff on equity[16]. Furthermore, other studies, e.g., Fama and French (1988b), find that the dividend-price ratio, which is related to the payoff on equity, has explanatory power for long-horizon returns. This suggests that the quarterly time intervals may be too short to confirm the predictions of the equilibrium model.

Overall, the analysis shows that the calibrated equilibrium models imply a substantial variation in the conditional risk-free rate, the conditional expected return on equity, the conditional equity premium and the conditional volatility of the return on equity. The direction of the movements of these conditional moments is consistent with the theoretical results. For the univariate Markov state-switching model the models' implications for the variation in the conditional moments are in most cases consistent with the empirically observed variation in the moments of subsequently realized returns. On the other hand, for the bivariate Markov state-switching model the observed and the implied variations in conditional moments are exactly contrary.

[16]The growth rate of the payoff on equity differs substantially across states, which is not the case for the growth rate of consumption (see table 8.7 on page 203).

9.3 Summary of Chapter 9

A selection of calibrated equilibrium models enclosing models with time-additive expected utility and models with recursive non-expected utility as well as models with the univariate exogenous process and models with the bivariate exogenous process has been assessed with respect to various properties of the implied rates of return. This assessment yields the following results.

The calibrated model with recursive non-expected utility based on the bivariate exogenous process is consistent with the empirically observed skewness, kurtosis and first-order autocorrelation of the return on equity and the equity premium. All models are consistent with the empirically observed correlation between the risk-free rate and the return on equity respectively the equity premium. The calibrated equilibrium models are therefore consistent with stylized facts of asset returns that cannot be explained with constant conditional expected returns and risks. However, none of the models generates the empirically observed second moments. The implied standard deviation of the risk-free rate is too low and that of the equity premium and the return on stocks is too high.

The calibrated equilibrium models imply as much negative serial correlation in multiperiod returns on equity and multiperiod equity premia as is empirically observed. Long-term mean reversion in stock prices is therefore consistent with the equilibrium models. Large sample simulations indicate that most of the mean reversion in both the observed rates of return and the implied equilibrium rates of return may be attributed to the small sample bias.

The calibrated equilibrium models imply a substantial variation in conditional expected returns and risks. The structure of this variation is strongly affected by the structure of the underlying endowment process. The models based on the univariate exogenous process imply that the values of stocks and risk-free assets move together, whereas the model based on the bivariate exogenous process implies that both move in opposite directions. For the equilibrium models based on the univariate process the models' implications for the variation in the moments of subsequent returns, i.e. the models' conditional forecasts for next period's returns, are consistent with subsequently realized returns. This is not the case for the equilibrium model based on the bivariate process. For this model the forecasts and the subsequently realized moments move in opposite directions between the states.

Chapter 10

Conclusion

The purpose of this study was to explain both the average levels and the co-movement of the German risk-free interest rate and the German equity premium over the 35 years from 1960 to 1994. In particular, the study analyzed whether intertemporal equilibrium asset pricing models can explain the empirically observed risk-free rates and equity premia in Germany, maintaining the assumptions of complete and frictionless markets and rational investors. The study contributes to the rapidly growing literature on the Equity Premium Puzzle and the Risk-free Rate Puzzle initiated by Mehra and Prescott (1985).

The comparison of the average risk-free rates and average equity premia of eight industrialized countries revealed that the real risk-free rate was relatively high and the real equity premium was relatively low in Germany in the period under consideration. The differences to the other countries' values, however, are statistically insignificant. The detailed investigation of historical German rates of return (chapter 6) yielded an average real risk-free rate of about 3.0 percent and an average real equity premium of about 3.3 percent per annum, both before taxation. Taking taxes into account, for private investors the average real risk-free rate declines to -.4 percent and the average real equity premium rises up to 5 percent per annum with increasing tax rates; for corporate investors the average real risk-free rate was -.3 percent and the average real equity premium was 1.4 percent per annum. Measured over quarterly time intervals and before taxation, the real risk-free rate and the real equity premium were slightly negatively correlated, whereas the correlation between the real risk-free rate and the real return on stocks was slightly positive.

To account for this evidence, different intertemporal equilibrium asset pricing models were derived and empirically tested by two approaches: the variance bound approach and the calibration and simulation approach. Chapter 2 started the analysis with an introduction to intertemporal asset pricing using the market pricing kernel framework that places very few restrictions a priori on the structure of asset pricing models. All existing asset pricing models, whether static or intertemporal, can be written in the form of the market pricing kernel. The existence of such a pricing kernel is guaranteed by the Law of One Price. The market pricing kernel is positive under the absence of arbitrage opportunities and uniquely specified if the market is complete. A linear conditional asset pricing model in terms of the market pricing kernel was derived exclusively assuming the Law of One Price.

The knowledge of the distribution of the market pricing kernel for a specific date enables the pricing of any payoff at that date. Chapter 3 surveyed different approaches attempting to reveal properties of the market pricing kernel without assuming a specific parametric model. Theoretical approaches have shown that there exist strong links between the properties of the market pricing kernel and the properties of asset prices. This has broad implications for models that restrict either the properties of the market pricing kernel or those of the asset prices. Empirical approaches either estimate the distribution of the market pricing kernel conditional on one date from cross-sectional asset prices or place restrictions on the moments of the unconditional distribution of the market pricing kernel using time series data. The variance bound approach of Hansen and Jagannathan (1991) is to place restrictions on the first and second unconditional moment of the market pricing kernel using the first and second unconditional moments of asset returns. This approach proved to be useful (1) for analyzing the information content of different asset return data for intertemporal asset pricing and (2) for comparing the potential of different parametric models of the market pricing kernel to be consistent with these data.

So far the analysis was concerned with intertemporal asset pricing without assuming a specific functional form of the market pricing kernel. Chapter 4 reviewed different approaches for specifying parametric models of the market pricing kernel. Wealth-based one-period models such as the Capital Asset Pricing Model (CAPM), when applied intertemporally, restrict the expected returns and risks to be constant over time, or at least to be perfectly predictable. Widely applied multifactor models allowing for variations in expected returns and risks have a disadvantage in that their factors are not theoretically specified so that these models are not easy to interpret. However, the main argument for using consumption-based equilibrium models derived from individuals' preferences was that these models determine the risk-free rate and the risk premium on the market portfolio endogenously, whereas the other approaches take both as exogenously given.

In the consumption-based equilibrium asset pricing framework asset prices are related to consumption. In equilibrium, asset prices and consumption are such that no individual can improve utility by intertemporally substituting consumption or by purchasing or selling assets. Within this framework the market pricing kernel equals the intertemporal marginal rate of substitution of a representative agent. The analysis started with a characterization of the individuals' preferences. It was demonstrated that in the intertemporal context time-additive expected utility implies strong behavioral restrictions such as the indifference to the timing of the resolution of uncertainty and the identity of risk aversion and aversion to intertemporal fluctuations in expected consumption levels. These restrictions may not be satisfied in reality. Therefore, the analysis was generalized to the less restrictive recursive non-expected preferences proposed by Kreps and Porteus (1978). For homothetic parametrizations of these preferences the conditions for the optimal consumption and investment choice were derived. After discussing the existence of a representative agent, the optimality conditions were applied to three alternative parametrizations of the recursive preferences, namely, time-additive expected utility with constant relative risk aversion, time-additive expected logarithmic utility and the recursive non-expected utility suggested by Epstein and Zin (1989). The first parametrization leads to the standard Consumption-based CAPM (C-CAPM). As a consequence of the myopic behavior of individuals with logarithmic utility, the second parametrization yields an asset pricing model closely related to the static CAPM. Numerical examples illustrated that the third parametrization allows to separate risk aversion and the elasticity of intertemporal substitution and that an early (late) resolution of uncertainty is preferred if risk aversion is larger (smaller) than the reciprocal of the elasticity of intertemporal substitution. This parametrization generates a two-factor CAPM encompassing both the C-CAPM and the static CAPM as special cases.

The main findings from the empirical analysis of the variance bounds placed on the market pricing kernel by German stock, bond and money market returns in the period 1968 to 1994 (chapter 7) are as follows.

(1) The joint analysis of one-period risk-free rates and aggregate stock returns places stronger bounds on the parametric models of the market pricing kernel than equity premia or aggregate stock returns taken separately. Adding a bond market index to the set of assets does not sharpen the variance bound, which indicates that aggregate bond returns contain no additional information about the distribution of the market pricing kernel.

(2) Similar to the results for the US it is found that a parametric model of the market pricing kernel derived from time-additive expected utility with constant relative risk aversion does not exhibit sufficient variability to satisfy the variance bounds for plausible parameters of risk aversion and time preference. The parameter of relative risk aversion must be larger than 15 for the market pricing kernel to be sufficiently variable to comply with the bound calculated from the equity premia. This finding illustrates the Equity Premium Puzzle for Germany. To satisfy the variance bound implied by risk-free rates and aggregate stock returns either the discount factor must be increased beyond one, given a relative risk aversion around 15, or rela-

tive risk aversion must be increased to values larger than 100. This points to the Risk-free Rate Puzzle. The occurrence of both puzzles in this general framework indicates that the puzzles do not result from special analysis tools such as estimation or calibration.

(3) Changing the length of the time intervals over which returns are measured, calculating returns from period average prices or using tax adjusted returns does not substantially alter these results, indicating that such modifications do not help to solve the puzzles.

(4) The parametric model of the market pricing kernel derived from recursive non-expected utility satisfies the variance bound calculated from risk-free rates and aggregate stock returns with a parameter of relative risk aversion around five and an elasticity of intertemporal substitution around .4. A time-additive expected utility model with relative risk aversion of five would have implied an elasticity of intertemporal substitution of .2. This indicates that German investors are less averse to intertemporal fluctuations in expected consumption than implied by time-additive expected utility and that the investors prefer an early resolution of uncertainty, a result that corresponds to the findings of experimental research in decision making. Thus, using recursive non-expected utility may help to solve the Equity Premium Puzzle and the Risk-free Rate Puzzle.

A comparison of the methodological issues of estimation and calibration approaches (chapter 5) revealed that the calibration approach can provide stronger evidence for or against a model's underlying economic theory and can indicate why a specific parametric model of the market pricing kernel fails to produce equilibrium rates of return that mimic observed data. Therefore, the calibration approach was applied in chapter 8. It is based on a general equilibrium framework with exogenous production. Stocks were alternatively modeled as (1) a claim to the exogenous consumption stream, (2) a levered claim to the exogenous consumption stream, or (3) a claim to an exogenous dividend stream that is different from the exogenous consumption stream. The exogenous consumption and dividend streams were assumed to follow Markov state-switching processes with two states. State-switching processes for the exogenous endowments are preferable because they allow to model time-varying conditional expectations, heteroscedasticity, and especially a time-varying one-period risk-free rate of return and still permit closed-form solutions for the endogenous equilibrium rates of return. The estimated bivariate Markov state-switching process for consumption and dividend growth identifies up- and down-states more clearly than the estimated univariate Markov state-switching process for consumption growth. Allowing the variances and the covariances to differ across the states yielded estimation results consistent with the empirical finding of increased volatilities during recessions. The evaluation of the estimated Markov state-switching models showed that only the bivariate switching model explains the data significantly better than a stationary i.i.d. process. However, both models capture features of the data that alternative models cannot capture.

The results from the calibration exercises can be summarized as follows:

(1) When stocks are modeled as a claim to the exogenous consumption stream and time-additive expected utility with constant relative risk aversion and a discount factor β below one are used, the implied equity premium does not exceed .052 percent per annum for risk-free rates less than four percent per annum. Increasing the parameter of risk aversion simultaneously raises the equity premium and the risk-free rate due to the interrelation of risk aversion and elasticity of intertemporal substitution. This interrelation can justify the use of discount factors beyond one as a compensation for the low elasticity of intertemporal substitution implicit in high levels of risk aversion. With β larger than one equity premia up to .36 percent per annum are obtained with relative risk aversion below ten. Choosing β beyond one in time-additive expected utility only yields the same results as models that separately specify risk aversion and elasticity of intertemporal substitution if the exogenous process is i.i.d.. If this is not true, using recursive non-expected utility gives more satisfying results from a theoretical point of view. With recursive non-expected utility a change in risk aversion affects the risk premium but has nearly no effect on the risk-free rate; a change in the discount factor or in the elasticity of intertemporal substitution affects the risk-free rate, leaving the risk premium almost unchanged. However, also with recursive non-expected utility the observed equity premium is not obtainable with plausible parameters of risk aversion. For a parameter of relative risk aversion of ten the implied equity premium is only about .28 percent per annum. Hence, in the specified economy both utility specifications do not allow to simultaneously match the first moments of the empirically observed one-period risk-free rate and equity premium.

(2) When stocks are modeled as a levered claim to the exogenous consumption stream, both time-additive expected utility and recursive non-expected utility imply higher equity premia. With a reasonable level of leverage the parameters of relative risk aversion required to generate the historical equity premium lies in a plausible range between two and six. Matching the historical risk-free rate with time-additive expected utility necessitates discount factors between 1.01 and 1.035. With recursive non-expected utility and a discount factor less than one, the necessary elasticity of intertemporal substitution of about .8 indicates a preference for an early resolution of uncertainty.

(3) When stocks are modeled as a claim to an exogenous dividend stream which differs from consumption the implied equity premium lies between that of the univariate model without leverage and that of the univariate model with leverage. Thus, allowing for the dividends to be different from

aggregate consumption does not substantially increase the implied equity premium. This follows from the fact that the larger variability of dividends as compared to consumption increases the equity premium, but, on the other hand, relaxing the assumption of a strong correlation between dividends and consumption lowers the equity premium. The empirically observed low correlation between dividends and aggregate consumption may arise because aggregate consumption may not be a good proxy for the consumption of equityholders. But even a higher correlation cannot account for the historical equity premium. Introducing leverage in addition increases the implied equity premium. However, only with recursive non-expected utility the historical equity premium can be generated in the bivariate model.

The further assessment of calibrated equilibrium models, i.e. models that match the first moments of the historical risk-free rate and equity premium, (chapter 9) yielded the following results:

(1) For all calibrated models the hypothesis that the empirically observed second moments are generated by the calibrated model and the underlying state-switching economy must be rejected. The implied standard deviation of the risk-free rate is too low and that of the equity premium and the return on stocks it too high.

(2) All calibrated models are consistent with the empirically observed first-order autocorrelation of the return on equity and the equity premium, the correlation between the risk-free rate and the return on equity and the correlation between the risk-free rate and the equity premium. The model with recursive non-expected utility based on the bivariate exogenous process is furthermore consistent with the empirically observed skewness and kurtosis of the return on equity and the equity premium.

(3) The calibrated models generate long-horizon predictability of stock index returns and equity premia. Long-term mean reversion in stock prices is therefore consistent with the equilibrium models and, hence, does not indicate market inefficiency.

(4) The calibrated equilibrium models imply a substantial variation in conditional expected returns and risks. The univariate models' implications for the variation in the moments of subsequent returns are consistent with subsequently realized returns. The bivariate model's implications are contrary to the subsequently realized returns.

Overall, using recursive non-expected utility, a reasonable level of leverage and plausible preference parameters, representative-agent intertemporal equilibrium asset pricing models can explain both the average levels and the co-movement of the German risk-free interest rate and the German equity premium. However, the models cannot account for the second moments of the risk-free rate and the equity premium. Especially the implied standard deviation of the equity premium is generally much higher than the empirically observed standard deviation. This observation does not raise a new puzzle stating that the observed volatility of the equity premium is too low to be explained, but can be attributed to the Equity Premium Puzzle. The first moment of the empirically observed equity premium can only be explained with reasonable levels of risk aversion if the volatility of the equity premium is unrealistically high. Hence, the historical equity premium is too high to be explained with reasonable levels of risk aversion, given its "low" volatility. Nevertheless, the calibrated models based on the state-switching endowment processes are consistent with stylized facts of asset returns such as skewness, kurtosis, autocorrelation, their long-term predictability and their correlation with the risk-free rate that cannot be explained with constant conditional expected returns and risks.

This study neither claims to solve the Equity Premium Puzzle and the Risk-free Rate Puzzle nor to produce a generally accepted model of the market pricing kernel. But since the analyzed representative-agent models are the basis of wide areas of theoretical research, it is worth searching for improvements of these models such that they account for at least economically important aspects of the data. Recursive non-expected preferences, leverage and state-switching economic growth, for example, proved to be helpful. The search for a better performing market pricing kernel can be carried on along many other directions than those followed here, e.g., by incorporating market frictions or market incompleteness. Also taking into account investor heterogeneity may provide valuable insights. In the general equilibrium framework for the calibration approach it may further be beneficial to include production into the model. This may account for empirical findings such as highly correlated stock returns and investment growth.

Altogether, this study provided insights into the basic mechanisms of intertemporal equilibrium asset pricing models derived from the consumption and investment choice of individuals. In general, hoping to explain all aspects of the data with models as simple and as aggregated as the analyzed representative-agent models is somewhat presumptuous. But it was shown that with reasonable and not very complicated modifications of the standard intertemporal equilibrium models important properties of historical German rates of return can be explained with plausible assumptions about individuals' preferences, maintaining the assumptions of complete and frictionless markets and rational investors.

Appendix

A.1 Expected Value of the Product of Jointly Lognormally Distributed Variables

For a lognormally distributed random variable $\tilde{x}$, i.e. $\ln \tilde{x}$ is normally distributed with mean $\mathrm{E}\{\ln \tilde{x}\}$ and variance $\mathrm{var}(\ln \tilde{x})$, the expected value is given by (Patel and Read 1982, p. 24)

$$\ln \mathrm{E}\{\tilde{x}\} = \mathrm{E}\{\ln \tilde{x}\} + \frac{1}{2}\mathrm{var}(\ln \tilde{x}).$$

A similar expression is now derived for the expected value of the product of K jointly lognormally distributed random variables $\tilde{x}_k$, $k \in \{1, \ldots, K\}$, $K \geq 1$. This expression allows to calculate the expected value of the product of jointly lognormally distributed random variables, given the expected values, variances and covariances of the individual variables. The following derivation is based on the fact that the product of lognormally distributed random variables is lognormally distributed itself.

$$\begin{aligned}
\ln \mathrm{E}\left\{\prod_{k=1}^{K} \tilde{x}_k\right\} &= \mathrm{E}\left\{\ln\left(\prod_{k=1}^{K} \tilde{x}_k\right)\right\} + \frac{1}{2}\mathrm{var}\left(\ln\left(\prod_{k=1}^{K} \tilde{x}_k\right)\right) \\
&= \mathrm{E}\left\{\sum_{k=1}^{K} \ln \tilde{x}_k\right\} + \frac{1}{2}\mathrm{var}\left(\sum_{k=1}^{K} \ln \tilde{x}_k\right) \\
&= \sum_{k=1}^{K} \mathrm{E}\{\ln \tilde{x}_k\} + \frac{1}{2}\sum_{k=1}^{K}\sum_{l=1}^{K} \mathrm{cov}(\ln \tilde{x}_k, \ln \tilde{x}_l).
\end{aligned}$$

The expected value of the product of two jointly lognormally distributed random variables is hence given by

$$\begin{aligned}
\ln \mathrm{E}\{\tilde{x}_1 \tilde{x}_2\} &= \mathrm{E}\{\ln \tilde{x}_1\} + \mathrm{E}\{\ln \tilde{x}_2\} \\
&\quad + \frac{1}{2}\Big(\mathrm{var}(\ln \tilde{x}_1) + \mathrm{var}(\ln \tilde{x}_2) + 2\mathrm{cov}(\ln \tilde{x}_1, \ln \tilde{x}_2)\Big).
\end{aligned}$$

For the product of three random variables it follows

$$\begin{aligned}\ln \mathrm{E}\{\tilde{x}_1\tilde{x}_2\tilde{x}_3\} &= \mathrm{E}\{\ln \tilde{x}_1\} + \mathrm{E}\{\ln \tilde{x}_2\} + \mathrm{E}\{\ln \tilde{x}_3\} \\ &+\frac{1}{2}\Big(\mathrm{var}(\ln \tilde{x}_1) + \mathrm{var}(\ln \tilde{x}_2) + \mathrm{var}(\ln \tilde{x}_3) \\ &+2\mathrm{cov}(\ln \tilde{x}_1, \ln \tilde{x}_2) \\ &+2\mathrm{cov}(\ln \tilde{x}_1, \ln \tilde{x}_3) \\ &+2\mathrm{cov}(\ln \tilde{x}_2, \ln \tilde{x}_3)\Big).\end{aligned}$$

A.2 Additional Tables and Figures

For further illustration additional tables and figures are given in this appendix. These tables and figures are referred to in chapter 7, chapter 8 and chapter 9.

Table A.1: Markov state-switching model with two states and state-dependent normal distributions for the logarithm of the quarterly real percentage growth of per capita consumption expenditures on nondurables and services (CNS-Q) ("first optimum").

Panel 1:	Maximum Likelihood Estimates[a]			
	$s = 1$		$s = 2$	
prob_{ss}	.98941	(.01449)[b]	.99075	(.01227)
prob_s[c]	.46631		.53369	
μ_{cs} [%]	.23152	(.09330)	.80339	(.10339)
σ^2_{cs} [$\%^2$]	.48081	(.10363)	.69748	(.12546)
Panel 2:	**Unconditional Moments**			
	Observed		Implied[d]	
$\text{E}\{\ln \tilde{\lambda}\}$ [%]	.55715		.53672	
$\text{var}(\ln \tilde{\lambda})$ [$\%^2$]	.68933		.67783	
Skewness	-.11986		.16014	
Excess Kurtosis	.73620		.03220*	
$\text{corr}(\ln \tilde{\lambda}_t, \ln \tilde{\lambda}_{t-1})$	-.08029		.11769	
Panel 3:	**LR[e] Statistic**			
LR	10.7024			

[a] prob_{ss} denotes the conditional probability for staying in state s, prob_s denotes the unconditional probability of state s, μ_{cs} [σ^2_{cs}] denotes the mean [variance] of the state-dependent normal distribution for logarithmic consumption growth $\ln \tilde{\lambda}$.

[b] Standard deviations in parentheses.

[c] The unconditional stationary probabilities of the states are calculated as $\text{prob}_1 = (1 - \text{prob}_{22})/(2 - \text{prob}_{11} - \text{prob}_{22})$ and $\text{prob}_2 = 1 - \text{prob}_1$.

[d] The calculation of the implied unconditional moments is described in the footnote of table 8.1.

*** (**, *) indicates that the hypothesis that the implied value is equal to the observed value can be rejected at a 1 (5, 10) percent level of significance based on a simulation exercise similar to that described in section 8.1.3.3.

[e] Likelihood ratio statistic: twice the difference of the maximized loglikelihood value of the Markov state-switching model (H_1) and that using a single normal distribution (H_0).

Table A.2: Markov state-switching model with two states and state-dependent bivariate normal distributions for the logarithm of the quarterly real percentage growth of per capita consumption expenditures on nondurables and services (CNS-Q) and the logarithm of the quarterly real percentage growth of per capita corporate dividends (D-Q).

Panel 1:	Maximum Likelihood Estimates[a]			
	$s=1$		$s=2$	
prob_{ss}	.73335	(.13168)	.93829	(.05129)
prob_s	.18794		.81206	
μ_{cs} [%]	.26094	(.26907)	.62269	(.09600)
μ_{es} [%]	-1.32158	(.58973)	1.10054	(.40343)
σ^2_{cs} [%^{2}]	1.14326	(.54120)	.55914	(.07828)
σ^2_{es} [%^{2}]	5.19941	(2.30646)	5.42258	(.91624)
σ_{ces} [%^{2}]	2.05805	(1.24991)	.26066	(.23721)
Panel 2:	**Unconditional Moments**			
	$\ln\tilde{\lambda}$		$\ln\tilde{\gamma}$	
	Observed	Implied[b]	Observed	Implied
E{·} [%]	.55715	.55470	.66175	.64533
var(·) [%^{2}]	.68933	.68889	6.29759	6.27598
Skewness	-.11986	-.17709	.23042	-.07035
Excess Kurtosis	.73620	.42170	.25341	-.00718
$\text{corr}(\cdot_t, \cdot_{t-1})$	-.08029	.01947	.08615	.09582
$\text{corr}(\ln\tilde{\lambda}_t, \ln\tilde{\gamma}_t)$	.34625	.35213		
Panel 3:	**LR[c] Statistic and Significance Level**			
LR	11.0250			
Significance level	.50	.10	.05	.01
UTF-LR[d]	7.9864	15.34362	17.7608	23.3043

[a] prob_{ss} denotes the conditional probability for staying in state s, prob_s denotes the unconditional probability of state s. The unconditional probabilities are calculated as $\text{prob}_1 = (1 - \text{prob}_{22})/(2 - \text{prob}_{11} - \text{prob}_{22})$ and $\text{prob}_2 = 1 - \text{prob}_1$. μ_{cs} [σ^2_{cs}] denotes the mean [variance] of the state-dependent normal distribution for logarithmic consumption growth $\ln\tilde{\lambda}$, μ_{es} [σ^2_{es}] denotes the mean [variance] of the state-dependent normal distribution for the logarithmic growth rate of corporate dividends $\ln\tilde{\gamma}$, σ_{ces} denotes the state-dependent covariance of both growth rates. Standard deviations are given in parentheses.

[b] The calculation of the implied unconditional moments is described in the footnote of table 8.1. The implied unconditional correlation of consumption growth and dividend growth is calculated as follows:
$\text{corr}(\ln\tilde{\lambda}_t, \ln\tilde{\gamma}_t) = \sum_{i=1}^{2} \text{prob}_i(\sigma_{cei} + (\mu_{ci} - \mu_c)(\mu_{ei} - \mu_e))/(\sigma_c\sigma_e)$. *** (**, *) indicates that the hypothesis that the implied value is equal to the observed value can be rejected at a 1 (5, 10) percent level of significance based on a simulation exercise similar to that described in section 8.4.4.3.

[c] Likelihood ratio statistic: twice the difference of the maximized loglikelihood value of the bivariate Markov state-switching model (H_1) and that using a single bivariate normal distribution (H_0).

[d] Upper tail fractiles of the simulated distribution of the LR statistic for the null of an i.i.d. bivariate normally distributed process.

Table A.3: Means and fractiles of the simulated distributions of the first four moments and the first-order autocorrelation of the one-period risk-free rate, the return on equity and the equity premium implied by the calibrated model BN1b. For the return on equity and the equity premium also the simulated distribution of the correlation with the one-period risk-free rate is given. The simulation consists of 10,000 time series with 139 periods.

		Simulation Fractiles						
	Mean	.5 %	2.5 %	5 %	50 %	95 %	97.5 %	99.5 %
Panel 1:	One-period risk-free rate							
$E\{\tilde{r}^f\}$ [%]	.7424	.6243	.6569	.6688	.7460	.8053	.8142	.8320
$std(\tilde{r}^f)$ [%]	.1761	.0902	.1201	.1316	.1805	.2041	.2050	.2055
Skewness $\tilde{r}^f$	-1.1560	-4.0504	-2.7482	-2.3732	-1.0773	-.2017	-.0861	.2309
Kurtosis $\tilde{r}^f$	-.1562	-2.0136	-1.9936	-1.9607	-.8453	3.6586	5.5935	14.5112
$corr(\tilde{r}^f_{t-1}, \tilde{r}^f_t)$	.7360	.4107	.5501	.5959	.7451	.8469	.8653	.9022
Panel 2:	Return on equity							
$E\{\tilde{r}^E\}$ [%]	1.3813	-2.4028	-1.4255	-.9709	1.4218	3.5677	3.9439	4.6889
$std(\tilde{r}^E)$ [%]	16.7283	11.9961	13.1465	13.7867	16.7639	19.5491	20.0610	20.9422
Skewness $\tilde{r}^E$	-.0704	-.9514	-.7266	-.6089	-.0631	.4483	.5464	.7535
Kurtosis $\tilde{r}^E$	2.4445	.8424	1.1148	1.2694	2.3174	4.0477	4.5046	5.5115
$corr(\tilde{r}^E_{t-1}, \tilde{r}^E_t)$	-.0096	-.2492	-.1903	-.1612	-.0087	.1381	.1626	.2154
$corr(\tilde{r}^f, \tilde{r}^E)$	-.0303	-.3273	-.2398	-.2011	-.0302	.1423	.1755	.2585
Panel 3:	Equity premium							
$E\{\tilde{p}^E\}$ [%]	.6390	-3.0621	-2.1095	-1.6602	.6809	2.7795	3.1413	3.8768
$std(\tilde{p}^E)$ [%]	16.7336	12.0005	13.1484	13.7801	16.7674	19.5751	20.0733	20.9692
Skewness $\tilde{p}^E$	-.0581	-.9411	-.7140	-.5987	-.0508	.4625	.5632	.7728
Kurtosis $\tilde{p}^E$	2.5083	.8955	1.1655	1.3235	2.3833	4.1233	4.5903	5.6089
$corr(\tilde{p}^E_{t-1}, \tilde{r}^E_t)$	-.0147	-.2537	-.1941	-.1662	-.0139	.1321	.1564	.2089
$corr(\tilde{r}^f, \tilde{p}^E)$	-.0408	-.3343	-.2480	-.2092	-.0409	.1301	.1635	.2460

Figure A.1: Bounds imposed on the standard deviation of the market pricing kernel by real **end-of-quarter returns before taxation** (Q1, S1) of different sets of assets. Mean-standard deviation pairs of a market pricing kernel derived from **time-additive expected utility** with constant relative risk aversion are shown for different levels of risk aversion α (a) and a discount factor $\beta = 1$. To illustrate the effect of large values of risk aversion, the lower graph shows the same variance bounds as the upper graph using a different scale. The mean-standard deviation pairs using the time series of consumption expenditures on nondurables and services (CNS) are represented by diamond symbols, those using the time series of total consumption expenditures (C) are represented by triangle symbols.

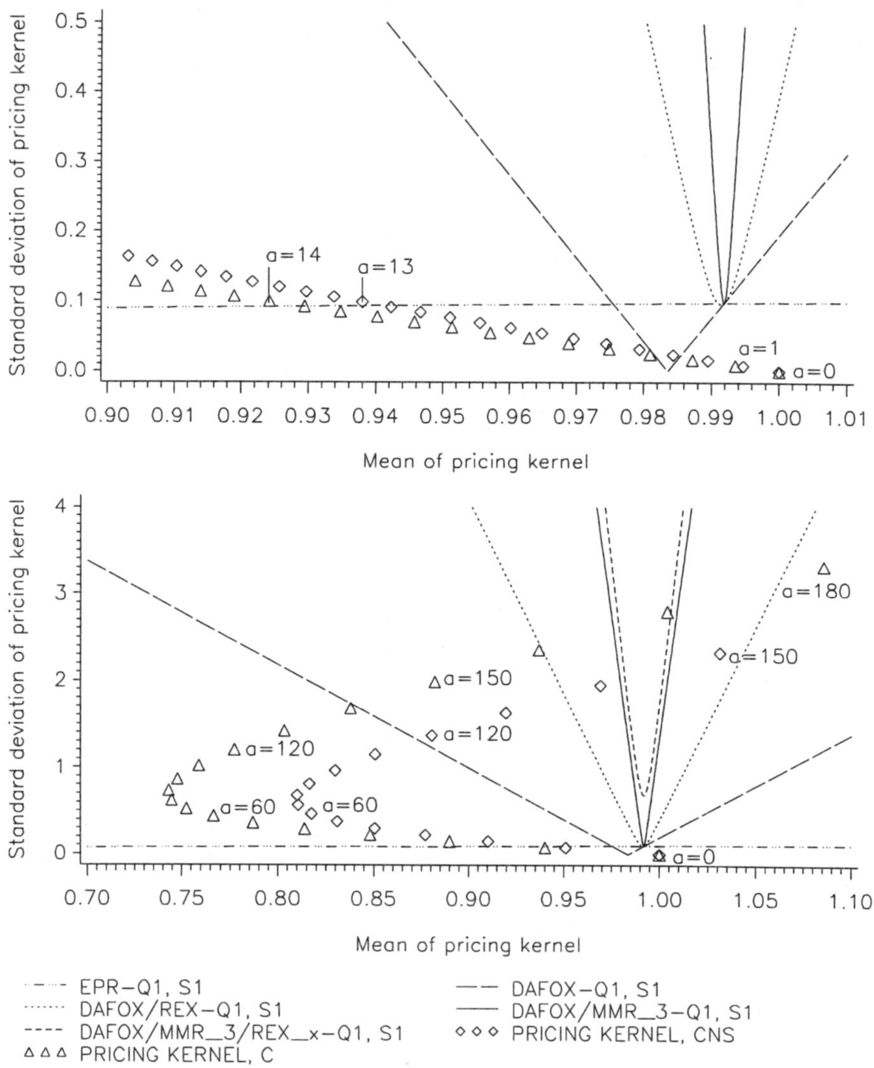

Figure A.2: Bounds imposed on the standard deviation of the market pricing kernel by real **tax adjusted quarterly average returns** (Q2, S2) of different sets of assets. Mean-standard deviation pairs of a market pricing kernel derived from **time-additive expected utility** with constant relative risk aversion are shown for different levels of risk aversion α (a) and a discount factor $\beta = 1$. To illustrate the effect of large values of risk aversion, the lower graph shows the same variance bounds as the upper graph using a different scale. The mean-standard deviation pairs using the time series of consumption expenditures on nondurables and services (CNS) are represented by diamond symbols, those using the time series of total consumption expenditures (C) are represented by triangle symbols.

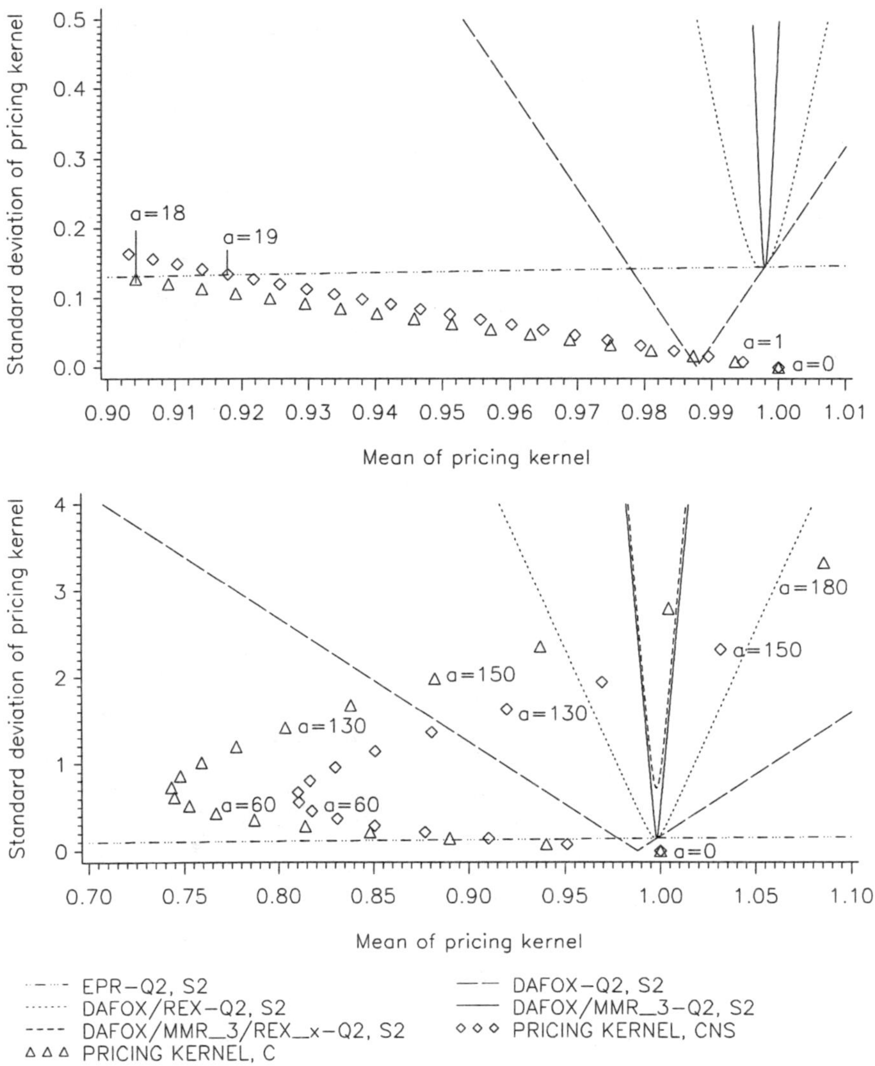

Figure A.3: Bounds imposed on the standard deviation of the market pricing kernel by real **end-of-quarter returns before taxation** (Q1, S1) of different sets of assets. Mean-standard deviation pairs of a market pricing kernel derived from **recursive non-expected utility** are shown for different levels of risk aversion α (a), different levels of elasticity of intertemporal substitution σ (s) and a discount factor $\beta = 1$. In the upper graph the time series of consumption expenditures on nondurables and services (CNS) is used; in the lower graph the time series of total consumption expenditures (C) is used.

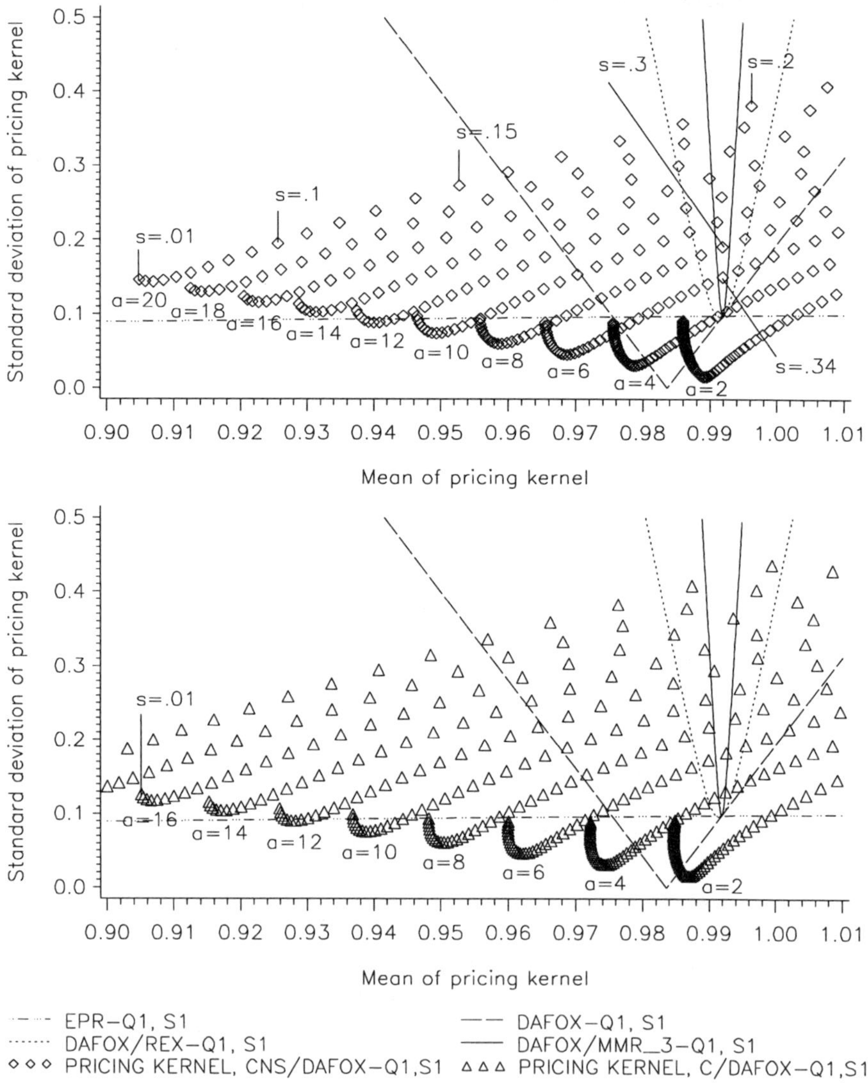

Figure A.4: Bounds imposed on the standard deviation of the market pricing kernel by real **tax adjusted quarterly average returns** (Q2, S2) of different sets of assets. Mean-standard deviation pairs of a market pricing kernel derived from **recursive non-expected utility** are shown for different levels of risk aversion α (a), different levels of elasticity of intertemporal substitution σ (s) and a discount factor $\beta = 1$. In the upper graph the time series of consumption expenditures on nondurables and services (CNS) is used; in the lower graph the time series of total consumption expenditures (C) is used.

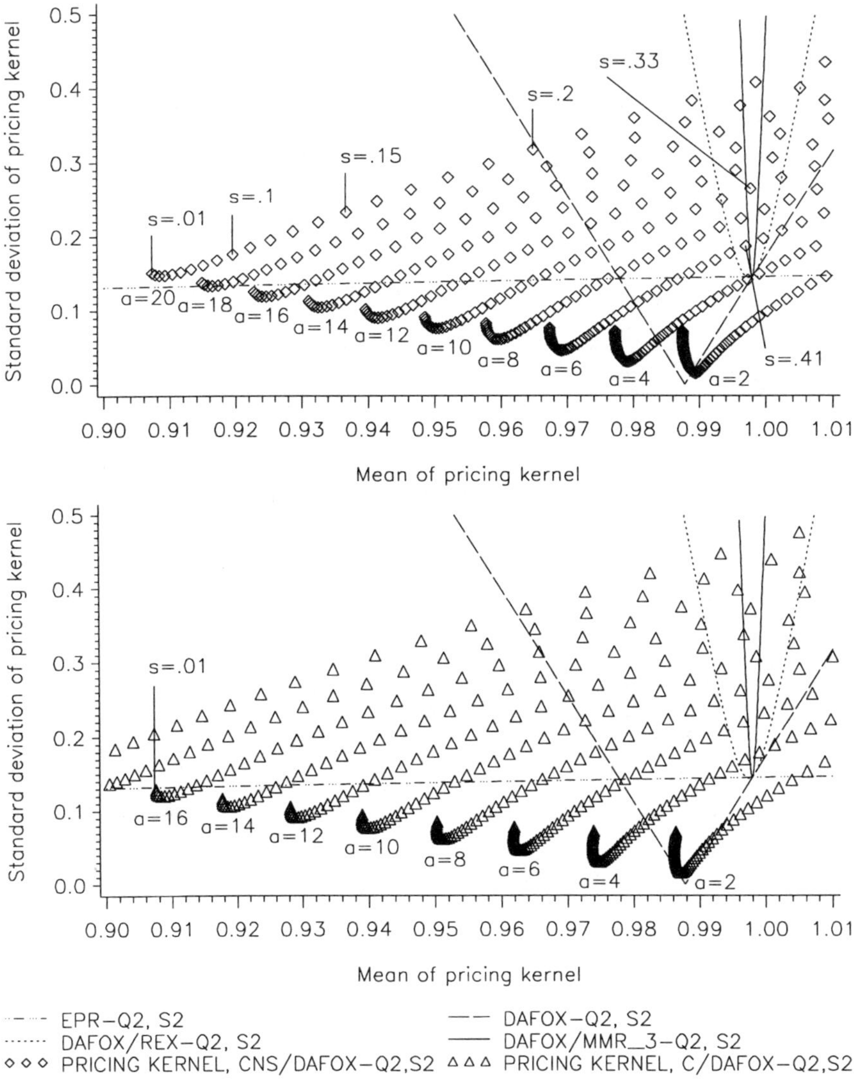

Figure A.5: The upper graph exhibits the time series of quarterly logarithmic real growth rates of per capita consumption expenditures on nondurables and services (CNS-Q). The lower graph shows the time series of the conditional smoothed probabilities of state 1, the down-state of the estimated Markov state-switching model for consumption growth ("first optimum").

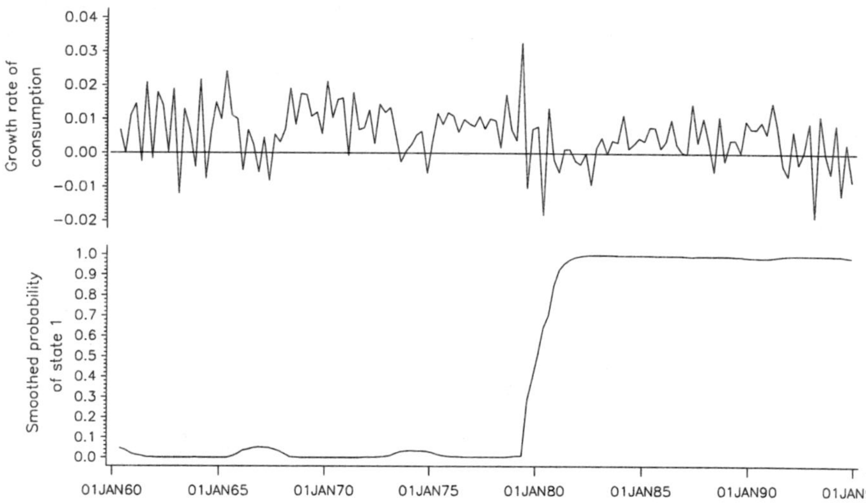

Figure A.6: Two randomly chosen time series of the 10,000 time series of the quarterly logarithmic real growth rates of consumption expenditures on nondurables and services (CNS-Q) which are simulated using the estimated Markov state-switching model.

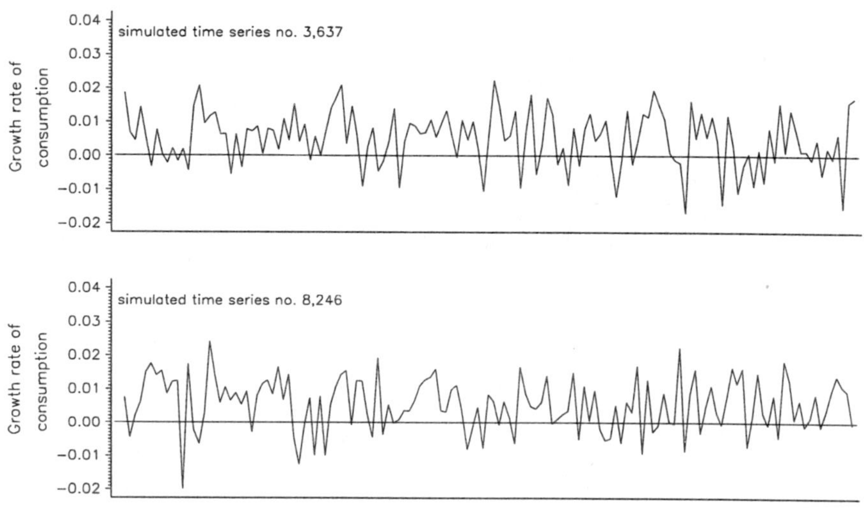

Figure A.7: The upper two graphs exhibit the time series of quarterly logarithmic real growth rates of per capita consumption expenditures on nondurables and services (CNS-Q) and of per capita corporate dividends (D-Q). The lower graph shows the time series of the conditional smoothed probabilities of state 1, the recession state of the estimated bivariate Markov state-switching model for consumption growth and growth of corporate dividends.

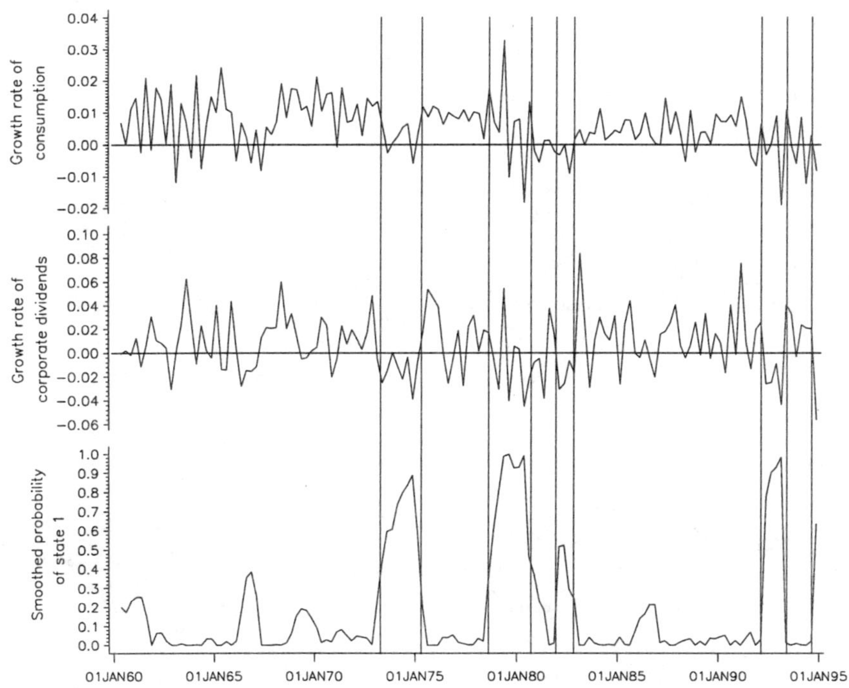

Figure A.8: The upper part exhibits the density functions of the two estimated state-dependent bivariate normal distributions of the Markov state-switching model for consumption growth and growth of corporate dividends. The lower part shows the density function of the mixture of the estimated state-dependent bivariate normal distributions in comparison with that of a single bivariate normal distribution fitted to the data.

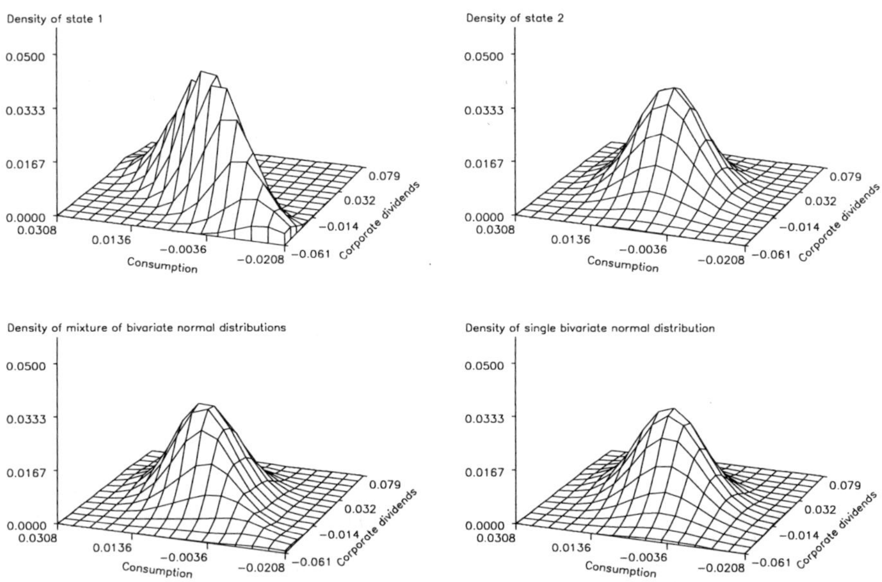

Figure A.9: Empirically observed serial correlation in multiperiod real returns on equity and in multiperiod real equity premia for various investment horizons in comparison with the serial correlation implied by the calibrated equilibrium model UN1b.

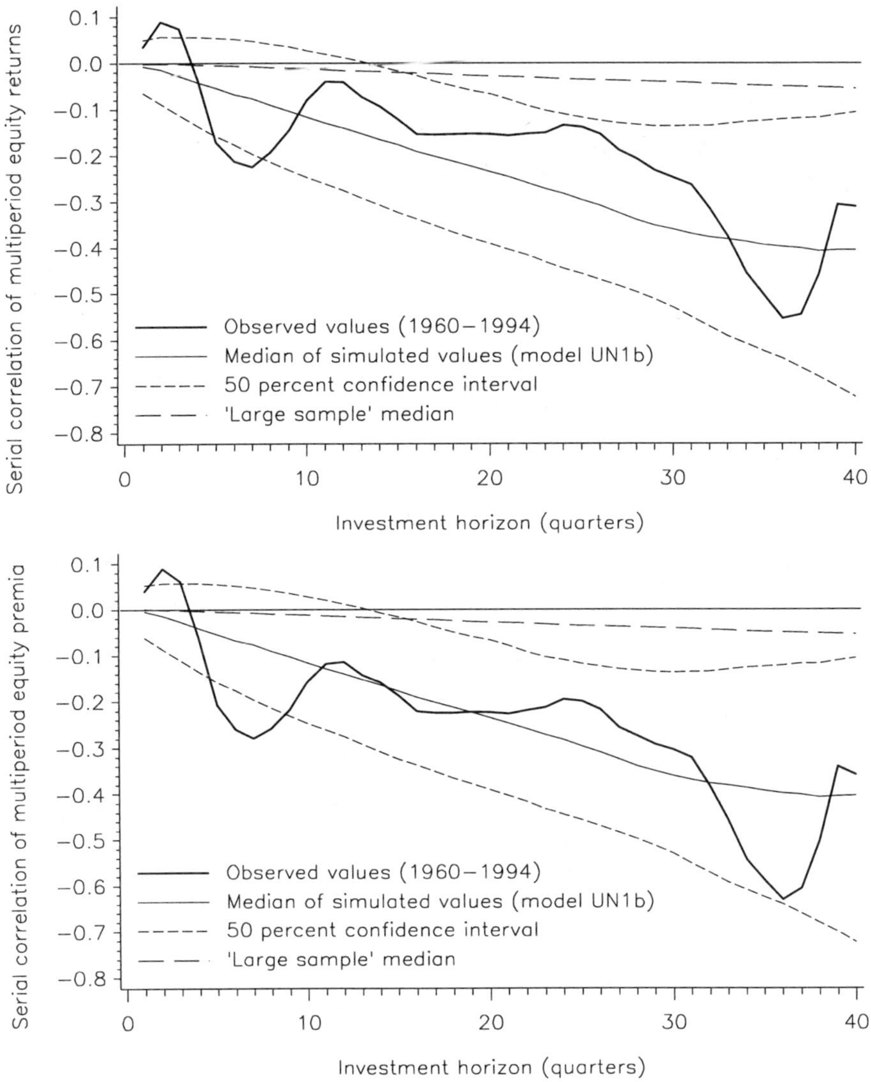

Figure A.10: Empirically observed serial correlation in multiperiod real returns on equity and in multiperiod real equity premia for various investment horizons in comparison with the serial correlation implied by the calibrated equilibrium model BN1b.

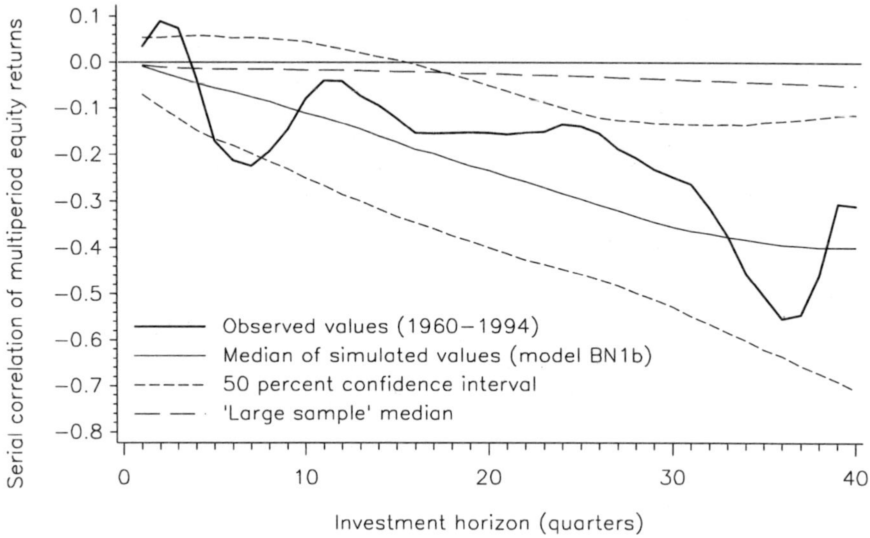

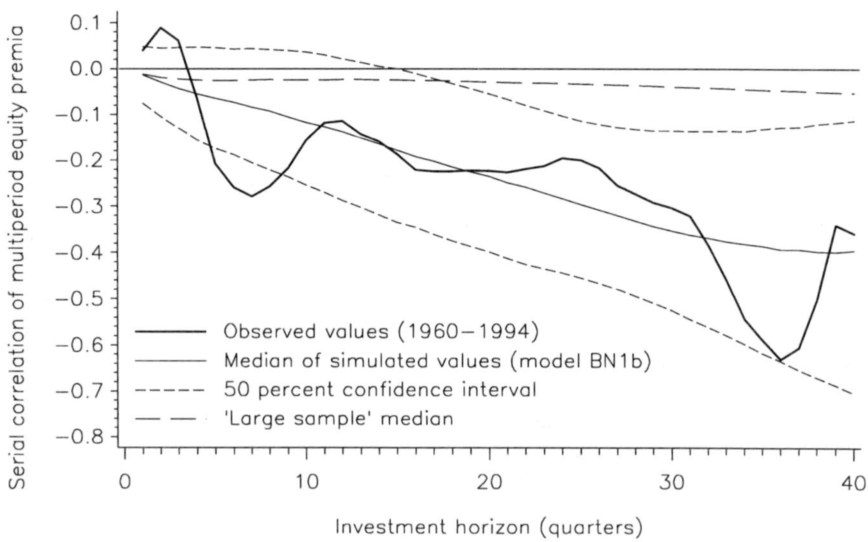

List of Symbols

Indices

I	number of assets
i	index of assets, $i \in \{1, \ldots, I\}$
f	index of the one-period risk-free asset
M	index of the aggregate wealth portfolio (market portfolio)
E	index of aggregate equity
z	index of an asset that satisfies $\text{cov}_t(\tilde{\phi}_{t+1}, \tilde{R}^z_{t+1}) = 0$ (first reference asset or zero-beta asset)
m	index of a second reference asset
ϕ	index of a second reference asset that satisfies $\lvert\text{corr}_t(\tilde{\phi}_{t+1}, \tilde{R}^\phi_{t+1})\rvert = 1$
P, P^*_t	index of a portfolio; index of the optimal portfolio from date t to date $t+1$
J	number of individuals in the economy
j	index of individuals, $j \in \{1, \ldots, J\}$
T	number of periods
t, τ	date at the end of period t, $t \in \{0, \ldots, T\}$; date at the end of period τ, $\tau \in \{t, \ldots, T\}$
S	number of states
s	index of states, $s \in \{1, \ldots, S\}$
X	number of instrumental variables

Variables and Parameters

$\tilde{\phi}_{t+1}$, $\phi_{t+1,s}$	random market pricing kernel at date t for pricing date $t+1$ payoffs; market pricing kernel at date t for pricing date $t+1$ payoffs in state s
$\pi_{t+1,s}$	date t price of a date $t+1$ payoff of one unit in state s and zero otherwise (price of a state-contingent claim) ($\pi_{t+1,s} = \phi_{t+1,s} \cdot \text{prob}_s$)

prob'_s	risk-neutral probability of state s (equivalent martingale measure) ($\text{prob}'_s = \phi_{t+1,s} \cdot \text{prob}_s \cdot R^f_{t+1}$)
prob_s, $\text{prob}_{s's}$	unconditional probability of state s; probability of state s conditional upon being in state s'
P^i_t, $\tilde{P}^i_t$, $P^i_{t,s}$	ex-dividend price of asset i, random ex-dividend price, resp. ex-dividend price in state s at date t
D^i_t, $\tilde{D}^i_t$, $D^i_{t,s}$	amount of any dividends, interest or other payments on asset i, random amount, resp. amount in state s at date t
d_t, $\tilde{d}_t$	aggregate dividend at date t; random aggregate dividend at date t
$e^i_{t,s}$	total payoff on asset i at date t in state s ($e^i_{t,s} = P^i_{t,s} + D^i_{t,s}$)
e_t, $\tilde{e}_t$	aggregate payoff on equity at date t; random aggregate payoff on equity at date t
$\tilde{\gamma}_{t+1}$	random gross growth rate of the aggregate payoff on equity from date t to date $t+1$ ($\tilde{\gamma}_{t+1} = \tilde{e}_{t+1}/e_t$)
R^i_{t+1}, $\tilde{R}^i_{t+1}$	gross return on asset i from date t to date $t+1$, resp. random gross return ($\tilde{R}^i_{t+1} = \tilde{r}^i_{t+1} + 1$)
$\tilde{R}^i_{s's}$	random gross return on asset i if the economy switches from state s' at date t to state s at date $t+1$
R^f_s	gross return on the one-period risk-free rate from date t to date $t+1$ if the economy is in state s at date t
$\tilde{\mathbf{R}}_{t+1}$	$[I \times 1]$ vector of random gross returns on the I assets from date t to date $t+1$
r^i_{t+1}, $\tilde{r}^i_{t+1}$	return on asset i from date t to date $t+1$, resp. random return
$\tilde{r}^i_{t,t+k}$	compound random return on asset i from date t to date $t+k$
$\tilde{p}^i_{t+1}$	random excess return on asset i over the one-period risk-free rate from date t to date $t+1$ ($\tilde{p}^i_{t+1} = \tilde{r}^i_{t+1} - r^f_{t+1} = \tilde{R}^i_{t+1} - R^f_{t+1}$)
$\tilde{p}^i_{s's}$	random excess return on asset i over the one-period risk-free rate if the economy switches from state s' at date t to state s at date $t+1$
$\tilde{u}^i_{t+1}$	model's forecast error concerning the return on asset i from date t to date $t+1$
v^i_t	fraction of the portfolio invested in asset i at date t
$\mathbf{v}$	$[I \times 1]$ vector of portfolio weights of the I assets
w_t, $\tilde{w}_t$	initial wealth at date t; random initial wealth at date t
c_t, $\tilde{c}_t$, $c_{t,s}$	consumption at date t; random consumption at date t; consumption in state s at date t
$\tilde{\lambda}_{t+1}$	random gross growth rate of aggregate consumption from date t to date $t+1$ ($\tilde{\lambda}_{t+1} = \tilde{c}_{t+1}/c_t$)
m_t, $\tilde{m}_t$	aggregate production at date t; random aggregate production at date t

α	coefficient of relative risk aversion
β, β_τ	constant one-period discount factor; discount factor from date t to date τ
σ	elasticity of intertemporal substitution
ρ	reciprocal of the elasticity of intertemporal substitution ($\rho = 1/\sigma$)
A^j, B^j	parameters of the HARA-class preference function of individual j
θ	parameter used at various occassions
a	constant term in regressions
b	slope coefficient in regressions
$\mathbf{b}$, $\mathbf{b}^*$	$[I \times 1]$ vectors of slope coefficients in multiple regressions
ε_t	regression error of time series regressions at date t
τ_{Priv}^{max}	maximum marginal income tax rate for private investors including supplementary tax but not church tax
τ_{Corp}	corporate income tax rate including supplementary tax
τ_{credit}	tax credit on dividend payments including supplementary tax
τ_{ID}	tax rate on interest and dividend payments
τ_{CG}	tax rate on capital gains
ID	nominal return from interest and dividend payments on an asset
CG	nominal return from capital gains on an asset
Θ, Φ	proportionality factors
Θ_t, $\tilde{\Theta}_t$	proportionality factor at date t; random proportionality factor at date t
Θ_s, Ψ_s	proportionality factors in state s
Ω_t	information available at date t
x_t	instrumental variable at date t, $x_t \in \Omega_t$

Functions

$\mathrm{E}\{\cdot\}$	expectation operator
$\mathrm{E}_t\{\cdot\}$	expectation taken at date t for a stochastic variable realized at date $t+1$, i.e. the expectation is conditional on the information available at date t
$\mathrm{var}(\cdot)$	variance operator
$\mathrm{var}_t(\cdot)$	variance of the realization of a stochastic variable at date $t+1$ conditional on the information available at date t
$\mathrm{std}(\cdot)$	standard deviation operator
$\mathrm{std}_t(\cdot)$	standard deviation of the realization of a stochastic variable at date $t+1$ conditional on the information available at date t

$\mathrm{cov}(\cdot,\cdot)$	covariance
$\mathrm{cov}_t(\cdot,\cdot)$	covariance of the realizations of two stochastic variables at date $t+1$ conditional on the information available at date t
$\mathrm{corr}(\cdot,\cdot)$	correlation coefficient
$\mathrm{corr}_t(\cdot,\cdot)$	correlation coefficient of the realizations of two stochastic variables at date $t+1$ conditional on the information available at date t
$\beta^{x,y}$	beta coefficient of y with respect to x ($\beta^{x,y} = \mathrm{cov}(x,y)/\mathrm{var}(x)$)
$\beta_t^{x,y}$	beta coefficient of y with respect to x at date $t+1$ conditional on information available at date t ($\beta_t^{x,y} = \mathrm{cov}_t(x,y)/\mathrm{var}_t(x)$)
$\Sigma_{\tilde{\mathbf{R}}}$	$[I \times I]$ covariance-matrix of gross returns
$\phi(\cdot)$	parametric model of the market pricing kernel, i.e. the market pricing kernel is a function of exogenous variables and parameters
$IMRS_t$	intertemporal marginal rate of substitution of consumption between dates $t-1$ and t
$U_t(\cdot)$, $U_t^*(\cdot)$, $\mathcal{U}_t(\cdot)$	functions giving the utility of a consumption path measured at date t
$u_t(\cdot)$, $u_t^*(\cdot)$, $u(\cdot)$	time-dependent, resp. time-independent functions giving the utility of current consumption
$J_t(\cdot)$	function giving the indirect utility of wealth, measured at date t, if the optimal consumption and investment strategy is followed subsequent to t
$W_t[\cdot,\cdot]$, $W[\cdot,\cdot]$, $V[\cdot,\cdot]$, $\mathcal{V}[\cdot,\cdot]$	time-dependent, resp. different time-independent aggregator functions
$V_{x,t}$	partial derivative of $V[\cdot,\cdot]$ with respect to its xth argument at date t
$\mu_t(\cdot)$	function giving the certainty equivalent of random future utility conditional on the information available at date t

Other Symbols

$\mathbf{1}$	$[I \times 1]$ vector of ones
$\forall$	"for all"
$\sim$	denotes random variables
T	denotes the transpose of a vector
μ	denotes sample means at various occasions

List of Tables

List of Figures

Bibliography

ABEL, A. B. (1988): "Stock Prices under Time-varying Dividend Risk - An Exact Solution in an Infinite-Horizon General Equilibrium Model," *Journal of Monetary Economics*, 22, 375–93.

——— (1990): "Asset Pricing under Habit Formation and Catching up with the Joneses," *American Economic Review*, 80(2), 38–42.

——— (1991): "The Equity Premium Puzzle," *Federal Reserve Bank of Philadelphia Business Review*, Sep./Oct.(2), 3–14.

——— (1994): "Exact Solutions for Expected Rates of Return Under Markov Regime Switching: Implications for the Equity Premium Puzzle," *Journal of Money, Credit & Banking*, 2(3), 345–61.

AHLBRECHT, M. (1996): *Entscheidungen unter Berücksichtigung des Zeitbezugs der Konsequenzen.* Peter Lang, Frankfurt am Main.

AHLBRECHT, M., AND M. WEBER (1996): "The Resolution of Uncertainty: An Experimental Study," *Journal of Institutional and Theoretical Economics*, 152(4), 593–607.

AMIHUD, Y., AND M. MURGIA (1997): "Dividends, Taxes, and Signaling: Evidence from Germany," *Journal of Finance*, 52(1), 397–408.

ATKESON, A., AND M. OGAKI (1996): "Wealth-varying Intertemporal Elasticities of Substitution: Evidence from Panel and Aggregate Data," *Journal of Monetary Economics*, 38, 507–34.

ATTANASIO, O. P., AND G. WEBER (1989): "Intertemporal Substitution, Risk Aversion and the Euler Equation for Consumption," *Economic Journal*, 99, 59–73.

BAKSHI, G. S., AND A. NAKA (1997): "An Empirical Investigation of Asset Pricing Models using Japanese Stock Market Data," *Journal of International Money and Finance*, 16(1), 81–112.

BANSAL, R., AND W. J. COLEMAN II (1996): "A Monetary Explanation of the Equity Premium, Term Premium, and Risk-free Rate Puzzles," *Journal of Political Economy*, 104(6), 1135–71.

BANZ, R. W. (1981): "The Relationship between Return and Market Value of Common Stocks," *Journal of Financial Economics*, 25, 3–18.

BANZ, R. W., AND M. H. MILLER (1978): "Prices for State-contingent Claims: Some Estimates and Applications," *Journal of Business*, 51(4), 653–72.

BARSKY, R. B. (1989): "Why Don't the Prices of Stocks and Bonds Move Together?," *American Economic Review*, 79(3), 1132–45.

BAUM, L. E., T. PETRIE, G. SOULES, AND N. WEISS (1970): "A Maximization Technique Occuring in the Statistical Analysis of Probabilistic Functions of Markov Chains," in *Annals of Mathematical Statistics*, vol. 41, pp. 164–71.

BAY, W. (1990): *Dividenden, Steuern und Steuerreform: ein internationaler Vergleich*. Deutscher Universitätsverlag, Wiesbaden.

BEKAERT, G., AND R. J. HODRICK (1992): "Characterizing Predictable Components in Excess Returns on Equity and Foreign Exchange Markets," *Journal of Finance*, 47(2), 467–510.

BEKAERT, G., R. J. HODRICK, AND D. A. MARSHALL (1997): "The Implications of First-order Risk Aversion for Asset Market Risk Premiums," *Journal of Monetary Economics*, 40(1), 3–39.

BENNINGA, S., AND A. PROTOPAPADAKIS (1990): "Leverage, Time Preference, and the Equity Premium Puzzle," *Journal of Monetary Economics*, 25, 49–58.

BICK, A. (1987): "On the Consistency of the Black-Scholes Model with a General Equilibrium Framework," *Journal of Financial and Quantitative Analysis*, 22(3), 259–75.

——— (1990): "On Viable Diffusion Price Processes of the Market Portfolio," *Journal of Finance*, 45(2), 673–89.

BITTLINGMAYER, G. (1995): "Output, Political Uncertainty, and Stock Market Fluctuations: Germany, 1890-1940," Discussion Paper 87, Center for Economic Studies, University of Munich.

BLACK, F. (1972): "Capital Market Equilibrium with Restricted Borrowing," *Journal of Business*, 45, 444–55.

BONOMO, M., AND R. GARCIA (1994): "Can a Well-Fitted Equilibrium Asset-Pricing Model Produce Mean Reversion?," *Journal of Applied Econometrics*, 9, 19–29.

——— (1996): "Consumption and Equilibrium Asset Pricing: An Empirical Assessment," *Journal of Empirical Finance*, 3, 239–65.

BOSSARD, A. (1988): *Konsum und Kapitalmarkt: Die intertemporale Konsum-Entscheidung der Haushalte und die Preisbestimmung auf dem Kapitalmarkt*. Haupt, Bern.

BOSWELL, M. T., S. D. GORE, G. P. PATIL, AND C. TALLIE (1993): "The Art of Computer Generation of Random Variables," in *Handbook of Statistics*, ed. by C. R. Rao, vol. 9, pp. 661–721. North Holland.

BREEDEN, D. T. (1979): "An Intertemporal Asset Pricing Model with Stochastic Consumption and Investment," *Journal of Financial Economics*, 7(3), 265–96.

——— (1989): "Intertemporal Portfolio Theory and Asset Pricing," in *The New Palgrave: Finance*, ed. by J. Eatwell, M. Milgate, and P. Newman, pp. 180–93. Macmillan Press, London.

Breeden, D. T., M. R. Gibbons, and R. H. Litzenberger (1989): "Empirical Tests of the Consumption-oriented CAPM," *Journal of Finance*, 44(2), 231–62.

Breeden, D. T., and R. H. Litzenberger (1978): "Prices of State-contingent Claims Implicit in Option Prices," *Journal of Business*, 51(4), 621–51.

Brown, D. P., and M. R. Gibbons (1985): "A Simple Econometric Approach for Utility-based Asset Pricing Models," *Journal of Finance*, 40, 359–81.

Bufman, G., and L. Leiderman (1990): "Consumption and Asset Returns under Non-Expected Utility," *Economics Letters*, 34, 231–35.

Burnside, C. (1994): "Hansen-Jagannathan Bounds as Classical Tests of Asset-Pricing Models," *Journal of Business and Economics Statistics*, 12(1), 57–79.

Campbell, J. Y. (1993): "Intertemporal Asset Pricing without Consumption Data," *American Economic Review*, 83(3), 487–512.

——— (1996): "Consumption and the Stock Market: Interpreting International Experience," Discussion Paper 5610, National Bureau of Economic Research (NBER).

Campbell, J. Y., and J. H. Cochrane (1995): "By Force of Habit: A Consumption-based Explanation of Aggregate Stock Market Behavior," Discussion Paper 4995, National Bureau of Economic Research (NBER).

Canova, F. (1994): "Statistical Inference in Calibration Models," *Journal of Applied Econometrics*, 9, 123–44.

Carhart, M. M., R. J. Krail, R. L. Stevens, and K. D. Welch (1996): "Testing the Conditional CAPM," Unpublished manuscript, University of Southern California.

Cecchetti, S. G., P.-S. Lam, and N. C. Mark (1990): "Mean Reversion in Equilibrium Asset Prices," *American Economic Review*, 80, 398–418.

——— (1993): "The Equity Premium and the Risk-free Rate: Matching the Moments," *Journal of Monetary Economics*, 31(1), 21–45.

——— (1994): "Testing Volatility Restrictions on Intertemporal Marginal Rates of Substitution Implied by Euler Equations and Asset Returns," *Journal of Finance*, 49(1), 123–52.

Cecchetti, S. G., and N. C. Mark (1990): "Evaluating Empirical Tests of Asset Pricing Models: Alternative Interpretations," *American Economic Review - Papers & Proceedings*, 80, 48–51.

Chew, S., and J. Ho (1994): "Hope: An Empirical Study of Attitude toward the Timing of Uncertainty Resolution," *Journal of Risk and Uncertainty*, 8, 267–88.

Cochrane, J. H., and L. P. Hansen (1992): "Asset Pricing Explorations for Macroeconomics," in *NBER Macroeconomics Annual*, ed. by O. J. Blanchard, and S. Fischer, vol. 11, pp. 115–82. Cambridge, Massachusetts: MIT Press.

Constantinides, G. M. (1982): "Intertemporal Asset Pricing with Heterogeneous Consumers and without Demand Aggregation," *Journal of Business*, 55(2), 253–67.

——— (1990): "Habit Formation: A Resolution of the Equity Premium Puzzle," *Journal of Political Economy*, 98(3), 519–43.

CONSTANTINIDES, G. M., AND D. DUFFIE (1996): "Asset Pricing with Heterogeneous Consumers," *Journal of Political Economy*, 104(2), 219–40.

D'AGOSTINO, R. B., AND M. A. STEPHENS (1986): *Goodness-of-fit techniques.* Marcel Dekker, INC., New York.

DANTHINE, J.-P., J. B. DONALDSON, AND R. MEHRA (1992): "The Equity Premium and the Allocation of Income Risk," *Journal of Economic Dynamics and Control*, 16(3), 509–32.

DEATON, A. (1992): *Understanding Consumption.* Oxford University Press, Oxford.

DEATON, A., AND J. MUELLBAUER (1980): *Economics and Consumer Behavior.* Cambridge University Press, Cambridge.

DEBONDT, W., AND R. THALER (1985): "Does the Stock Market Overreact?," *Journal of Finance*, 40(1), 793–805.

DEJONG, D. N., B. F. INGRAM, AND C. H. WHITEMAN (1996): "A Bayesian Approach to Calibration," *Journal of Business & Economic Statistics*, 14(1), 1–9.

DEUTSCHE BUNDESBANK (1996): "Reaktionen der Geldmarkt- und kurzfristiger Bankzinsen auf Änderungen der Notenbanksätze," *Monthly Report*, October, 33–49.

——— (1997): "Die Aktie als Finanzierungs- und Anlageinstrument," *Monthly Report*, January, 27–41.

DEUTSCHES AKTIENINSTITUT (ed.) (1996): *DAI-Factbook 1996.* Deutsches Aktieninstitut e.V., Frankfurt am Main.

DIXIT, A. K. (1990): *Optimization in Economic Theory.* Oxford University Press, New York, 2nd edn.

DOTHAN, M. U. (1990): *Prices in Financial Markets.* Oxford University Press, Oxford.

DUFFIE, D. (1988): *Security Markets – Stochastic Models.* Academic Press, San Diego.

——— (1996): *Dynamic Asset Pricing Theory.* Princeton University Press, Princeton, 2nd edn.

DUFFIE, D., AND L. G. EPSTEIN (1992): "Asset Pricing with Stochastic Differential Utility," *Review of Financial Studies*, 5(3), 411–36.

DYBVIG, P. H., AND S. A. ROSS (1989): "Arbitrage," in *The New Palgrave: Finance*, ed. by J. Eatwell, M. Milgate, and P. Newman, pp. 57–71. Macmillan Press, London.

EICHENBAUM, M. S., AND L. P. HANSEN (1990): "Estimating Models with Intertemporal Substitution Using Aggregate Time Series Data," *Journal of Business & Economic Statistics*, 8(1), 53–69.

EPSTEIN, L. G. (1992): "Behavior under Risk: Recent Developments in Theory and Applications," in *Advances in Economic Theory*, ed. by J. Laffont, vol. 2, pp. 1–63. Cambridge University Press, Cambridge.

EPSTEIN, L. G., AND S. E. ZIN (1989): "Substitution, Risk Aversion, and the Temporal Behavior of Consumption and Asset Returns: A Theoretical Framework," *Econometrica*, 57(4), 937–69.

——— (1990): "'First-order' Risk Aversion and the Equity Premium Puzzle," *Journal of Monetary Economics*, 26, 387–407.

——— (1991): "Substitution, Risk Aversion, and the Temporal Behavior of Consumption and Asset Returns: An Empirical Analysis," *Journal of Political Economy*, 99(2), 263–86.

FAMA, E. F., AND K. R. FRENCH (1988a): "Permanent and Temporary Components of Stock Prices," *Journal of Political Economy*, 96(2), 246–73.

——— (1988b): "Dividend Yields and Expected Stock Returns," *Journal of Financial Economics*, 22, 3–25.

——— (1992): "The Cross-section of Expected Stock Returns," *Journal of Finance*, 47, 427–65.

——— (1993): "Common Risk Factors in the Returns on Stocks and Bonds," *Journal of Financial Economics*, 33, 3–56.

FERSON, W. E., AND G. M. CONSTANTINIDES (1991): "Habit Persistence and Durability in Aggregate Consumption," *Journal of Financial Economics*, 29, 199–240.

FERSON, W. E., AND S. R. FOERSTER (1994): "Finite Sample Properties of the Generalized Method of Moments in Tests of Conditional Asset Pricing Models," *Journal of Financial Economics*, 36, 29–55.

FERSON, W. E., AND C. R. HARVEY (1992): "Seasonality and Consumption-based Asset Pricing," *Journal of Finance*, 47(2), 511–52.

FERSON, W. E., AND R. W. SCHADT (1996): "Measuring Fund Strategy and Performance in Changing Economic Conditions," *Journal of Finance*, 51(2), 425–61.

FISHER, S. J. (1994): "Asset Trading, Transaction Costs and the Equity Premium," *Journal of Applied Econometrics*, 9, S71–S94.

FRANKE, G. (1984): "Conditions for Myopic Valuation and Serial Independence of the Market Excess Return in Discrete Time Models," *Journal of Finance*, 39(2), 425–42.

FRANKE, G., AND H. HAX (1994): *Finanzwirtschaft des Unternehmens und Kapitalmarkt.* Springer-Verlag, Berlin, Heidelberg, 3rd edn.

FRANKE, G., R. C. STAPLETON, AND M. G. SUBRAHMANYAM (1996): "Why are Options Expensive?," Unpublished manuscript, University of Konstanz.

FRIEDMAN, M. (1953): "The Methodology of Positive Economics," in *Essays in Positive Economics.* Chicago University Press, Chicago.

FUNKE, M. (1992): "Asset Prices and Aggregate Consumption in West Germany - Multi-Input Transfer Function Model Analysis," *Jahrbücher für Nationalökonomie und Statistik*, 210, 193–202.

GALLANT, R. A., L. P. HANSEN, AND G. E. TAUCHEN (1990): "Using Conditional Moments of Asset Payoffs to Infer the Volatility of Intertemporal Marginal Rates of Substitution," *Journal of Econometrics*, 45, 141–79.

GARCIA, R., AND P. PERRON (1996): "An Analysis of the Real Interest Rate Under Regime Shifts," *Review of Economics and Statistics*, 78(1), 111–25.

GIOVANNINI, A., AND P. WEIL (1989): "Risk Aversion and Intertemporal Substitution in the Capital Asset Pricing Model," Discussion Paper 2824, National Bureau of Economic Research (NBER), Cambridge, Mass.

GOLLIER, C., AND H. SCHLESINGER (1997): "Changes in Risk and Asset Prices," Unpublished manuscript, University of Alabama.

GÖPPL, H., T. LÜDECKE, C. SCHLAG, AND H. SCHÜTZ (1995): "The German Equity Market: Risk, Return, and Liquidity," Discussion Paper 183, Institut für Entscheidungstheorie und Unternehmensforschung, University of Karlsruhe.

GÖPPL, H., AND H. SCHÜTZ (1993): "The Design and Implementation of a German Stock Price Research Index (Deutscher Aktien-Forschungsindex DAFOX)," in *Mathematical Modelling in Economics - Essays in Honor of Wolfgang Eichhorn*, ed. by D. Diewert, F. Spremann, and F. Stehling, pp. 506–19. Springer Verlag, Berlin.

GORDON, S., L. SAMSON, AND B. CARMICHAEL (1995): "Finite-sample Inferences about Mean-standard Deviation Bounds for Stochastic Discount Factors," *Economics Letters*, 49, 295–300.

GRAY, S. F. (1996): "Modeling the Conditional Distribution of Interest Rates as a Regime-switching Process," *Journal of Financial Economics*, 42, 27–62.

GREGORY, A. W., AND G. W. SMITH (1990): "Calibration as Estimation," *Econometric Reviews*, 9(1), 57–89.

GREGORY, A. W., AND G. W. SMITH (1991): "Calibration as Testing: Inference in Simulated Macroeconomic Models," *Journal of Business & Economic Statistics*, 9(3), 297–303.

GROSSMAN, S. J., A. MELINO, AND R. J. SHILLER (1987): "Estimating the Continuous-Time Consumption-based Asset-Pricing Model," *Journal of Business and Economic Statistics*, 5(3), 315–27.

GROSSMAN, S. J., AND R. J. SHILLER (1981): "The Determinants of the Variability of Stock Market Prices," *American Economic Review - Papers & Proceedings*, 71, 222–7.

HAKANSSON, N. H. (1989): "Financial Markets," in *The New Palgrave: Finance*, ed. by J. Eatwell, M. Milgate, and P. Newman, pp. 135–44. Macmillan Press, London.

HALL, R. E. (1988): "Intertemporal Substitution in Consumption," *Journal of Political Economy*, 96(2), 339–57.

HAMILTON, J. D. (1989): "A new Approach to the Economic Analysis of Nonstationary Time Series and the Business Cycle," *Econometrica*, 57(2), 357–84.

——— (1990): "Analysis of Time Series Subject to Changes in Regime," *Journal of Econometrics*, 45, 39–70.

——— (1993): "Estimation, Inference and Forecasting of Time Series Subject to Changes in Regime," in *Nonstationary Time Series Analysis and Cointegration*, ed. by C. Hargreaves, Oxford.

——— (1996): "Specification Testing in Markov-switching Time Series Models," *Journal of Econometrics*, 70, 127–57.

HANSEN, L. P., J. HEATON, AND E. G. LUTTMER (1995): "Econometric Evaluation of Asset Pricing Models," *Review of Financial Studies*, 8(2), 237–74.

HANSEN, L. P., AND J. J. HECKMAN (1996): "The Empirical Foundations of Calibration," *Journal of Economic Perspectives*, 10(1), 87–104.

HANSEN, L. P., AND R. JAGANNATHAN (1991): "Implications of Security Market Data for Models of Dynamic Economies," *Journal of Political Economy*, 99(2), 225–62.

——— (1997): "Assessing Specification Errors in Stochastic Discount Factor Models," *Journal of Finance*, 52(2), 557–90.

HANSEN, L. P., AND K. J. SINGLETON (1982): "Generalized Instrumental Variables Estimation of Nonlinear Rational Expectations Models," *Econometrica*, 50(5), 1269–86.

——— (1983): "Stochastic Consumption, Risk Aversion, and the Temporal Behavior of Asset Returns," *Journal of Political Economy*, 91(2), 249–65.

——— (1984): "Errata," *Econometrica*, 52(1), 267–8.

HARDOUVELIS, G. A., D. KIM, AND T. A. WIZMAN (1996): "Asset Pricing Models with and without Consumption Data: An Empirical Evaluation," *Journal of Empirical Finance*, 3, 267–301.

HE, H., AND D. M. MODEST (1995): "Market Frictions and Consumption-based Asset Pricing," *Journal of Political Economy*, 103(1), 94–117.

HEATON, J., AND D. LUCAS (1992): "The Effects of Incomplete Insurance Markets and Trading Costs in a Consumption-based Asset Pricing Model," *Journal of Economic Dynamics and Control*, 16(3), 601–20.

HERRMANN, M. (1996): *Unternehmenskontrolle durch Stiftungen: Untersuchung der Performancewirkung.* Gabler Verlag, Wiesbaden.

HOOVER, K. D. (1995): "Facts and Artifacts: Calibration and the Empirical Assessment of Real-Business-Cycle Models," *Oxford Economic Papers*, 47(1), 24–44.

HUANG, C.-F., AND R. H. LITZENBERGER (1988): *Foundations for Financial Economics.* North Holland, New York, 2nd edn.

HUNG, M.-W. (1994): "The Interaction between Nonexpected Utility and Asymmetric Market Fundamentals," *Journal of Finance*, 49(1), 325–43.

——— (1995): "On Mean-Standard Deviation Frontier of Stochastic Discount Factor in the Presence of Regime Shifts," in *Research in Finance*, ed. by A. H. Chen, vol. 13, pp. 143–59. JAI Press Inc.

JACKWERTH, J. C., AND M. RUBINSTEIN (1996): "Recovering Probability Distributions from Option Prices," *Journal of Finance*, 51(5), 1611–31.

JACOBS, K. (1996): "Incomplete Markets and Security Prices: An Investigation Using Household Data," Unpublished manuscript, McGill University, Montreal.

JAGANNATHAN, R., K. KUBOTA, AND H. TAKEHARA (1995): "The CAPM with Human Capital: Evidence from Japan," Unpublished manuscript, University of Minnesota.

JAGANNATHAN, R., AND Z. WANG (1996): "The Conditional CAPM and the Cross-Section of Expected Returns," *Journal of Finance*, 51(1), 3–53.

JARROW, R. A. (1988): *Finance Theory.* Prentice-Hall, New Jersey.

JORION, P., AND A. GIOVANNINI (1993): "Time Series Tests of a Non-Expected-Utility Model of Asset Pricing," *European Economic Review*, 37(5), 1083–100.

KANDEL, S., AND R. F. STAMBAUGH (1990): "Expectations and Volatility of Consumption and Asset Returns," *Review of Financial Studies*, 3(2), 207–32.

——— (1991): "Asset Returns and Intertemporal Preferences," *Journal of Monetary Economics*, 27(1), 37–71.

KAROLYI, G. A., AND B. KHO (1996): "Time-varying Risk Premia and the Return to Buying Winners and Selling Losers: Caveat Emptor et Venditor," Unpublished manuscript, The Charles A. Dice Center for Research in Financial Economics, Ohio State University.

KIMBALL, M. S. (1990): "Precautionary Saving in the Small and in the Large," *Econometrica*, 58(1), 53–73.

——— (1991): "Precautionary Motives for Holding Assets," Discussion Paper 3586, National Bureau of Economic Research (NBER).

——— (1993): "Standard Risk Aversion," *Econometrica*, 61(3), 589–611.

KIMBALL, M. S., AND P. WEIL (1992): "Precautionary Saving and Consumption Smoothing Across Time and Possibilities," Discussion Paper 3976, National Bureau of Economic Research (NBER).

KIRMAN, A. P. (1992): "Whom or What Does the Representative Individual Represent?," *Journal of Economic Perspectives*, 6(2), 117–36.

KISCHKA, P., AND C. PUPPE (1992): "Decisions Under Risk and Uncertainty: A Survey of Recent Developments," *Zeitschrift für Operations Research*, 36, 125–47.

KOCHERLAKOTA, N. R. (1990a): "Disentangling the Coefficient of Relative Risk Aversion from the Elasticity of Intertemporal Substitution: An Irrelevance Result," *Journal of Finance*, 45(1), 175–90.

——— (1990b): "On the 'Discount' Factor in Growth Economies," *Journal of Monetary Economics*, 25(1), 43–7.

——— (1990c): "On Tests of Representative Consumer Asset Pricing Models," *Journal of Monetary Economics*, 26(2), 285–304.

——— (1996): "The Equity Premium: It's Still a Puzzle," *Journal of Economic Literature*, 34(1), 42–71.

KREPS, D. M. (1988): *Notes on the Theory of Choice.* Westview Press, Boulder, Colorado.

——— (1990): *A Course in Microeconomic Theory.* Princeton University Press.

KREPS, D. M., AND E. L. PORTEUS (1978): "Temporal Resolution of Uncertainty and Dynamic Choice Theory," *Econometrica*, 46(1), 185–200.

——— (1979): "Temporal von Neumann-Morgenstern and Induced Preferences," *Journal of Economic Theory*, 20, 81–109.

KUGLER, P. (1988): "Intertemporal Substitution, Taste Shocks and Cointegration," *Economics Letters*, 26, 235–9.

KURZ, M., AND A. BELTRATTI (1997): "The Equity Premium is No Puzzle," in *Endogenous Economic Fluctuations*, ed. by M. Kurz, pp. 283–315. Springer-Verlag, Berlin.

KYDLAND, F. E., AND E. C. PRESCOTT (1982): "Time to Build and Aggregate Fluctuations," *Econometrica*, 50(6), 1345–70.

——— (1996): "The Computational Experiment: An Econometric Tool," *Journal of Economic Perspectives*, 10(1), 69–85.

LAIRD, N. (1993): "The EM Algorithm," in *Handbook of Statistics*, ed. by C. R. Rao, vol. 9, pp. 509–20. North Holland.

LEWBEL, A. (1989): "Exact Aggregation and a Representative Consumer," *Quarterly Journal of Economics*, 104, 622–33.

LINTNER, J. (1965): "The Valuation of Risky Assets and the Selection of Risky Investments in Stock Portfolios and Capital Budgets," *Review of Economics*, 47, 13–37.

LO, A. W., AND A. C. MACKINLAY (1988): "Stock Market Prices Do Not Follow Random Walks: Evidence from a Simple Specification Test," *Review of Financial Studies*, 1, 41–66.

LONG, J. (1974): "Stock Prices, Inflation, and the Term Structure of Interest Rates," *Journal of Financial Economics*, 1, 131–70.

LONGSTAFF, F. (1995): "Option Pricing and the Martingale Restriction," *Review of Financial Studies*, 8, 1091–124.

LUCAS, D. J. (1994): "Asset Pricing with Undiversifiable Income Risk and Short Sales Constraints – Deepening the Equity Premium Puzzle," *Journal of Monetary Economics*, 34(3), 325–41.

LUCAS, R. E. J. (1978): "Asset Prices in an Exchange Economy," *Econometrica*, 46(6), 1429–45.

——— (1980): "Methods and Problems in Business Cycle Theory," in *Studies in Business-Cycle Theory.* Blackwell, Oxford.

LUND, J., AND T. ENGSTED (1996): "GMM and Present Value Tests of the C-CAPM: Evidence from the Danish, German, Swedish and UK Stock Markets," *Journal of International Money and Finance*, 15(4), 497–521.

LYNCH, A. W. (1996): "Decision Frequency and Synchronization across Agents: Implications for Aggregate Consumption and Equity Return," *Journal of Finance*, 51(4), 1479–97.

MACHINA, M. (1987): "Choice under Uncertainty: Problems solved and unsolved," *Journal of Economic Perspectives*, 1(1), 121–54.

——— (1989): "Dynamic Consistency and Non-Expected Utility Models of Choice under Uncertainty," *Journal of Economic Literature*, 32, 1622–68.

MANKIW, N. G. (1986): "The Equity Premium and the Concentration of Aggregate Stocks," *Journal of Financial Economics*, 17(1), 211–19.

MANKIW, N. G., J. J. ROTEMBERG, AND L. H. SUMMERS (1985): "Intertemporal Substitution in Macroeconomics," *Quarterly Journal of Economics*, 100, 225–51.

MANKIW, N. G., AND S. D. ZELDES (1991): "The Consumption of Stockholders and Non-Stockholders," *Journal of Financial Economics*, 29(1), 97–112.

MAURO, P. (1995): "Stock Markets and Growth - A Brief Caveat on Precautionary Savings," *Economics Letters*, 47, 111–6.

MEHRA, R., AND E. C. PRESCOTT (1985): "The Equity Premium - A Puzzle," *Journal of Monetary Economics*, 15(2), 145–61.

——— (1988): "The Equity Premium Puzzle: A Solution?," *Journal of Monetary Economics*, 22(1), 133–36.

MELLA, F. (1988): *Dem Trend auf der Spur.* Verlag Börsen-Zeitung, Frankfurt am Main.

MERTON, R. C. (1973): "An Intertemporal Capital Asset Pricing Model," *Econometrica*, 41(5), 867–87.

MEYER, B. (1994): *Der Overreaction-Effekt am deutschen Aktienmarkt - Einordnung und empirische Untersuchung der langfristigen Überreaktion.* Fritz Knapp Verlag, Frankfurt am Main.

——— (1995): "Die langfristige Performance von 'Gewinner'- und 'Verlierer'-Aktien am deutschen Aktienmarkt," *Finanzmarkt und Portfoliomanagement*, 9(1), 61–80.

——— (1996): "Are German Stock and Bond Returns Consistent with Equilibrium Asset Pricing? - A Calibration Exercise Using Recursive Non-Expected Utility," Discussion Paper 300, SFB 178, University of Konstanz.

——— (1997): "Variance Bounds for Pricing Kernels in Intertemporal Asset Pricing Models: The German Case," in *Geld, Finanzwirtschaft, Banken und Versicherungen (1996)*, ed. by C. Hipp, pp. 459–91. Verlag Versicherungswirtschaft e.V., Karlsruhe.

MISSONG, M. (1994): *Intertemporale Substitution und stochastische Eulergleichungsmodelle: eine theoretische und empirische Untersuchung.* Haag und Hercher, Frankfurt am Main.

——— (1996): "Euler Equations and Linear Asset Pricing," Discussion Paper 97/1996, Institute of Statistics and Econometrics, Christian Albrechts University at Kiel.

MOORE, B., AND H. SCHALLER (1996): "Learning, Regime Switches, and Equilibrium Asset Pricing Dynamics," *Journal of Economic Dynamics and Control*, 20, 979–1006.

MORAWIETZ, M. (1994): *Rentabilität und Risiko deutscher Aktien- und Rentenanlagen seit 1870: unter Berücksichtigung von Geldentwertung und steuerlichen Einflüssen.* Deutscher Universitätsverlag Gabler, Wiesbaden.

MOSSIN, J. (1966): "Equilibrium in a Capital Asset Market," *Econometrica*, 34, 768–83.

NEELY, CHRISTOPHER J., A. R., AND C. H. WHITEMAN (1996): "Identification Failure in the Intertemporal Consumption CAPM," Discussion paper, Federal Reserve Bank of St. Louis.

NEUMANN, M., AND C. SCHLAG (1996): "Martingale Restrictions and Implied Distributions for German Stock Index Option Prices," Discussion Paper 196, Institut für Entscheidungstheorie und Unternehmensforschung, University of Karlsruhe.

NI, S. X. (1993): "Habit Formation as a Resolution to the Equity Premium Puzzle: What is in the Data, what is not," *Southern Economic Journal*, 59(4), 749–59.

OERTMANN, P. (1994a): "Size Effect und Performance von deutschen Aktien," *Finanzmarkt und Portfolio Management*, 8, 197–211.

——— (1994b): "Firm-Size-Effekt am deutschen Aktienmarkt," *Zeitschrift für betriebswirtschaftliche Forschung*, 46(3), 229–59.

PAGAN, A. R. (1994): "Calibration and Econometric Research: An Overview," *Journal of Applied Econometrics*, 9, S1–S10.

PATEL, J. K., AND C. B. READ (1982): *Handbook of the Normal Distribution.* Marcel Dekker, New York.

PEARSON, E. S., AND H. O. HARTLEY (1966): *Biometrika Tables for Statisticians - Volume 1.* Cambridge University Press, Cambridge.

POTERBA, J. M., AND L. H. SUMMERS (1988): "Mean Reversion in Stock Prices - Evidence and Implications," *Journal of Financial Economics*, 22, 27–59.

QUAH, D. T. (1995): "Business Cycle Empirics: Calibration and Estimation," *The Economic Journal*, 105(433), 1594–6.

RIETZ, T. A. (1988): "The Equity Premium Puzzle: A Solution," *Journal of Monetary Economics*, 22(1), 117–31.

ROLL, R. (1977): "A Critique of the Asset Pricing Theory's Tests: Part I: On Past and Potential Testability of the Theory," *Journal of Financial Economics*, 4, 129–76.

ROSS, S. (1976): "The Arbitrage Theory of Capital Asset Pricing," *Journal of Economic Theory*, 13, 341–60.

RUBINSTEIN, M. (1974): "An Aggregation Theorem for Securities Markets," *Journal of Financial Economics*, 1, 225–44.

——— (1976): "The Valuation of Uncertain Income Streams and the Pricing of Options," *Bell Journal of Economics*, 7, 407–25.

——— (1994): "Implied Binomial Trees," *Journal of Finance*, 49(3), 771–818.

SAUER, A. (1991): "Die Bereinigung von Aktienkursen - Ein kurzer Überblick über Konzept und praktische Umsetzung," Discussion paper, Institut für Entscheidungstheorie und Unternehmensforschung, Universität Karlsruhe.

SAUER, A., AND A. MURPHY (1992): "An Empirical Comparison of Alternative Models of Capital Asset Pricing in Germany," *Journal of Banking and Finance*, 16, 183–96.

SAVAGE, L. (1972): *Foundations of Statistics.* Dover, New York.

SCHÜLER, K., AND V. SPIES (1991): "Einkommen aus Unternehmenstätigkeit und Vermögen in den Volkswirtschaftlichen Gesamtrechnungen," *Wirtschaft und Statistik*, 10/91, 653–66.

SCHWERT, W. G. (1989a): "Business Cycles, Financial Crises, and Stock Volatility," in *Carnegie-Rochester Conference Series on Public Policy*, vol. 31, pp. 83–126, Amsterdam. North-Holland.

——— (1989b): "Why Does Stock Market Volatility Change Over Time?," *Journal of Finance*, 44, 1115–45.

SELDEN, L. (1978): "A New Representation of Preferences over 'Certain × Uncertain' Consumption Pairs: The 'Ordinal Certainty Equivalent' Hypothesis," *Econometrica*, 46(5), 1045–60.

SHARPE, W. F. (1964): "Capital Asset Pricing: A Theory of Market Equilibrium under Conditions of Risk," *Journal of Finance*, 19, 425–42.

SHILLER, R. J. (1981): "Do Stock Prices Move Too Much to be Justified by Subsequent Changes in Dividends?," *American Economic Review*, 71(3), 421–35.

——— (1982): "Consumption, Asset Markets and Macroeconomic Fluctuations," in *Carnegie-Rochester Conference Series on Public Policy*, vol. 17, pp. 203–38, Amsterdam. North-Holland.

SIEGEL, J. J. (1992): "The Equity Premium: Stock and Bond Returns since 1802," *Financial Analysts Journal*, Jan.-Feb., 28–38.

SINGLETON, K. J. (1990): "Specification and Estimation of Intertemporal Asset Pricing Models," in *Handbook of Monetary Economics*, ed. by B. Friedman, and F. H. Hahn, vol. 1, pp. 583–626.

——— (1994): "Econometric Analysis of Representative Agent Intertemporal Asset Pricing Models," in *Advances in Econometrics*, vol. 2, pp. 277–311.

SNOW, K. N. (1991): "Diagnosing Asset Pricing Models Using the Distribution of Asset Returns," *Journal of Finance*, 46(3), 955–84.

SOLNIK, B. (1996): *International Investments.* Addison-Wesley, Mass., 3rd. edn.

STAPLETON, R. C., AND M. G. SUBRAHMANYAM (1990): "Risk Aversion and the Intertemporal Behavior of Asset Prices," *Review of Financial Studies*, 3, 677–94.

STEHLE, R. (1997): "Der Size-Effekt am deutschen Aktienmarkt," *Zeitschrift für Bankrecht und Bankwirtschaft*, 9(3), 237–260.

STEHLE, R., AND A. HARTMOND (1991): "Durchschnittsrenditen deutscher Aktien 1954-1988," *Kredit und Kapital*, 24(3), 371–411.

STEHLE, R., R. HUBER, AND J. MAIER (1996): "Rückrechnung des DAX für die Jahre 1955 bis 1987," *Kredit und Kapital*, 29(2), 277–304.

TELMER, C. I. (1993): "Asset Pricing Puzzles and Incomplete Markets," *Journal of Finance*, 48(5), 1803–32.

TURTLE, H., A. BUSE, AND B. KORKIE (1994): "Tests of Conditional Asset Pricing with Time-varying Moments and Risk Prices," *Journal of Financial & Quantitative Analysis*, 29(1), 15–29.

VON NEUMANN, J., AND O. MORGENSTERN (1953): *Theory of Games and Economic Behavior.* Princeton University Press, Princeton, New Jersey, 3rd edn.

WATSON, M. W. (1993): "Measures of Fit for Calibrated Models," *Journal of Political Economy*, 101(6), 1011–41.

WEIL, P. (1989): "The Equity Premium Puzzle and the Risk-free Rate Puzzle," *Journal of Monetary Economics*, 24(2), 401–21.

——— (1990): "Nonexpected Utility in Macroeconomics," *Quarterly Journal of Economics*, 105(1), 29–42.

——— (1992): "Equilibrium Asset Prices with Undiversifiable Labor Income Risk," *Journal of Economic Dynamics and Control*, 16(3/4), 769–90.

WILHELM, J. (1986): "Zum Verhältnis von Höhenpräferenz und Risikopräferenz - eine theoretische Analyse," *Zeitschrift für betriebswirtschaftliche Forschung*, 38, 467–92.

WOLTERS, J. (1992): "Der Zusammenhang zwischen Konsum und Einkommen: Alternative ökonometrische Ansätze," *RWI-Mitteilungen*, 43, 115–32.

WORKING, H. (1960): "Note on the Correlation of First Differences of Averages in a Random Chain," *Econometrica*, 28(4), 916–8.

Contributions to Economics

Hagen Bobzin
Indivisibilities
1998. ISBN 3-7908-1123-8

Helmut Wagner (Ed.)
Current Issues in Monetary Economics
1998. ISBN 3-7908-1127-0

Peter Michaelis/Frank Stähler (Eds.)
Recent Policy Issues in Environmental and Resource Economics
1998. ISBN 3-7908-1137-8

Jessica de Wolff
The Political Economy of Fiscal Decisions
1998. ISBN 3-7908-1130-0

Georg Bol/Gholamreza Nakhaeizadeh/
Karl-Heinz Vollmer (Eds.)
Risk Measurements, Econometrics and Neural Networks
1998. ISBN 3-7908-1152-1

Joachim Winter
Investment and Exit Decisions at the Plant Level
1998. ISBN 3-7908-1154-8

Printing: Weihert-Druck GmbH, Darmstadt
Binding: Buchbinderei Schäffer, Grünstadt